M000286315

CUTTING GREEN TAPE

Independent Studies in Political Economy

THE ACADEMY IN CRISIS
The Political Economy of Higher Education
Edited by John W. Sommer
Foreword by Nathan Glazer

AMERICAN HEALTH CARE
Government, Market Processes, and the Public Interest
Edited by Roger D. Feldman
Foreword by Mark V. Pauly

CUTTING GREEN TAPE
Toxic Pollutants, Environmental Regulation and the Law
Edited by Richard L. Stroup and Roger E. Miners
Foreword by W. Kip Viscusi

MONEY AND THE NATION STATE
The Financial Regulation, Government, and
the World Monetary System
Edited by Kevin Dowd and Richard H. Timberlake Jr.

PRIVATE RIGHTS & PUBLIC ILLUSIONS
Tibor R. Machan
Foreword by Nicholas Rescher

TAXING CHOICE
The Predatory Politics of Fiscal Discrimination
Edited by William F. Shughart II
Foreword by Paul W. McCracken

WRITING OFF IDEAS
Taxation, Foundations, and Philanthropy in America
Randall G. Holcombe

CUTTING GREEN TAPE

Toxic Pollutants, Environmental Regulation and the Law

Edited by Richard L. Stroup
and Roger E. Meiners
Foreword by W. Kip Viscusi

Transaction Publishers
New Brunswick (U.S.A.) and London (U.K.)

Copyright © 2000 by The Independent Institute, Oakland, Calif..

All rights reserved under International and Pan-American Copyright Conventions. No part of this book may be reproduced or transmitted in any form or by any means, electronic or mechanical, including photocopy, recording, or any information storage and retrieval system, without prior permission in writing from the publisher. All inquiries should be addressed to Transaction Publishers, Rutgers—The State University, 35 Berrue Circle, Piscataway, New Jersey 08854-8042.

This book is printed on acid-free paper that meets the American National Standard for Permanence of Paper for Printed Library Materials.

Library of Congress Catalog Number: 00-020661
ISBN: 1-56000-429-0 (cloth); 0-7658-0618-5 (paper)
Printed in the United States of America

Library of Congress Cataloging-in-Publication Data

Cutting green tape : toxic pollutants, environmental regulation, and the law / Richard L. Stroup and Roger E. Meiners, editors ; with a forword by W. Kip Viscusi.
    p. cm.—(Independent studies in political economy)
    Includes bibliographical references and index.
    ISBN 1-56000-429-0 (alk. paper) —ISBN 0-7658-0618-5 (pbk. : alk. paper)
    1. Toxic torts—United States. 2. Liability for hazardous substances pollution damages—United States. 3. Liability for environmental damages—United States. 4. Law and politics. I. Stroup, Richard. II. Meiners, Roger E. III. Series
KF1299.H39 C88 2000
344.73'04633—dc21   00-020661

The INDEPENDENT INSTITUTE

The Independent Institute is non-profit, non-partisan, scholarly research and educational organization that sponsors comprehensive studies on the political economy of critical social and economic problems.

The politicization of decision making in society has largely confined public debate to the narrow reconsideration of existing policies. Given the prevailing influence of partisan interests, little social innovation has occurred. In order to understand both the nature of and possible solutions to major public issues, The Independent Institute's program adheres to the highest standards of independent inquiry and is pursued regardless of prevailing political or social biases and conventions. The resulting studies are widely distributed as books and other publications, and publicly debated through numerous conference and media programs.

Through this uncommon independence, depth, and clarity, The Independent Institute pushes at the frontiers of our knowledge, redefines the debate over public issues, and fosters new and effective directions for government reform.

FOUNDER & PRESIDENT
David J. Theroux

RESEARCH DIRECTOR
Alexander T. Tabarrok

SENIOR FELLOWS
Bruce L. Benson
Robert Higgs
Richard K. Vedder

ACADEMIC ADVISORS
Stephen E. Ambrose
University of New Orleans
Martin Anderson
Hoover Institution
Herman Belz
University of Maryland
Thomas E. Borcherding
Claremont Graduate School
Boudewijn Bouckaert
University of Ghent, Belgium
James M. Buchanan
George Mason University
Allan C. Carlson
Rockford Institute
Robert W. Crandall
Brookings Institution
Arthur A. Ekirch, Jr.
State U. of New York, Albany
Richard A. Epstein
University of Chicago
B. Delworth Gardner
Brigham Young University

George Gilder
Discovery Institute
Nathan Glazer
Harvard University
Ronald Hamowy
University of Alberta
Steve H. Hanke
Johns Hopkins University
Ronald Max Hartwell
Oxford University
H. Robert Heller
International Payments Institute
Lawrence A. Kudlow
Schroder and Company, Inc.
Deirdre N. McCloskey
University of Iowa
J. Huston McCulloch
Ohio State University
Forrest McDonald
University of Alabama
Merton H. Miller
University of Chicago
Thomas Gale Moore
Hoover Institution
Charles Murray
American Enterprise Institute
William A. Niskanen
Cato Institute
Michael J. Novak, Jr.
American Enterprise Institute
Charles E. Phelps
University of Rochester

Paul Craig Roberts
Inst. for Political Economy
Nathan Rosenberg
Stanford University
Simon Rottenberg
University of Massachusetts
Pascal Salin
University of Paris, France
Arthur Seldon
Inst. of Econ. Affairs, England
William F. Shughart II
University of Mississippi
Joel H. Spring
State University of New York Old Westbury
Richard L. Stroup
Montana State University
Thomas S. Szasz
State U. of New York, Syracuse
Robert D. Tollison
University of Mississippi
Arnold S. Trebach
American University
Gordon Tullock
George Mason University
Richard E. Wagner
George Mason University
Sir Alan A. Walters
AIG Trading Corporation
Carolyn L. Weaver
American Enterprise Institute
Walter E. Williams
George Mason University

THE INDEPENDENT INSTITUTE
100 Swan Way, Oakland, CA 94621-1428, U.S.A.
Telephone: 510-632-1366 • Fax 510-568-6040
E-mail: info@independent.org • Website: http://www.independent.org

Contents

Foreword

Mass toxic torts are the most salient policy problem in the risk area. The stakes involved are enormous, the potential for error is huge, and social institutions have yet to deal effectively with such risks. Mass toxic torts have received the greatest attention with respect to our judicial system. From the mid-1980s to the early 1990s, the majority of product liability cases in the U.S. Federal courts involved lawsuits over asbestos-related illnesses. This experience is not unique. Other mass tort actions, such as those pertaining to DES, also involved thousands of claimants. The breast implant litigation potentially could involve even more claimants than the asbestos litigation. The prospect of incurring the monumental litigation costs spurred the companies to fashion an administrative settlement to the breast implant litigation. The settlement was initially rejected, although the increasing medical evidence questioning the link between breast implants and many of the more serious alleged health effects has apparently induced plaintiffs to accept a more recent settlement offer.[1]

Government regulators also have not had great success in addressing the kinds of risks associated with mass toxic torts. In the case of policies such as the Superfund program that addressed the risks posed by hazardous wastes, the Congress imposed ex post liability in much the same way as the courts did for asbestos. Companies must now incur obligations that they could not anticipate at the time of the waste disposal action because the current legal requirements for hazardous waste cleanup did not exist then.

The difficulties caused by mass toxic torts can be traced to several distinctive elements. Here I will explore four of the critical characteristics associated with toxic liability and indicate how they affect the functioning of our various risk institutions. I will then outline several principles for responsible risk management that represent a substantial departure from society's current approach to toxic liability.

Difficulties Posed by Toxic Liability[2]

Perhaps the most distinctive characteristic of toxic risks is the gestation period before the risk outcomes of toxic exposures become apparent. In the case of acute injuries, such as an automobile accident, the adverse health outcome is immediate. For dimly understood hazards, such as those posed by asbestos, the links occur after a long period of time, perhaps decades. This time period, coupled with the small probabilities involved, makes inferences difficult. The existence of a health risk will not be known immediately and, even when it is well established, the presence of multiple potential causes will make assessing the risk difficult. Although there are a few exceptional ailments that are signature diseases, such as the mesothelioma risks associated with asbestos, other hazards such as the lung cancer risks associated with asbestos can rise from multiple causes. We would not know for example, whether a cigarette smoking asbestos worker contracted lung cancer from asbestos exposures, or from smoking behavior, or some combination of the two.

In part because of the tremendous uncertainties arising from the substantial gestation period, the workers' compensation system did not function effectively in compensating asbestos workers for their ailments. Thousands of unsuccessful workers' compensation claimants subsequently filed cases to recover through tort liability. Causality issues, however, also pose substantial difficulties for the courts in asbestos and other mass toxic tort cases. In the case of the Agent Orange litigation, Judge Weinstein ultimately fashioned a settlement that provided modest compensation because he believed that causality was too difficult to prove.[3] Breast implants pose a similar problem. Based on fragmentary evidence of a link between implants and serious ailments, claimants were successful in receiving multimillion dollar punitive damages awards, but now the scientific evidence seems to indicate that the graver risks claimed in these cases were overstated. Claimants have since been faring less well, although a multibillion dollar settlement fund has been negotiated. Even though the breast implant product itself did not change with respect to its underlying risk characteristics, the courts have shifted in their treatment of this potentially risky product because of the shifting state of scientific knowledge. This knowledge will remain imprecise because the gestation period between receiving the breast implants and the onset of any illnesses makes distinguishing potential causalities a hazardous undertaking.

This gestation period and the fact that companies will be liable *ex post* have a variety of major implications. First, if companies do not anticipate the liability, there will be no incentive effects. This theme is articulated by Donald Dewees in his paper dealing with deterrence issues. The absence of incentives with respect to the hazardous waste cleanup program Superfund and the consequent failure of this effort to have a major risk reduction effect are continuing themes throughout this volume. The *ex* post character of Superfund and related efforts is of concern to Bruce Yandle, Jo-Christy Brown, Roger Meiners, Donald Dewees, and David Fractor. *Ex* post liability also undermines the insurance function that could otherwise take place because it is impossible to contract *ex* ante for unanticipated liabilities and to spread these costs in the usual manner.

A telling case that dramatizes the problems posed for insurance companies is the experience of Lloyd's of London. Throughout the period in which it wrote asbestos insurance coverage and even through the 1980s Lloyd's consistently underestimated the potential number of court cases that would arise from asbestos. Its estimates of the worst case scenarios of asbestos-related suits was 81,000 in 1982, which they then increased to 180,000 in 1990, but even these estimates remained far below the tally by the American Bar Association that indicated that the suits would number 340,000.[4]

The difficulty of *ex* post liability either with respect to Superfund, asbestos, or any other hazard is that the unanticipated nature of such costs undermines many of the critical market mechanisms that regulate risk. Firms will not experience a deterrence effect if the legal sanctions and the expected penalties for hazardous behavior are not known at the time they make their decisions. Similarly, insurance firms will not charge appropriate premium levels to the extent that future legal regimes cannot be reasonably anticipated. Firms cannot charge premiums retroactively given the passage of new laws, and even if contracts were structured in this manner the financial insolvency resulting from mass toxic torts would often prevent recouping costs. For analogous reasons the usual economic parables in which firms spread the costs of insurance across all product purchasers will not hold true because these are costs for actions taken in the past, not costs associated with current products. Thus, it will not be feasible for firms to pass these costs on to consumers because other firms without the history of such liabilities will be able to undercut them.

The second distinguishing characteristic of mass toxic torts is their mass nature. A large number of people are often exposed to these risks. Coupled with the lengthy gestation period before the ailments become apparent, this characteristic leads to a situation in which there will be waves of claimants once the risks become known. For example, suppose that breast implants begin in year one at 100,000 implants per year and continue through year ten. If problems posed by breast implants were apparent immediately, then presumably before even the 100,000 women had received the implants in the first year, these problems would be identified and companies would cease selling this product. If, however, the risk outcomes are not apparent until after a ten-year lag, there will be 1,000,000 claimants. There is consequently an explosive growth in the claimant pool that arises because companies will continue to sell these risky products during this lag period unaware of the ultimate risks. When the risks become apparent, they will be hit by successive waves of purchasers of the product.

The fact that these lawsuits will arrive on mass in the courts is not a benign outcome. The large number of suits, whether it is for asbestos, breast implants, or some other toxic liability situation, will clog the courts. The large number of suits against individual firms also will impose huge unanticipated litigation costs on the firms. Firms may settle such suits to avoid the litigation expense even if they have a product they believe to be safe. If the costs are too great, firms will simply reorganize, declaring bankruptcy because of the costs generated by this unanticipated wave of suits. Insurers will respond in ways that will not foster risk management. Lloyd's tottered on the brink of bankruptcy, sending a strong lesson to the insurance industry at large. It is now much more difficult to obtain the kind of coverage for environmental or toxic liability that firms were previously able to purchase. Insurance companies have become wary of the long-term risks posed by these open-ended policies. The result of imposing liability retroactively has been to lead insurance companies to exit the market for mass tort risks. The entire scenario in terms of company and insurance firm responses for toxic liability contrasts quite starkly with that for manufacturing defects. For these hazards, the firm can identify the problem quickly, react in a way that remedies the defect without an undue lag, and incur the kinds of costs that are potentially insurable.

The uninsurability of the risks associated with toxic liability stems from a single characteristic relevant to insurance firms, which is that in addition to being unanticipated, these risks are highly correlated. Insurance companies are very successful in insuring independent risks that, in effect, cancel themselves out in the insurance company's portfolio. Thus, insuring famous athletes' health and the hands of concert pianists is well within the abilities of insurance firms such as Lloyd's of London. However, if all such risk ailments were not independent but would all occur simultaneously to all of the firm's insured parties, then insurance would not be financially feasible. It is just that type of problem that arises with respect to mass toxic torts in which there are waves of correlated risks rather than independent risks that do not move together.

A third distinguishing characteristic of toxic risks is that they often involve small probabilities. The small magnitude of the risks is often well below a level that many would regard to be a de minimis hazard. For example, R. L. Stroup observes that the air toxics policy of the U.S. Environmental Protection Agency addresses risks that are comparable in magnitude to the chance of being hit by an airplane. There is little doubt that most people would agree that such risks do not merit our attention, much less the expenditure of billions of dollars. Unfortunately, the failure to provide for tradeoffs in the legislation governing risk management makes examples such as this the norm rather than the exception. Although the risks tend to be small, the number of people exposed to these hazards may be substantial. It is the presence of very large populations exposed to relatively small risks that gives rise to the mass toxic tort nature of these hazards.

Because toxic hazards are often low probability events, it becomes more difficult for individuals to learn about the probabilities involved. In situations of multiple causes it may be hard to distinguish several decades after the time of the exposure what the cause of a particular ailment was. The small probabilities coupled with a long gestation period and problems of multiple causation compound the difficulties of a risk assessor.

A consequence of the small magnitude of the probabilities and the difficulty of learning about the risk is that the risks will tend to be ambiguous. There may be a fairly broad range of scientific uncertainty regarding the magnitude of the risk. The appropriate guide should be the mean value of the risk, whereas government risk regulation agencies often focus on various upper bound values, such as the 95th percentile of the distribution. In practice, this bias is even more extreme as government risk analysts often use the 95th percentile of each particular parameter to calculate the overall risk, such as the exposure level to the chemical. The result is that there is a compounding of conservatism that leads to an overall degree of conservatism with respect to the risks posed by hazardous wastes that goes well beyond the 99th percentile of the overall risk distribution. Often, there is less than a one percent chance that the risks are as large as government estimates. This distortion in the probabilities shifts our priorities from the more substantial risks that are better known, to imprecisely understood but small risks that should merit less attention.

The government and the courts have both failed us in addressing toxic liability issues. In each case, liability imposed by these institutions is *ex* post. Because this liability is after the fact, at the time the risks are generated companies cannot anticipate the subsequent liability costs, eliminating any deterrent effects. It is also not feasible for companies to contract sensibly for such risks. Moreover, it is particularly striking that the most significant perpetrator of many of these mass torts—the U.S. Government—is often exempt from liability. The paper by Bruce Benson on "Toxic Torts by Government" delves into both the asbestos and Agent Orange experiences. In the case of asbestos, the largest exposure occurred during World War II to shipyard workers who installed asbestos insulation in Navy ships. However, this decision was not one that resulted from corporate behavior but instead was the result of specifications for ship construction mandated by the U.S. Navy. Similarly, the herbicide Agent Orange that was the target of massive litigation was a defoliant used in Vietnam. Although the U.S. Government exerted all control over the decision to apply this powerful herbicide, the chemical companies producing the herbicide, such as Dow Chemical Company, were the subsequent targets in the lawsuits.[5]

Toward a Sound Policy

Although there are substantial errors in the way that our social institutions have addressed toxic liability issues, it is possible to formulate a sensible policy in this area. Here I will distinguish four principles for a rational risk policy involving toxics. Many of these guidelines are interrelated, as they each reflect an efficiency orientation.

First, all policies should have social benefits in excess of social costs. This is the basic principle of benefit-cost analysis discussed by R. L. Stroup and is the social analog of the market test discussed by Bruce Yandle. The difference from a market situation is that rather than asking whether the individual purchasers of a product would find it attractive to do so we instead are asking the question of whether society is willing to pay an amount for a particular policy that exceeds its cost.

What would be the implications of such an efficiency-oriented policy? In the case of courts, the focus should be on whether the risk level is efficient. Thus, a negligence test would rule rather than strict liability in which all ailments are compensated irrespective of whether the risk level reflects a suitable balancing of costs and benefits. The infeasibility of strict liability in the case of mass toxic torts arises because of the large exposed population. If one wished, for example, to compensate all victims of asbestos exposures, then the magnitude of this compensation would bankrupt not only all asbestos producers but also all their insurance companies. A continuing difficulty in mass toxic torts situations is identifying causality. Thus, it is not feasible to distinguish which lung cancers of asbestos workers were due to asbestos exposures as opposed to other causes. In the case of asbestos, this inability to distinguish causality and the need to compensate all lung cancers of asbestos workers irrespective of cause boost the potential compensation costs by roughly an order of magnitude.

A negligence-type approach focuses on deterring inefficient behavior. In that regard, the test should be relative to the state of information at the time of the exposure rather than to standards of information and liability that may develop decades after the exposure period. *Ex* post liability does not play any constructive efficiency role in fostering safer decisions.

The government should likewise apply a benefit-cost test to its policies. At present no such tests are applied to environmental and occupational exposures to toxics. The government instead pursues more uncompromising objectives, which often leads to inordinate attention to insignificant risks and diversion of resources from the truly substantial hazards. In the case of the hazardous waste cleanup effort under the Superfund program, the cost per case of cancer prevented by this effort is well in excess of $10 billion per case. This amount is more than 1,000 times greater than the figure that the U.S. Department of Transportation is willing to spend per accident prevented through improved highway and airline safety. Enormous dividends could be reaped by shifting our attention from dimly understood toxics with very small risks to more fundamental hazards such as those posed by highway risks.[6]

A second principle for sound policy is that there should be honest risk assessment based on the mean level of the risk. Current policies focus on upper bound values that distort our risk priorities. In some cases, these distortions arise from extrapolating the results from animal tests to humans, as is noted by Aaron Wildavsky. There is also a tendency to rely on "junk science" in situations of scientific uncertainty, which is a danger discussed by David Bernstein, Kenneth Foster and Peter Huber. Even if we foster accurate risk assessments, individuals may misperceive these risks. Daniel Benjamin discusses the potential for risk overestimation in the case of very small risks, which is the level that most toxics pose. He also indicates that many of these patterns of overestimation may not stem from irrationality but rather from lack of accurate risk information. His essay highlights the importance of the government as a purveyor of risk information. But for such information to be useful it must promote sound risk judgments rather than alarm because of excessive attention to worst case scenarios.

A third principle for a sound risk policy is that the size of the exposed population matters. Surprisingly, the number of potential exposures often does not enter government risk assessments. Rather, the focus is on a hypothetically exposed individual. However, if no one is actually exposed to the risk then surely we should devote less attention to it than a risk for which there may be very substantial exposed populations.

In our analysis of 200 Superfund sites, James T. Hamilton and I found that the EPA did in fact adhere to its former policy of ignoring the size of exposed populations.[7] Indeed, the agency gives no preference whatsoever to hazardous waste sites for which there are large exposed populations as opposed to sites where there are no people exposed to the hazard. Shifting focus to the total benefits and total costs of policy efforts will help eliminate this inattention to the role of human exposures.

The recognition of exposed populations is also pertinent when courts attempt to judge a company's performance within the context of corporate liability. The company places products on the market but generally cannot distinguish which individuals will buy the product. Thus, the appropriate test for whether a product or product-related risk is inefficient should be from the standpoint of the entire market. From this market perspective, would the benefits of reducing the risk exceed the costs? The focus of individual court cases on particular exposed individuals rather than on the entire market narrows the jury's attention in a way that does not foster such a comprehensive perspective. In contrast, regulatory agencies could potentially address such issues from a broader perspective, but a successful regulatory approach requires that current regulatory mandates incorporate the kind of benefit-cost balancing and sound risk analysis that has been advocated here.

The fourth and final issue pertaining to a sound policy approach is that sensible risk management policy is better for human health than the current profligate approach. Squandering resources on risk reduction efforts that achieve very little is not simply wasteful. In doing so we divert resources from real risks that could be eliminated. Such efforts take funds out of our consumer market basket that could have been spent on better food, housing, medical care, and other commodities relating to health. The result is that ineffective regulations impose real health costs. Indeed, estimates suggest that every time we spend from $10 million to $50 million on risk regulation we sacrifice one statistical life that could have been saved through the improved standard of living that would have resulted from such expenditures on a normal mix of consumer goods (Viscusi 1998, ch. 5).

The most powerful force improving human health throughout this and earlier centuries has been the improvement in our standard of living. Many of these improvements can be traced to a variety of social innovations as well as to our increased affluence. As discussed in the paper by David Addock and Daniel Polsby, it is somewhat ironic that judicial and policy efforts designed to enhance human health may stymie the innovation and economic progress that is necessary to promote our well-being. Sensible judicial and regulatory policies will do more to promote human health than will seemingly uncompromising, but misdirected risk reduction efforts. The papers in this volume will help establish the framework for more sensible management of the risks associated with toxic liability.

— W. Kip Viscusi
Cogan Professor of Law and Economics
Director, Program on Empirical Legal Studies
Harvard Law School

Notes

1. In July 1998 the U.S. silicone breast implant manufacturer Dow Corning announced a $3.2 billion dollar compensation package to settle the estimate 177,000 joint cases against the company. Shortly thereafter an independent panel of experts commissioned by the British government concluded that "Silicone gel breast implants are not associated with any greater health risk than other surgical implants"; and "In particular, there is no evidence of an association with an abnormal immune response or typical or atypical connective tissue diseases or syndromes." Similar conclusions had been reached earlier by two British reviews, a Canadian and a French study, and a review by the American Medical Association. The most recent British study (Silicone Gel Breast Implants: The Report of the Independent Review Group. 1998. London: Department of Health) references previous reviews and the voluminous scientific literature. It can be found on the web at http://www.silicone-review.gov.uk. Angell (1997) is an accessible yet scientifically accurate account of the breast implant controversy.
2. I provide a broader discussion of many of these issues in Viscusi (1991).
3. See the excellent volume by Peter Schuck (1986).
4. See Viscusi (1998) for documentation of this and related risk issues.
5. The potential for serious risk should have been evident at the time of Agent Orange's use since this herbicide was powerful enough to be a defoliant in areas with dense vegetation.
6. For a review of such tradeoffs, see Viscusi (1998).
7. See Viscusi and Hamilton (1996).

References

Angell, Marcia. (1997). *Science on Trial: The Clash of Medical Evidence and the Law in the Breast Implant Case.* New York: W. W. Norton & Company.

Schuck, Peter. 1986. *Agent Orange on Trial.* Cambridge, MA: Harvard University Press.

Viscusi, W. Kip. 1991. *Reforming Products Liability.* Cambridge, MA: Harvard University Press.

———. 1998. *Rational Risk Policy.* Oxford: Oxford University Press.

Viscusi, W. Kip, and James T. Hamilton. 1996. "Cleaning Up Superfund." *The Public Interest* 124 (Summer): 52-60.

Introduction

The Toxic Liability Problem:
Why is It Too Large?

Richard L. Stroup and Roger E. Meiners

> *Anyone can raise a fire alarm, but it takes a lot of money to send the fire engines out.*
> — *New York Governor Hugh Carey,*
> New York Times *(May 19, 1980)*

I. Introduction

If, in recent months, unprecedented numbers of fire trucks had roared, sirens wailing, through the streets of a city, a question would naturally arise: Why are there so many more fires? Or if there are not, then why so many new false alarms? In either case, responding to the added alarms is costly. It is also dangerous. Fire trucks speeding through the city not only consume resources: the trucks themselves present a danger. Accidents happen as trucks answer alarms.

Faced with this situation, the city's fire authorities would try to learn what could be done to reduce the number of costly and dangerous responses to alarms. The goal would be to reduce both the number of fires and the number of false alarms. A similar approach should guide our thinking about risks from chemicals and chemical wastes that are toxic or possibly toxic.

1

Governor Carey was not speaking of fire trucks per se in the remark quoted above. He was referring by analogy to U.S. Environmental Protection Agency actions in regard to the Love Canal landfill in Niagara Falls, New York. Once the alarms had been sounded, it became extremely costly to cope with them, as Carey discovered as governor. We are still learning just how costly those responses were.

This introduction explains what we know about the size of the toxic liability problem and explores why these costs are so large. It also puts forth some criteria to be considered for reducing the toxic liability burden. Specifically, section II provides some historical background on the laws that create toxic liabilities. Section III provides estimates of the size of the direct costs associated with some components of the toxic liability problem. Section IV explains the importance of the indirect costs, which are comprised of distorted incentives, perverse penalties on productivity, and the chilling effect on investment caused by large penalties assigned with little regard to fault or to effectiveness of costly remediation. Section V offers an explanation of how and why the liability system has gone wrong. Section VI provides a set of criteria to help evaluate alternative institutions for determining legal liability.

II. The Historical Background

The problems at Love Canal were real. Chemical wastes had leaked in the late 1970s, producing "purple lawns [and] multicolored basement walls" in homes built on the landfill.[1] But an isolated problem quickly developed into a national emergency. A public health emergency was declared to give officials state funding to study the situation.[2] One unanticipated side effect of the strong "emergency" language was that Love Canal became national news.

Raising the alarm frightened and angered residents, who organized for political action. "Mustering the tools of dramatization and protest perfected by an earlier generation of peace and civil rights activists, the residents focused worldwide attention on what they perceived to be their plight,"[3] reported Marc Landy et al. in their policy history and analysis, *The Environmental Protection Agency: Asking the Wrong Questions*.

The Homeowners Association, as part of their strategy, held two EPA officials hostage for two days at their headquarters. They also hired a scientist who did a door-to-door survey and reported that residents claimed to suffer a large number of ailments which they attributed to chemical leaks from Love Canal. The EPA, seeking to justify its action, commissioned a quick study that reported chromosome damage among Love Canal residents. Neither study was peer reviewed, but both garnered press attention. Both studies were discredited by the scientific community and the EPA subsequently disavowed its own study. But the alarmist announcements had been page one news, while the later statements discrediting the studies were not. National press coverage was intense.

The EPA recommended that 700 families be moved, at a cost of $3 to $5 million. Governor Carey's remark about the fire trucks came as he criticized this recommendation. He estimated that while moving the families would cost $3-$5 million, providing them with new housing would cost $30 million to $60 million. He recognized that such a costly step should not be taken if this was a false alarm.

Love Canal reached its highest visibility during the 1980 presidential campaign. President Jimmy Carter recognized the political significance of Love Canal. The White House assumed control of negotiations on the issue the day after the Governor's statement. Carter ordered the families moved. Hundreds of families were uprooted without gaining any clear health benefits. And taxpayers were liable for bills running into the many millions of dollars.

Yet there was then, and there is today, no real evidence of a serious health problem, much less a health emergency.[4] "Then, as now, no reliable epidemiological studies showed that area residents were subject to greater health risks than the population at large,"[5] states Landy, citing articles in *Science* and the report from a blue ribbon commission appointed by Governor Carey.

Despite the lack of evidence on health effects, dramatic announcements about health emergencies continued to be made by government agencies seeking to further their agenda, including the EPA. A few months later, after more publicity and bureaucratic and political maneuvering, Congress passed the Comprehensive Environmental Response and Liability Act (CERCLA or, as commonly known, Superfund) to deal with cleaning up sites such as Love Canal. Politicians and others asserted that other "Love Canals" around the nation were "ticking time bombs."

The Superfund program meant that tens of billions of dollars in resources would be expended on cleanups of sites that supposedly presented health hazards. There would be no need for any reliable evidence that a health (or environmental) problem existed at any of them. Some risks, real or imagined, would be reduced by the cleanups. But other risks would be introduced by the investigation and remediation efforts themselves.

The Superfund process, and the stigma associated with Superfund sites, caused many families to worry about their health, and about whether they might have to move from their homes at a time when no one else would buy them. A Superfund "cleanup" meant disturbing buried chemicals, and bringing in heavy machinery to dig and transport enormous amounts of earth. Air pollution would be produced from all the fuel burned to do this, from burning polluted dirt in some cases, and to pump and treat groundwater. On balance, would the program increase or decrease health and environmental damage? No one really knows. Aaron Wildavsky's chapter in this book makes a strong case that because the EPA so badly skews its data, Superfund may well do more harm than good.[6]

The same process, an apparent crisis exploited to generate political benefits, has given us a hodge-podge of government programs that address risks from mishandled toxic chemicals. The hazardous waste provisions of the Resource Recovery and Conservation Act (RCRA), the Underground Storage Tank (UST) program, and the application of similar requirements at federal facilities dealing with hazardous wastes, were enacted in much the same environment that gave us Superfund. Environmental groups, in particular, raise real or alleged health alarms, and the press often amplifies the alarms. That political and bureaucratic environment also gave us federal drinking water standards, provisions of the Clean Air Act that deal with air toxics, restrictions on paint containing lead, and other laws that require high-cost measures to deal with tiny chemical exposures, often with little evidence that the exposures represent serious risks or risks that could not have been resolved in a much more cost-effective manner. As Foster, Bernstein and Huber discuss in their chapter in this volume, because of poor science, many regulations are based on phantom risk.

Most laws, and the regulations issued under them, are aimed at risks that, if shown to be large, would justify efforts at prevention and remediation. But in most cases, science gives little support to claims that the risks are large. Indeed, as we shall see, it was often the weak scientific evidence that led to demands for new statutes and regulations, since the courts refused to take action on the basis of weakly supported suspicions. Under the statutes, the mere suspicion of harm, if shown to be plausible under the most extreme assumptions, can be enough for agencies to force others to pay for very costly attempts to restore air, water, and soil to pristine conditions. In what follows, Superfund is used to illustrate the problems that, to a large degree, are shared by the EPA's other programs to control toxic chemicals.

The tens (and potentially hundreds) of billions of dollars spent on the Superfund program represent liabilities to individuals, firms (that is, shareholders, workers, and customers), and government agencies (that is, taxpayers) who are billed for the costs. To the extent that remediation projects rectify substantial damages, or reduce large risks, and to the extent that the remediation is accomplished in a cost-effective manner, it is money well spent. And to the extent that polluters are forced to pay in proportion to harms they perpetrated on others, such liabilities represent a fair placement of the burdens and a proper warning and incentive to all who consider mismanaging wastes. Such an optimistic view of Superfund is not supported by the evidence. Bruce Yandle's chapter in this volume documents how the change to political and bureaucratic control from common law control of hazardous sites has led to overkill on waste sites that the EPA thinks might, even hypothetically, pose a danger. The fact that we do not know whether there are benefits, while we do know that the costs are enormous, should make us question both Superfund and the policy process that put the program in place.

However great or small are the actual toxic dangers at the center of this rising tide of activity, some of the problems result from institutional failures. Actions by some users and disposers of toxic materials may not have been sufficiently constrained by rules that existed to help force accountability. The weaknesses of institutional constraints caused some dangerous situations to develop.

In a parallel way, where the dangers in fact are smaller than claimed, a different set of institutional failings are taking place. These failings are permitting exaggerated claims to cause large amounts of resources to be squandered in remediation and litigation. The liability system is not functioning properly as federal environmental laws have expanded liability to a host of parties. While much of the rest of this chapter focuses on the case of Superfund, the issue is much the same with respect to air toxics, as Stroup discusses in his chapter later in this volume.

III. The Size of the Liability Problem in the United States

The liabilities facing citizens from the remediation of dangers from toxic materials is conservatively estimated to be in the hundreds of billions of dollars. A comprehensive estimate of the size of the problem comes from a study by Russell et al.[7] The study estimated that under current policy, remedying existing hazardous waste problems over thirty years would require about $750 billion worth of real resources. The estimates of the cost ranged from $375 billion to $1.7 trillion, reflecting plausible changes in cost assumptions and small changes in program emphases.

These estimates are for remediation programs required under three federal programs (Superfund sites listed on the National Priority List [NPL], the RCRA Corrective Action program, and the UST program), cleanup of federal facilities, and state-required and voluntary private cleanups. The costs are estimated on an "as built" basis in 1990 dollars. They are neither discounted nor adjusted for inflation over the thirty years. They are incomplete since they deal with resource requirements for remediation activities only. They explicitly exclude litigation and other transaction costs, the costs of preventing future contamination, and the future costs of eliminating the wastes that are assumed to be contained rather than destroyed in the remediation process.

The transaction costs excluded in the Russell estimates also seem likely to range into the hundreds of billions of dollars. These are primarily litigation costs, representing resources expended in apportioning liability among potentially responsible parties (PRPs). Adding them to remediation costs increases the total cost estimates substantially. A study cited by Russell as complementary to their own work, in that it covers transactions costs, is by the Institute for Civil Justice.[8] That study found that for a representative sample of insured cleanups in the late 1980s, litigation and related transaction costs for remediation efforts averaged 88 percent of total expenses. That is, for every one dollar spent on remediation, more than seven additional dollars were spent on transaction costs. For expenses paid by large PRPs without insurer involvement, the transaction costs were smaller, but still significant, averaging 21 percent. In those cases, for every four dollars spent on remediation, another dollar was spent on apportioning liability. As early litigation results in additional precedents, uncertainty about the outcomes should decline, reducing future litigation costs. Costly litigation is seldom pursued when the outcome is nearly certain. But even a substantial reduction leaves large transaction costs to be borne, along with remediation expenditures.

Adding transaction costs and the excluded future costs of dealing with existing problem sites under current policy to the $750 billion "best guess" estimate of Russell, it is reasonable to think of the liability problem at existing hazardous waste sites as a trillion-dollar problem over the next thirty years. Russell points out that relatively minor changes in policy could make large differences in remediation costs. Moving from current policy to one that allows greater use of waste containment and requires less destruction of contaminants could, with no change in health risks, save $268 billion. However, moving away from containment and toward requiring more destruction of contaminants would cost $425 billion more than current policy, again with no change in expected health risks. The difference between cost savings and cost add-ons sums to $693 billion.

If we add in transactions costs, which vary directly with remediation costs, it is reasonable to estimate that the trillion-dollar liability could be raised or lowered by half simply by changing policy on what forms of remediation will be required. In other words, choosing between alternative policies, a choice that Russell thinks will not affect the health risks, represents the difference between spending $0.5 trillion and $1.5 trillion over the next thirty years. A difference of about one trillion dollars in real resources hangs in the balance *with no resulting change in human health risks or benefits expected.*

If risk-reduction methods are allowed to be freely chosen, even greater savings are possible. Russell reported on a study prepared by Colglazier et al.[9] based on EPA decision documents for 231 Superfund NPL sites. The estimated costs of eliminating environmental and human health risks from these sites over the next thirty years could range from $352 billion if "best available technology" were required, to $150 billion if the technologies selected for the sites under EPA requirements are followed, to as little as $15 billion if institutional controls (such as fences and deed restrictions to keep people from contacting the contamination) were allowed when feasible. The $15 billion option would require future generations to continue those controls past thirty years, which adds to the cost. The point of these studies is that the technologies and techniques chosen to reduce risks can make an enormous difference in the cost of reaching a given risk reduction goal. Will the responsible parties be allowed to choose the most cost-effective way to meet the risk reduction goals at each site? Political and bureaucratic decisions will provide the answer.[10]

Beyond Cost-Effectiveness: Which Sites Pose Important Risks?

A major reason for the enormous costs projections is that remediation projects are all treated as if they pose an extremely serious risk and justify sending all the fire engines. But the evidence indicates that some, perhaps most, of the remediation projects required under current policy are responses to "false alarms," or to greatly exaggerated claims of danger, just as the response to Love Canal was excessive.

Prior to the existence of EPA and extensive regulation of hazardous wastes, liability for harms done or threatened was enforced primarily through common law, as explained by Brown and Meiners and Donald Dewees in their chapters. Parties are liable for harm inflicted on others. Complainants have to meet a burden of proof before courts act on their behalf. Injunctive relief can be sought if harm is threatened. But there, too, a burden of proof must be met under the rules of evidence, so that each case might be judged in an unbiased manner. If the weight of the evidence supports the plaintiff, then the court can hold for them. Parties are not to have costs imposed upon them for unproven harms or risks.

Toxic chemical laws and regulations expanded liability to satisfy political and bureaucratic requirements. Superfund, as Bruce Yandle explains in another chapter of this book, sought to reduce risks that did not meet common law standards of nuisance or strict liability. The links between the harm (or potential harm) done and the risk reduction requirement have been largely dissolved in the political process. Congress required the EPA to protect human health and the environment. Agency personnel must use their judgment as to what levels of air, water, and soil pollution are acceptable. The EPA is not instructed to balance costs and benefits or to base decisions on substantive scientific evidence.

The evidence indicates that these conditions have led to vastly exaggerated risk estimates, and thus to false alarms. What we see in the case of Superfund, including the way in which Superfund sites are chosen for inclusion on the National Priority List (NPL), is a process that *appears* to be reasonable but is highly political and not subject to strong scientific standards. Sites are examined, and numerical values are assigned to various indicators of danger and potential for harm to health and the environment. Variables include the size of the site, the levels and types of toxic substances present, whether air, water or soil have been contaminated, and proximity to aquifers, population centers, and surface water. But many of the indicators are subjective, rather than objectively observable, so that political factors are important. This is no surprise. The EPA is political by definition.

Many analysts question the entire process. For example, Curtis Travis, director of the Center for Risk Management of Oak Ridge National Laboratory, writes that:

> First, there is little correlation between a site being on Superfund's National Priority List and the risk it poses to human health. Second, once a site is on the list, remediation usually is selected without turning to risk assessment as a priority-setting tool. Finally, most remedial alternatives are selected without adequate evidence of their effectiveness and permanence.[11]

Travis cites evidence that "Many sites on the list pose little or no risk to human health," and that "Conversely, many sites with significant health and environmental risk are not on the list."[12]

Similarly, in a report to the Congress on Superfund program management, the General Accounting Office (GAO) reported: "The federal government does not have an effective way to measure the relative risk of these sites across agency lines or to assign priorities to these cleanups, which could cost hundreds of billions of dollars."[13] Superfund, most observers agree, is a program that is using billions of dollars in real resources, with precious little to show in the way of health benefits, even as other, more risky problem sites persist. As the GAO reports:

> An effort as costly as our nation's hazardous waste cleanup problem should be justified on evidence that expenditures will result in commensurate benefits to human health and the environment. However, Superfund expenditures have not been based on an adequate comparison of the sites' risks with other environmental problems.[14]

The problems of combining good science, good analytic decision-making, and good politics have understandably proved to be difficult in the Superfund program. The same sorts of problems affect EPA environmental programs in general.[15]

Turning a problem over to regulatory authorities does not in itself produce the information required for rational decision-making. Since government authorities must respond to political considerations, an additional complexity is introduced. With hundreds of billions of dollars at stake, potential bankruptcies, and serious disruptions at the community level, the political pitfalls are large and many. To direct and administer a program to regulate the production, handling, disposal, and treatment of toxic substances in a way that is perceived to be effective, efficient, and just is a complex and politically difficult task. Agency goals regarding "turf" and budget add still more complexities.

Cleanups are accomplished only by diverting resources from other worthy missions, including the avoidance of other health risks. For example, putting the $6 billion per year being spent on Superfund toward cancer research would quadruple cancer research spending. But for those who administer Superfund and its cleanups, or the EPA's air toxics program, other health risks are not a part of the decision set under their control. Supreme Court Justice Stephen Breyer has pointed out that agency professionals naturally have "tunnel vision." They see the good they can do without looking at, or taking into account, that good things must be sacrificed to further their own mission.[16] Just as any bureaucracy blows its own horn, we should not be surprised that EPA rules to evaluate the risks it oversees are skewed toward overstatement.[17] More cleanup of potential risk, at the expense of projects not of concern to the EPA, can be "justified" by overstating risks.

Indeed, economists Albert Nichols and Richard Zeckhauser have pointed out that the EPA's Guidelines for Carcinogen Risk Assessment

> make no pretense of seeking a best estimate of actual cancer risks, but rather strive to find a 'plausible upper bound' for those risks....The cumulative effect of following the upper bound path, using a long series of conservative assumptions, can be monumental overestimates of health risks. The result is more stringent and costly regulation of at least some types of risk than if policymakers were more realistically informed.[18]

To see this at work, consider how the EPA calculates current and future risks. The distinction is not based on time periods, but on land-use conditions. For example, possible land uses, such as building houses on industrial land that is now a Superfund site, are considered to be future risks.

A study (funded by the EPA) of the risk assessments at seventy-eight Superfund sites conducted by James Hamilton and Kip Viscusi revealed that most risks the EPA calculates occur under future land-use conditions. Their study calculated a total of 1,430 pathways (ways that contaminants can potentially reach people to cause health damages) — 70 percent of the cancer pathways, 79 percent of the non-cancer pathways, and 72 percent of the total risk pathways — arise from future possible uses rather than current land uses.[19] Furthermore, 59 percent of all risk pathways occur to hypothetical future residents living on top of what is now a Superfund site. When cancer pathways are weighted for risk (that is, when both the frequency and severity of risk are considered), future scenarios account for 91 percent of a site's total cancer risks.[20]

Hamilton and Viscusi report:

> Easily the most important risk pathway driving the risk analysis is the set of scenarios in which there will be many more future on-site residents than there are at present, where these future on-site residents do not represent growth in current residential developments but rather departures from current population patterns.[21]

They conclude that "preventing future development of the site or use of the site for other purposes would...eliminate the most severe risks that arise,"[22] and that many of the risks that would likely remain after containment and land use restrictions, such as that to trespassers, are very low even without adopting policies to reduce those risks, such as fencing. The EPA, however, does not normally consider such proposals, preferring "permanent" solutions, however costly, dangerous, and disruptive to people they might be.

To an analyst seeking efficiency in government action, two basic criteria seem appropriate for deciding whether a situation that drew an alarm actually calls for remediation. First, the physical risks introduced by the proposed remediation effort itself, such as accidents in the transportation of people, equipment and materials, and in the manipulation and processing of materials, must be considered. If they offset, or more than offset, the safety gains produced by remediation, the proposed remediation project is not efficient. Second, if the remediation project does produce net reductions in dangers associated with the site, will the gains be offset, or more than offset, by harm caused by diversion of resources from other uses. In other words, are the cutbacks in other possible projects—cutbacks caused by the need to allocate resources to a site—losing more benefits than are provided by this project? If so, the project is not a good way to allocate the nation's resources. These simple questions are not asked by the EPA (or Congress) as they set the rules for programs such as Superfund. Each agency, and indeed those who head each program within an agency, display the "tunnel vision" described by Justice Breyer.[23]

As Haddock and Polsby explain in their chapter, private parties, faced with liability for a site, have an incentive to balance the cost of greater cleanup with the benefits of reduced future liability produced by the additional cleanup. That is, a private decision-maker has to pay both for the planned costs of remediation, but is also held accountable, by common law, for residual damages that may occur if the cleanup is insufficient, as well as the unintended negative effects of remediation. Such accountability is not always complete, since it requires that those potentially harmed have enough information about their harms to sue the injurer in court. The real world is imperfect but when liability rules are enforced, parties have incentives not to violate others' rights.

IV. The Calculated Costs Greatly Understate the Liability Problem

The cost estimates referenced above, as large and as misdirected as they may be, are only part of the overall toxic liability problem, which also includes the negative impacts of the current toxic liability policies on the economy generally and on the environment. The uncertainty that arises from a potential burden of regulatory requirements, especially when they can be as large as those imposed by Superfund and similar programs, generates significant economic harm to parties affected by the new rules.

Decision-makers whose ownership of capital is clouded by large potential liability claims may be deterred from investing in some projects. That is not always a bad thing. If liability claims are expected to be valid only when those facing them exercise bad judgment, deterrence is optimal. But the current regulatory liability scheme often does not hobble the makers of faulty decisions; instead, it extracts money from investors with "deep pockets" (that is, having wealth developed by previously successful investment decisions.) In these cases, the wrong decision-makers are being handicapped. When acting responsibly fails to protect wealth, projects where accountability is most important to society will be shunned. Similarly, projects that create no harm but that might expose investors to liabilities for harms caused by others, will be inappropriately avoided. An example is the now-familiar "brownfields-greenfields" problem in many cities.

Investors are wary of building new projects on "brownfields," land used previously for industrial purposes that is available for redevelopment. If contamination is found on the brownfield, a new owner can be held liable for a frighteningly expensive cleanup, even if the investor has polluted nothing and may have acted only to decrease potential harm. Under Superfund rules, no harm by the new owner need be shown, nor any strong evidence of future harms at all, to assess liability on a new developer.[24]

Hence, investors tend to seek "greenfields" where no industry has previously been and no contamination is likely to be found. Distortions of local economic development patterns result. Inner cities have less hope for renewal since they are brownfields. Furthermore, the policy keeps developers of new projects from the kind of containment or partial cleanup that would make contaminated land suitable and safe for new industrial use.

The taxes used to finance Superfund produce similarly perverse results. They are based on the output of firms, primarily chemical and petroleum producers. There is no relationship between the amount of tax a firm pays and the environmental harm it generates. Taxes levied by the Superfund law fall on the production of economic goods, not on the production of economic or environmental "bads." Good works are penalized, while harmful outputs are not. No firm can reduce its Superfund tax burden by reducing its waste emissions; rather, to reduce its tax burden, it must reduce its production of valuable products.

When the net is cast wide, and many who did no wrong are caught in it and subjected to large costs via joint-and-several, retroactive, and strict liability, as in the case of Superfund, extensive litigation is guaranteed. Further, since so many of the rules are politically determined, they are also subject to possible change each legislative session. When the liability climate is uncertain and subject to change with the political winds, the business climate is chilled. Property, including land with toxic substances that may bring on remediation liability *and* all property of firms facing such liabilities, is placed at risk. Property owners and their creditors cannot be sure that all property will not be confiscated to settle liability. Willingness to invest in such properties declines. Entrepreneurs become absorbed in the battle to protect what they own, rather than being able to focus on the creation of new wealth. Investment declines in favor of defensive maneuvering.

Economic Freedom Requires Effective Liability Laws

A key purpose of liability law, whether developed in statutes or by common law in precedents, has been to make parties accountable for harms resulting from their actions. Doing so serves justice and efficiency. Various aspects of these issues are discussed in the chapters by Dewees, Brown and Meiners, and by Haddock and Polsby.

Justice demands that individuals be accountable for results of their decisions, and that compensation be paid, when possible, by wrongdoers to those harmed. Accountability also has an efficiency role: those who are accountable have freedom of action. Those who might have to pay for mistakes naturally want veto power over any action. But when only those who invested and had a voice in decisions are held accountable for the decision, then those who decide can be given the freedom they need to innovate and to act. The entrepreneur who discovers a better way to fill a consumer's want will be more likely to be given the freedom to try the idea, if the investment risks only investors' capital. Investors will invest in attractive projects, if net gains from a successful project can be pocketed by those who backed it.

Freedom fosters economic growth. Development under common law, in England and the U.S. during the last half of the nineteenth century, and the first half of the twentieth, are examples of what can happen in such an environment. Proper liability rules see that wrongdoers compensate their victims, and that those who do well may reap rewards and not be subject to political and bureaucratic requirements based on unsubstantiated and dubious claims.

No liability system is perfect. When accountability is not achieved by liability, the incentive to use resources efficiently and to avoid harm to others is weakened. This is especially true when the law weakens the relationship between a faulty decision and payment of resulting costs by the decision-maker. The result is a poorer economy. Similarly, when non-polluters are made to pay for cleanups due to false alarms or for cleanups for which they cannot reasonably be found liable for having caused, resources are wasted and the economy is poorer.

Wasting resources so that economic growth is harmed has important ramifications for the goal of protecting society against toxic risks. A prosperous economy has at least two favorable impacts on health risks, including toxic risks. The first is that richer nations tend to be both healthier and cleaner. The second is that the willingness and ability to pay for environmental quality rises with income.

It is often asserted that we must choose between economic growth and environmental quality. Yet evidence from around the world suggests that in most respects, growth and quality are complements, not substitutes. Less-developed nations suffer from many ills, ranging from waterborne diseases to deforestation. With growth and technological advances, clean-burning fuels replace wood and animal dung; "slash and burn" agriculture is replaced by careful cultivation using far less land per unit of output; and sewage treatment reduces waterborne disease. Once people have enough income so that they are not struggling to put food on the table, they become more likely able to take actions to reduce or avoid environmental damage.

The relationship between economic growth and environmental protection is especially strong among industrialized nations. The public concern that became visible on Earth Day 1970 was in part a reflection of the increasing affluence of U.S. citizens. Members of environmental groups tend to be high-income individuals. Consider the Sierra Club whose magazine, *Sierra,* goes to every club member (though subscriptions can be purchased separately). The 1992 *Sierra* reader survey revealed an average reader household income of $79,400, more than double the U.S. average. People with higher incomes are more willing to devote more income, directly and through the political process, to environmental quality.

Economist Donald Coursey finds that in industrial nations, citizen support for measures to improve environmental quality is statistically sensitive to income changes.[25] Willingness to pay for environmental measures, such as pollution regulations, is highly elastic with respect to income. He estimates that the income elasticity of demand for environmental quality is about 2.5. That means that a 10 percent increase in income leads to a 25 percent increase in citizens' willingness to pay for environmental measures. Similarly, a 10 percent decline in a community's income leads to a 25 percent decline in that community's support for costly environmental measures. Coursey points out that the demand for BMW, Mercedes-Benz, and other luxury automobiles has the same sensitivity to income changes. The environment, in that sense, is a BMW. In a similar vein, other studies confirm that policies which reduce output, or the growth of output over time, reduce the health of citizens.[26]

Economic growth and investment make the adoption of new technologies easier and less costly than they would be in a stagnant economy. The same property right protections (including effective liability mechanisms), freedoms, and market processes that encourage growth also encourage technological advances. In the United States, advancing technology was itself cleaning the environment well before major environmental laws were passed. There is evidence that air quality was improving faster before the Clean Air Act was passed than afterwards,[27] at least in part because citizens were suing in court for relief from pollution, local governments were taking action, and because in a market system, profits rise when materials waste is reduced.

V. Why the Liability System Has Gone Wrong

What has changed to bring about such costly toxic liabilities? In simple terms, the public demanded changes. This happened because public perceptions about toxic risks changed, and because the public's perception of who pays, when extra caution is exercised, is incorrect.

Public perception of the dangers from toxic chemicals has shifted through a process that is generally traced to Rachel Carson's 1962 book, *Silent Spring*. This book used vivid and moving writing to depict nature (and humans) suffering deadly assaults from man-made chemicals, especially pesticides such as DDT. Environmental groups, such as the Environmental Defense Fund and the Sierra Club, seized on this well-written book to drum into the public's consciousness the claim that human beings are killing natural creatures and natural systems by chemical means, and that humans are very much at risk. At fault, in particular, were the greedy corporations which try to manipulate nature in search of profits.

That Carson's book was so influential is indicative of the problem we are discussing. Carson was wrong about some asserted "facts" and was biased in reporting other things. DDT was one of the most valuable chemicals in human history. It saved millions of lives and saved millions more from malaria and other diseases, yet Carson's book was key in stimulating the removal of DDT from the market. Malaria currently kills between 1 and 3 million people, mostly in third world countries.[28] Carson implied that the misuse of DDT was the fault of chemical makers when in fact most of its misuse can be attributed to the Department of Agriculture which, under orders from Congress, sprayed DDT and other chemicals willy-nilly, even over the objections of farmers who did not appreciate their property being bombarded.[29] Like the Superfund program today, which forces billions to be spent with little scientific justification, in the 1950s Congress authorized billions to be spent spraying chemicals but allocated almost nothing to study the effects of the chemicals.

As the chapter by Bruce Benson on "Toxic Tort by Government" discusses, one of the greatest ironies of all is that there is a natural inclination to look to government to solve problems that may have been caused by government. Unlike private parties who are liable for their mistakes, the government is not liable for its mistakes unless it volunteers liability.

While popular sentiment for a clean, safe environment is sensible, in recent decades voter support for dramatic environmental measures is often based on overestimates of the degree to which problems exist or are being addressed effectively. On any issue, misperceptions gradually give way to better, more complete scientific understanding. However, often the policy damage has been done. Policies chosen in response to narrowly focused[30] interest groups and rationally ignorant voters can be very expensive and counter-productive.[31] As a result of the electorate's susceptibility to thinly supported claims of environmental crisis, political results do not promote economic efficiency, equity, social cooperation, personal freedom, or even environmental quality in some cases.

Rational ignorance is not a stable situation. As Dan Benjamin discusses in his chapter in this book, people are rational about risk when they have incentives to become informed about risks. But the personal benefit to the voter for gaining added knowledge on most issues is so small that political battles can be won and lost, and policies put into place, well before enough voter ignorance is displaced by objective analysis.[32]

There is another dimension here to rational voter ignorance. Voter attitudes toward proposed political action depend not only on the benefits they perceive, but also the costs. The rationally ignorant voter is likely to be ignorant about who pays for costly regulations. Advocates of dramatic action nearly always say that "industry" should be regulated, and thus should be required to foot the bill. "Chemicals" should be restricted (of course, the regulation is imposed upon individuals, not upon chemicals). Part of the public's willingness to support very costly regulation is related to a misconception about who pays, and, indeed, even about who owns corporations. The fact that pension funds and small investors hold a large portion of corporate stock (perhaps approaching half) is not well understood by most voters.

In a constantly changing world, with a political process ultimately under the control of rationally ignorant voters, influenced by narrowly focused interest groups, there is good reason to expect that investments in the creation of new crisis perceptions will be profitable for various groups. For others who may know better, it is easy to cooperate, while it would be costly for them to dig more deeply to discredit the crisis claims.

Crisis, Incentives, and Environmental Politics

A "crisis" catches the public's eye, and politicians compete to provide "remedies," especially when the remedies are spending programs and regulations that benefit their political clients.[33] But why has the media gone along with the environmental activists' claims? And what has been the role of scientists, science laboratories, and government bureaus? Why have they not provided the public with a more balanced picture of the activists' claims?

Activist organizations benefit from "crisis." It helps keep them together,[34] and keeps them funded. Almost every fund-raising letter from environmental organizations features an environmental crisis described in vivid detail. The message is that if they do not receive donations to help them fight the crisis, then the establishment—what European Greens call "the Big Machine"—will do irreparable harm. With "crises" verified daily in the media, this is an effective technique. In the U.S., the ten largest environmental activist organizations raised more than $500 million in 1990.[35]

When a crisis is claimed, and drastic regulatory measures are called for, a producer group often stands to lose by bearing higher costs, at least temporarily.[36] Thus we can usually count on industry to fight new regulations. One political problem is that business has little credibility in such matters. Everyone recognizes that in seeking profit, businesses are likely to fight costly regulation that may be beneficial for society as a whole. Since company jobs are linked to profitability, both a narrow mission focus and personal financial rewards provide businesses with incentives to present information in a way calculated to minimize regulatory cost burdens. Lack of credibility with the electorate limits the effectiveness of even honest educational efforts by corporations—for perfectly logical reasons.

The media and most voters seem not to recognize that environmental groups are not objective and are not liable for making false assertions. These groups are nonprofit, but their financial health depends on donations and government grants, both of which are enhanced if they can convince the public that crises loom. Their jobs are enhanced when their organizations grow. It should be no surprise that the head of the National Wildlife Federation earned about a quarter-million dollar salary in 1990. The NWF budget was about $90 million that year. While business is logically tagged with a credibility problem, environmental groups are not perceived to have the same conflict of interest in making claims about the environment. For the media, any crisis, especially one that strikes close to home, is a good story. The publisher seeking readers and the reporter competing for space recognize that what is startling catches readers' attention. Information that reduces fears of a crisis do not make for an exciting story, unless the information supports a scandal on the part of those claiming crisis, or unless one scientist is flatly contradicting the others. The same seems to be true in other realms, too. Financial crisis and doomsday stories sell in the financial press.

The incentives facing the media to utilize crisis as a story line seem generalizable. Less clear is the role of media preferences and ideological predilections. There is evidence, however, that environmental crisis stories also appeal to journalists because journalists support a so-called pro-environmental agenda. Stanley Rothman and Robert Lichter studied how the media interpret science in the case of the danger from nuclear power. They found that journalists paid more attention to scientists who opposed nuclear power than those who supported it. (The risk posed by nuclear power is virtually a litmus test for environmental activists.)[37]

Important members of the media clearly signed on in support of the environmental activist view. *Time* magazine, for example, featured, in place of its normal "Man of the Year" issue a "Planet of the Year" issue (January 2, 1989) picturing the earth as under attack by humans. No attempt at balanced reporting was evident. Indeed, *Wall Street Journal* writer David Brooks reported[38] that the editor in charge of that issue, Charles Alexander, stated at an environmental conference: "As the science editor at *Time* I would freely admit that on this issue [environment] we have crossed the boundary from news reporting to advocacy." Brooks proceeds: "After a round of applause from the gathered journalists and scientists, NBC correspondent Andrea Mitchell told the audience that 'clearly the networks have made that decision now, where you'd have to call it advocacy.'"

Since media portrayal of science is often skewed, and since this is helping to drive environmental and risk management policy in directions unsupported by clear scientific evidence, why do scientists themselves not speak up to rectify the skewed perceptions of risk portrayed by the media and held by the public? One reason that the views of ideologically leftist scientists appear to prevail was provided by the Rothman and Lichter study. They found that anti-nuclear sentiment among scientists was positively correlated with a willingness to publish articles without peer review; pro-nuclear scientists were more reluctant to go against the traditional scientific practice; thus scientists holding anti-nuclear sentiments were quoted and cited by the press far more often than their numbers would suggest.

Elizabeth Whelan contends[39] that environmental science depends for its funding on public concern about environmental problems—unlike other sciences such as physics. Many groups, private and public, are thriving precisely because private foundations, the public, and some governmental bureaus are motivated by fear of environmental catastrophe. Even the honest environmental scientist is reluctant to speak loudly against the importance, much less the existence, of an environmental problem. Most scientists report their study results honestly: their interpretation of those findings is likely to concede the possibility of a serious problem, even if the results themselves show no reason to expect one. In any event, the press does not pick up on the large number of studies published that fail to imply a crisis.

Over time we should expect the production of accurate scientific evidence. Providing a strongly supported alternative theory opposing the conventional wisdom is a time-honored way to earn an academic reputation.[40] The scientific process in free countries not only provides peer review, to keep data presentation honest, but it also rewards those who emphasize views (models with good empirical support) that oppose the previously established mainstream thinking. This process, however, does not usually operate in political "real time." Acceptance, challenge, and replacement of the conventional wisdom is slow by media standards. By the time the challenge succeeds, policies based on the false alarm are in place, complete with a constituency, including a bureaucracy to work tirelessly to defend it.[41]

Government agencies that handle risk have two good reasons to lean toward interpreting the issues they deal with in crisis terms. The most obvious reason is that a crisis loosens the purse strings and paves the way for increased power or "turf." The EPA has massive funding only because of fears of the sort propounded by environmental activists. Robert Higgs studied the relationship of crisis to the growth of government more generally in *Crisis and Leviathan*.[42] With a larger budget, an agency can better serve whatever its managers see as its mission. It can also better serve its political supporters—those who can lobby and at budget time provide political support to obtain funds and preserve or extend its turf.

Agencies such as the EPA provide grant monies to environmental groups, which help beat the drum in support of further EPA funding. The more scare stories the groups provide to the media, the larger the demand for politicians and the EPA to protect us from these hazards that are difficult to evaluate. Support for any agency is not restricted to the non-business sector. Some firms owe much of their business to certain regulations. Asbestos removal from buildings, another "crisis" unsupported by evidence, provided billions of dollars worth of business to asbestos removal and construction firms that rushed in to fill this new market niche.

Crisis also helps an agency take a tough stand against possibly dangerous substances. Agency personnel, if they sensibly view their narrowly focused jobs as important, will be biased toward tough stands. The cost to them of failing to constrain an activity which turns out to be harmful (call this a "Type I" error) can be severe—opprobrium and professional disgrace. But the cost (to them) of over-regulating or banning a substance that turns out to be harmless (a "Type II error") may be virtually nil. After all, they were simply being cautious and we often cannot know of potential benefits that were never allowed to come into existence.

Since Type II errors impose only small costs on regulatory decision-makers, agency officials can be expected to accept them in order to minimize the risk of Type I errors. Regulatory delay resulting from Type II errors results in lost output, and even lost lives as when people die because new medicines or pesticides are kept off the market for "safety" reasons; but these are seldom blamed on the cautious bureaucrat. Tragically, this kind of pressure is constantly on regulators, and their "conservative policies" are applauded by narrowly focused environmental activists.[43]

VI. Democracy, Witches, and Environmental Crisis

Errors occur in any human system. Yet political control of environmental risks seems biased toward overreaction to perceived crises.[44] It is difficult to think of examples of similar magnitude on the other side, where massive reaction was called for but rejected or long delayed to the great regret of the general public later. Political control has further weaknesses: political battle among rivals for resource allocation tends to breed rancor as each faction tries to delegitimize the other's goals and demands; unlike markets or the common law, wherein the power of the state is brought to bear against the freedom of an individual only after a burden of proving damages to another are met (in a court with rules of evidence), the procedural protections for individuals and organizations against political attacks and regulatory harassment are not extensive in the U.S. Congress.

The latter point is an important one for environmental policy, and is dramatized by Harvard ecologist William Clark on the political handling of risk.[45] Clark compares environmental policy in the current era to the Inquisitions four centuries ago in Europe, when societal problems often were blamed on witches. A person accused of being a witch was brought before the authorities (church and state officials) and tried. Since there was no burden of proof, and no rules of evidence (torture was common), accusation was typically tantamount to conviction. Innocence could of course not be proven, just as today, a chemical cannot be conclusively proven not ever to be harmful. There was no stopping rule which would exonerate the accused, but there was pressure on the authorities to "do something" about crop failures, mysterious illnesses and so on. They typically did. As Clark points out, hundreds of thousands (the exact number is disputed by historians) of persons were burned at the stake. The public may have obtained some satisfaction knowing that "guilty" parties were punished and that protection had been obtained against presumed risks. But crop failures are not solved by burning witches.

In the final analysis, the Inquisition created very great costs in pursuit of policies with few objective benefits. Much of the evidence presented in this book suggests that the toxic liability system is having the same effect.

Notes

1. Byong Hyoung Lee, "Shifting Gears in a Dynamic Environment: Changing Strategies in the Formulation of Superfund," Ph.D. diss., State University of New York, 1991, p. 50.
2. This fact and much of what follows in this and the next paragraph are based on materials in chapter 5, "Passing Superfund" in Marc K. Landy, Marc J. Roberts, and Stephen R. Thomas, *The Environmental Protection Agency: Asking the Wrong Questions* (New York: Oxford University Press, 1990), p. 135. This book provides interesting detail and insightful analysis of EPA and its programs.
3. Landy, et al.
4. For the history of Love Canal and the evidence of harm faced by residents, see the lengthy court decision (denying punitive damages due to lack of evidence of reckless disregard for safety by Hooker) in *U.S. v. Hooker Chemicals*, 850 F. Supp. 993 (W.D. NY, 1994).
5. Landy, et al., p. 133, and citations therein.
6. See Aaron Wildavsky in this volume.
7. See Milton Russell, E. William Colglazier, and Mary R. English, *Hazardous Waste Remediation: the Task Ahead* (Knoxville, TN: Waste Management Research and Education Institute, University of Tennessee, December 1991).
8. Jan Paul Acton and Lloyd S. Dixon, *Superfund and Transactions Costs: The Experiences of Insurers and Very Large Industrial Firms* (Santa Monica, CA: RAND, 1992), p. xi.
9. E. W. Colglazier, T. Cox, and K. Davis, *Estimating Resource Requirements for NPL Sites* (Knoxville: University of Tennessee, 1991), p. 65, cited in Milton Russell, "Cost Implications of Alternative Superfund Configurations," paper presented January 7, 1995, at the annual meeting of the American Economic Association in Washington, DC.
10. As of 1997 the Clinton administration was floating proposals about lowering clean-up standards for industrial "brownfields." The crushing economic burden of the trillion dollar liability is understood, but it remains to be seen if some of the less costly clean-up options will be allowed.
11. Travis, Curtis C., "Waste Remediation: Can We Afford It?," *Forum for Applied Research and Public Policy* (Spring 1993), p. 57.
12. Ibid., p. 58.
13. GAO, "Superfund Program Management," (Washington, DC: GAO, December 1992; GAO/HR-93-10), p. 19.
14. Ibid., p. 7.
15. See Landy and Michael Greve and Fred L. Smith, eds. (1992), *Environmental Politics: Public Costs, Private Rewards* {New York: Praeger)for a series of revealing episodes about how and why flawed decisions emerge from political and bureaucratic processes. For some quantification and further explanations of why resources are wasted in government regula-

tion of the environment, see Robert Crandall, *Why Is the Cost of Environmental Regulation So High?* (St. Louis, MO: Center for the Study of American Business, 1992). See also, Roger Meiners and Bruce Yandle (eds.), *Taking the Environment Seriously* (Lanham, MD: Rowman & Littlefield, 1992).

16. See Stephen Breyer, *Breaking the Vicious Circle: Toward Effective Risk Regulation* (Cambridge, MA: Harvard University Press, 1993), pp. 11-19.

17. For an excellent and readable summary of the issues involved in risk assessment, and the conservative nature of EPA's approach, see "Choices in Risk Assessment" (Washington, DC: Regulatory Impact Analysis Project, Inc., 1994). The non-profit study was funded by the U.S. Department of Energy.

18. Albert Nichols and Richard Zeckhauser, "The Perils of Prudence," *Regulation*, 10, 2 (November/December 1986), p. 13.

19. James Hamilton and W. Kip Viscusi, "Human Health Risk Assessments for Superfund," Paper prepared for NYU School of Law Conference on Superfund Reauthorization, December 3-4, 1993, p. 18, table I. (Available from the Center for Risk Analysis, Harvard School of Public Health.)

20. Hamilton and Viscusi, Tables 2, 3, 10.

21. Ibid., p. 21.

22. Ibid., p. 26.

23. See n. 14.

24. See David R. Allardice, Richard H. Matoon, and William A. Testa, "Brownfield Redevelopment and Urban Economies," *Chicago Fed Letter*, Federal Reserve Bank of Chicago, 93 (May 1995), pp. 1-3. The EPA recognizes the terrible loss generated by this policy and has floated ideas for modified standards for brownfields.

25. Coursey discussed this topic in "The Demand for Environmental Quality," a paper presented January 1993 at the annual meeting of the American Economic Association in Anaheim, CA.

26. See Wayne B. Gray, "The Cost of Regulation: OSHA, EPA, and the Productivity Slowdown," *American Economic Review*, 77, 5 (December 1987), pp. 998-1006; Aaron Wildavsky, *Searching for Safety* (New Brunswick, NJ: Transactions Press, 1988); Ann Gibbons, "Does War on Cancer Equal War on Poverty?" *Science*, 253 (1991), p. 260.

27. See Robert W. Crandall, *Controlling Industrial Air Pollution: The Economics and Politics of Clean Air* (Washington, DC: Brookings Institution, 1983), p. 19.

28. For an in-depth review of the political history of DDT, see Christopher J. Bosso, *Pesticides and Politics* (Pittsburgh, PA: University of Pittsburgh Press, 1987).

29. The chemical misuse that Carson reported was not news to those affected by it; some southern states objected to heavy chemical spraying by USDA before *Silent Spring* "exposed" the problem and most of the destructive practices had been halted. See Bosso, pp. 100 ff.

30. We must distinguish between narrow interests and selfishness. Even if someone is truly selfless (altruistic) in concern for the environment (or anything else), one can have a narrow view that favors diverting resources from other highly worthy human goals, to an extent which few others may approve. A wilderness advocate, while perhaps selfless, might think that spending more to save the lives in Zaire is less worthy than saving more wilderness, and work zealously to achieve that goal. While selfless, this is a narrow interest. Heads of businesses or government bureaus may be equally narrowly focused and passionate about their missions. It seems likely that only when faced with the sort of tradeoffs and incentives inherent in a system of property ownership and voluntary exchange, where such narrow zeal is harnessed to benefit the strongly held desires of all (expressed in offers to trade), will persons with narrow interests be prevailed upon to consider competing desires of others in society, rather than trying publicly to denigrate and delegitimize them.

31. For a more complete explanation of rational voter ignorance, see James D. Gwartney and Richard L. Stroup, *Economics, Private and Public Choice*, 7th ed. (New York:The Dryden Press, 1995), chapter 4.

32. But "educating" the public is not easy because information is costly and voters have little reason to become educated about issues. Vice President Gore has made a successful career, in part, by calling for ever more government "solutions" to alleged environmental problems and by bashing opponents who suggest any modification in current environmental policies that may be economically wasteful and environmentally unsound.

33. For accounts of how environmental legislation is used to confer political or economic advantage on favored firms, industries, or regions at the expense of the public interest, see Bruce Ackerman and William T. Hassler, *Clean Coal/Dirty Air* (New Haven, CT: Yale University Press, 1981; and Robert W. Crandall, "Economic Rents as a Barrier to Deregulation," *Cato Journal*, 6, 1 (Spring/Summer 1986), pp. 186-189.

34. See Mary Douglas and Aaron Wildavsky, *Risk and Culture* (Berkeley: University of California Press, 1983), especially chapters 2-6, for the combined explanations of an anthropologist and a political scientist on why it is necessary for groups such as environmentalists, internally, to perceive great risks—preferably crisis—imposed on them by the establishment.

35. See Wolcott Henry, "Letter from the Publisher," *Conservation Digest*, 2, 4 (September 1990), p. 2.

36. Over time, capital leaves less profitable industries until reduced output and rising prices are sufficient to provide the same rate of return that is available from other investments. But in the short run, capital will share at least part of the burden of new regulation.

37. See Stanley Rothman and S. Robert Lichter, "Elite Ideology and Risk Perception in Nuclear Energy Policy," *American Political Science Review*, 8, 2 (June 1987), pp. 387-404.

38. Brooks, David, "Journalists and Others for Saving the Planet," *Wall Street Journal* editorial, October 5, 1989.

39. Elizabeth Whelan, *Toxic Terror* (Ottawa, IL: Jameson Books, 1985), pp. 292-93.

40. Some popular beliefs, such as visits of UFOs, cannot be "disproved" and are largely ignored by scientists because they are non-issues, just as biologists do not care to have to spend time responding to fundamentalists who deny evolution.

41. Acid rain was asserted to be a major environmental crisis in the 1970s. Over a decade, scientists spent about $600 million studying the issue and decided it was not much of an issue. Congress, not wanting discussion of the fact that there was little evidence to support expensive policies long in place, literally suppressed publication of the final report. Acid rain is still commonly referred to as if it were an environmental issue.

42. See Robert Higgs, *Crisis and Leviathan* (New York: Oxford University Press, 1987).

43. An empirical study of an analogous situation in drug regulation is Sam Peltzman, "An Evaluation of Consumer Protection: The 1982 Drug Amendments," *Journal of Political Economy*, 81 (1973), pp. 1049-91. On the analysis of the FDA more generally, see Robert Higgs (ed)., *Hazardous to our Health? FDA Regulation of Health Care Products* (Oakland, CA: Independent Institute, 1995).

44. While the key features of the situation in the U.S. are not difficult to understand, it is less clear why the American public, and therefore politicians, are more concerned about many risks than is the case in, say, European countries, where American "obsession" about chemicals seems to strike Europeans as odd. Europeans have plenty of their own economically irrational policies, so it is not because they have superior understanding of how the world works. Why different issues arise in different countries that have much the same economy (and environment) is simply not well understood.

45. William C. Clark, "Witches, Floods, and Wonder Drugs: Historical Perspectives on Risk Management," in Richard C. Schwing and Walter Albers, Jr., eds., *Societal Risk Assessment: How Safe Is Safe Enough?* (New York: Plenum Press, 1980), pp. 287-313.

1

Superfund and Risky Risk Reduction

Bruce Yandle

1. Introduction

Hazardous Waste and Changing Regimes

In 1980, Congress decided to abandon the centuries-old procedures of common law and community action for dealing with the damages and risks caused by hazardous waste sites. With the passage of the 1980 Comprehensive Environmental Response, Compensation, and Liability Act, better known as Superfund, common law rules and state and local statutes were largely displaced by federal statutes and regulation. Actions to be taken on behalf of concerned citizens to minimize the likelihood of real harm from risky waste sites were assigned to the U.S. Environmental Protection Agency (EPA). Love Canal and events like it had raised the specter of imminent catastrophe. That event alone seemed reason enough for the change.[1]

Perhaps some thought the regime shift would be temporary, that once catastrophes like Love Canal were avoided or managed, common-law rules and control by state and local governments would return. After all, there was theoretically a fixed number of orphan sites to be fixed.

After that, other federal programs and community common sense would prevent future Superfund sites. The fact that more than 3,500 emergency cleanup actions would be taken under Superfund buttresses this viewpoint. [2] But sober reflection on past interventions would hardly support such conclusions. Once community procedures are pushed aside by higher legislative bodies, local control somehow seems gone forever. [3] Special interest rent-seeking at higher levels of government pours figurative concrete around newly discovered federal authority. And what to local communities has all the appearance of free money pours from federal coffers to fix local trouble spots. Those assured that Superfund cleanups were removing serious risks had little reason to tally the costs of the federal program or to question the extent to which real risks existed in the first place. Fewer still had reason to care about the costs.

Long before there was anything called the U.S. Environmental Protection Agency, the common law of torts focused on public and private nuisance and rules regarding trespass.[4] Under common law, individual landowners, tenants, and affected citizens can seek remedies for damages they suffer from hazardous waste seepage and related losses. But to receive a favorable settlement, the plaintiff must first show damages, or an imminent threat of harm, and then show evidence that the party charged was actually the cause of damages. Traditional common law remedies for environmental harms involve damages paid to the victim of the harm, not the state, and an injunction. That approach was largely replaced by the Superfund.[5]

In contrast, violation of the Superfund statute itself is a civil offense irrespective of a demonstration of damages or clear evidence that connects an offending party to damaged property. As will be explained more fully later, hundreds of otherwise innocent parties can find themselves liable for cleaning a site, even though they behaved legally when the site was formed. From the standpoint of a community that seeks to clean an abandoned waste site, Superfund offers far brighter prospects than common law action. Instead of having to rely on the empty hope of prevailing in a court battle against a bankrupt waste site operator, innocent victims of harm and concerned bystanders can call on the EPA and harness Superfund's power to recover resources from practically anyone who participated in the formation of the waste site.

Sweeping aside the notions of damages that link victims of harm with those who caused it, the Superfund statute simply states that the presence of a risky and abandoned hazardous site, once certified as a

Superfund site, is a per se violation of law. If the risk, somehow calculated, is high enough, the site will be cleaned up. Those responsible for having created the site, irrespective of their guilt or innocence under common law, can be sued by the federal government for the full cost of remediating the site. Risk is to be reduced by regulation.

Where common law may seem cumbersome, arbitrary, and somewhat unpredictable, Superfund may be viewed optimistically as rational and effective, at least in theory. Unfortunately, the Superfund cure for common law weaknesses may be worse than the disease.[6] While consuming huge amounts of plaintiff and defendant resources, Superfund has simply failed to bring much in the way of risk reduction. After some nineteen years, fewer than 500 sites have been mitigated. Indeed, the lessons of Superfund support the well-established theory that regulatory risk reduction carries risks itself.[7]

How This Chapter Is Organized

This chapter examines the Superfund program as an attempt to reduce risk by statute, instead of doing so through the use of incentives and rules of property, common law, and community action. The chapter proceeds as follows: The next section uses insurance principles as a template to evaluate Superfund. While Superfund outwardly shares some insurance features, few would describe the program as an effort to underwrite losses.[8] However, applying the insurance template provides a way to identify dynamic elements that affect risk reduction.

The next section examines the question of risk reduction. If common law remedies are replaced by regulation, it is useful to see how risk is evaluated and how demonstrable future losses are prevented. This is accomplished by reporting elements of risk assessments conducted on two Superfund sites in South Carolina. Finally, the last section reports the results of empirical work that estimated the effects of the 1980 passage of Superfund legislation on the riskiness of firms in the chemical and petroleum industry. A short summary concludes the chapter.

2. Superfund as a Risk Reducer

Superfund is a dynamic program that first identifies candidate sites for entry into a slow and costly process. The process ends when a former hazardous site has been mitigated. Since sites are continually being added and completed, reporting a current count of where the program stands is problematic. For example, in June 1999, EPA was reporting FY1997 data on the agency's Superfund webpage.[9] At the end of FY1997, there were 1,405 sites on the EPA's national priority list for cleanup, and construction had been completed on 498 sites. By comparison, at the end of 1995, there were 1,374 sites on the national priority list, some 91 had been completed, and work was underway on another 700 sites.[10] As an indicator of the potential magnitude of the program's undertaking, more than 30,000 candidate sites are in line to be evaluated, and as many as 425,000 additional sites have been proposed for review.[11] Through 1995, emergency and other removal actions had been completed on some 40 percent of the priority sites.

If the untreated waste that remains on the more than 1,400 sites poses risk to human life and the environment, those risks are still present. Would tort action under common law and community action have been slower and less effective? Or has Superfund perpetuated risks that might otherwise have been reduced? A generous reading of Superfund might lead one to argue that the complex federal program simply needs time to mature. At some point, cleanups will accelerate. The pipeline will become unclogged, and clean sites will emerge in increasing number. But there is more to the problem than might be explained by bureaucratic inertia. The $15.2 billion pool of government money now tagged for use in cleaning sites is short by half the amount needed to cover the scheduled cleanups.[12] At an average cost of $25 million each, cleanup of the current priority sites will cost $31 billion. The EPA estimates the bill will come to $40 billion.

If 30,000 other candidate sites are taken on and average cost remains the same, another $750 billion is added to the bill. The reserves are simply not there to fund completion. The shortage of funds tells us just how important it is to recover funds from any and all who had any connection with a Superfund site, no matter what their guilt in the matter. Viewed as a social insurance scheme, where funds are accumulated to cover catastrophic losses, Superfund is bankrupt.

Is Superfund an Insurance Scheme?

Just what kind of insurance scheme is Superfund? The classic case for insurance calls on the law of large numbers to generate an expected level of losses that occur randomly among exposure units. The exposure units are identifiable activities, such as driving automobiles, producing chemicals, or operating boilers and waste disposal sites. When a large number of units are exposed to a classified hazard, risk can be estimated and expected levels of loss can be predicted with sufficient accuracy to establish prepayment for losses.

Prices for insurance are established so that the expected value of losses are funded by insurance premiums. The scheme is based on risk, which can be identified before losses occur, not on uncertainty, which cannot be estimated a priori.[13] An effective insurance scheme is one where claims paid are equal to the earlier estimates of losses.

For insurance to work effectively, the scheme itself must not generate additional risk, which is to say that the moral hazard problem must be minimized. Those covered by insurance are required to coinsure, to bear part of the risk to which they are exposed so that they will have an incentive to maintain an appropriate level of care. In addition, exposure units are carefully selected for inclusion in the risk pool to avoid including units that will impose unexpected costs on the insurance fund. Moral hazard and adverse selection are the bane of insurance.

If moral hazard and adverse selection are the bane of insurance, uncertainty, which by definition cannot be estimated before events occur and therefore cannot be rationally priced, is the hobgoblin. When uncertainty confounds risk, insurance fails. But what are the exposure units and who are the insured in Superfund? And how do the taxes that fund the program and premiums paid by chemical companies relate to the level of risk?

Superfund Communities as Exposure Units

At first blush, one might think that communities of ordinary people are the exposure units that might seek insurance to cover losses generated by randomly occurring hazardous waste sites. Superfund revenues, funded partly by personal income taxes, can be thought of as the community insurance premium. But the taxes paid have nothing whatsoever to do with risky behavior. Not a living soul can identify the part of his income taxes that goes to fund Superfund. There is no linkage between the tax-premium and waste generation.

In addition, ordinary people in communities do not coinsure. All of the costs of cleanup are borne by others. People with mismanaged landfills in their communities have little incentive to reduce the future occurrence of the same mischief. Logic dictates just the reverse. Communities have an incentive to find unpleasant sites that can be cleaned by Superfund at no apparent cost to themselves.[14] Cleanup cost is shifted outside the community; benefits remain. Moral hazard enters the scheme.

Because of the poor incentive structure, thousands of pages of docket and federally inspired rules and regulations form the legal environment for community landfill operations. The incentives are identical to those found where health care is funded by third parties. From the standpoint of community leaders who seek to clean away hazardous waste, no level of care is too good so long as the cost is externalized. Like all other programs where demand is subsidized and linkages between benefits and costs are severed, Superfund will be perpetually short of funds.

A rationing system is needed to constrain the number of sites that might enter the process. But Congress has taken just the reverse tack. Superfund legislation included $17.7 million for a grant program that provides up to $50,000 each to Superfund communities. The funds can be used to improve their involvement in the process.[15] While woefully short of funds to actually address hazardous sites, Congress nonetheless, took steps to lengthen the list of candidate sites.

The Responsible Parties as Exposure Units

A different picture emerges if Superfund is viewed as an insurance scheme for those who somehow participated in the development of hazardous waste sites. Under the statute, any firm, organization, or government unit that contributed any amount of waste to a site, transportation firms that carried the waste, former operators or owners of sites, and, until modified by statute, even financial institutions that became owners through default of a borrower can be held to a standard of retroactive, strict, joint and several liability. Every one of the words in the standard is important. The standard is retroactive. Though completely law-abiding at the time, a firm that may have contracted with a waste hauler to dispose of waste on a legal site twenty years ago can be held liable for the full cost of cleaning the site today. The same is true for the waste hauler. The standard is strict. If evidence is presented indicating that a particular firm, individual, or organization did, in fact, participate in the development of a Superfund site, then the party can be held liable. There is no defense based on negligence or standard of care. The liability is joint and several. Any one party or group of parties can be held fully liable for the cost of mitigating a site, no matter how many other parties are involved. Of course, once one party receives notice, then that party can immediately sue all others to recover costs. And they can countersue. And their insurance carriers can sue, and be sued.

The collection of potentially responsible parties associated with Superfund sites in 1990 reveals an interesting mix. Some 39 percent are in the manufacturing sector.[16] Another 17 percent are cities and counties that operate landfills. U.S. Department of Energy sites and military bases, which deserve special notice, make up 21 percent of the sites. In October 1995, a federal interagency taskforce reported that as much as $389 billion would be needed over the next seventy-five years to address the problem of hazardous and radioactive waste on federal property.[17] Finally, 23 percent of sites are a broad collection of banks, transportation companies, hospitals, universities, and other miscellaneous institutions. There can be hundreds of potentially responsible parties associated with one site. Any one and all of them, along with their providers of general line insurance coverage, can be required to pay the cost of cleanup. If nothing is gained from the resulting suits, Superfund can be tapped.

It is here that insurers encounter an extreme case of uncertainty. Consider the insurance companies first. When contracts for the environmental insurance were written, long before the invention of Superfund, the premiums were based on the expected losses associated with the exposure units contained in the insurer's particular risk pool. Individual insurance companies estimated risk, set prices, and wrote contracts, expecting to encounter losses associated with environmental accidents. Individual insurance companies did not and could not calculate the cost associated with changing rules of liability and new federal statutes. Retroactive strict, joint and several liability exposed insurance companies to losses that had nothing to do with their clients' behavior. Environmental insurance ceased to be a financially attractive enterprise.

Now consider the firms and organizations caught in the Superfund web. To partially fund Superfund, chemical companies pay a tax on chemical feedstocks, and petroleum companies pay a tax on imported crude oil. But these taxes, like the income taxes paid by ordinary citizens, have nothing whatsoever to do with the behavior that creates a Superfund site. The most careful and cleanest chemical firm pays the same tax rate as the most careless. In short, Superfund prices contain no economic information. There are no price incentives that might predictably induce environment-conserving behavior.

The combination of strict, joint and several liability, insurance contracts, and efforts by firms caught in a Superfund web to stay afloat generates a set of perverse incentives where litigation is preferred to mitigation. Instead of reducing all risks associated with hazardous waste sites, Superfund removes some risk and extends the life expectancy of remaining low-level risks. While accomplishing this task, the program introduces additional uncertainty into the market, and converts some risk to uncertainty. Each of these activities tends to increase the number of transactions required to produce the same level of output, and that is costly, too.

Estimates of Superfund litigation or transaction costs now incurred by thousands of litigants rise to extraordinary levels. One indicator of this cost comes from a much-cited RAND study that obtained claims-cost information from a sample of national insurance companies that provide the largest share of environmental liability insurance.[18] The study reports that 88 percent of total insurance expenditures involving 13,000 Superfund claims were related to litigation and other transactions costs. Just 12 percent of the total cost of claims went to clean up sites. When extrapolated to the entire insurance industry, litigation and transaction costs amounted to $410 million for the three years, 1986 through 1989.

The insurance dynamics lead to a sad result. The demand for Superfund sites from communities is unbounded, yet very sensitive to price. The supply of remediation funds is restricted and inelastic. The market broker, EPA, has exhausted large amounts of scarce resources in the struggle to bring supply and demand together. Through 1991, the agency had used $4.4 billion, some 48 percent of Superfund appropriations, to cover administrative costs.[19]

Superfund as Reinsurance

Superfund does not fit the mold of an effective insurance scheme for communities and those who find themselves liable for cleaning abandoned sites. Perhaps the regulatory program should be evaluated as reinsurance. But reinsurance for whom?

The reinsurance market develops when private insurers seek to diversify the risk pool that develops among their clients. The law of large numbers comes to their rescue. Given a small number of exposure units, individual primary insurers (those who write the initial contracts) sell some of the coverage to organizations that pool the contracts with others they hold. In that way, the reinsurers are able to diversify risk for specialized contracts. For example, major hurricanes occur infrequently in a given coastal region of the U.S., but occur more frequently when all coastal regions are considered. Competing insurance companies can each write a few property insurance contracts and sell part of the liability to reinsurers, who pool the risk.

In a similar way, unknown hazardous waste sites can be discovered at long forgotten abandoned locations. New scientific knowledge can declare substances to be toxic that were previously deemed harmless. Sudden occurrence of toxic seepage in water systems is hard to predict for a small number of locations; so is new scientific learning. If communities are willing to self-insure for losses they can estimate, and commercial underwriters are willing to write contracts for predictable risks, it seems reasonable that Superfund could be an uncertainty backup for written and unwritten contracts that cover large-risk exposure. But instead of assuming some proportional part of the risks and assigning that proportion to the carriers or communities, Superfund works differently, and perversely.

First off, Superfund's rule of liability has basically eliminated the market for commercial insurance.[20] Instead of supplementing the private market, the program has almost wiped it out. The retroactive rule of strict, joint and several liability means that commercial insurers who wrote contracts decades ago can be hit with claims for losses they did not intend to insure. Commercial insurance contracts cover specified risks; they seek to make their clients whole. A typical contract is not intended to make some other client whole, which happens when joint and several liability enters the process.

Then, those who wish to tap Superfund must work through a political process that in no way resembles the reinsurance market. Superfund sites do not emerge after private insurance claims have been settled. They emerge before these events. The reinsurance analogy does not work here. The analogy may better fit communities that find it necessary to engage in public works projects to clean up old waste sites. Bad things do happen in towns, cities, and counties. Reserve funds are maintained to defray the cost of unexpected events. Asbestos found in city auditoriums has to be removed. Storm damage has to be repaired. Sometimes reserve funds (self-insurance) are inadequate.

Superfund offers a pool of funds to draw on when the town treasury is depleted. But the analogy goes just so far. Reinsurance pools are funded by those who face the risk. Superfund seeks to gain funds any way possible, by taxes, fees, and suits. There is no attempt to link costs to benefits. There is another problem. Concerned communities do not seek Superfund relief after having exhausted all other available remedies. The Superfund option typically emerges before any other cleanup action has occurred.

Common Law and Transaction Costs

The common law process largely replaced by Superfund was said to suffer because it was slow, cumbersome, and sometimes arbitrary. To bring a common law action, a plaintiff had to be tied to the soil and to prove damages, either individually or collectively. Statutes of limitation sometimes interfered with the process. Jurisdictional disputes often complicated the process. Plaintiffs and defendants could search for a friendlier forum. Sometimes common-law judges made outlandish awards and stretched the normal meaning of contract language when doing so. And sometimes, those who created hazards were no longer in existence. Plaintiffs had to suffer or turn to the state for assistance. If the Superfund process offers some advantage relative to common law, that advantage has to be found in the program's ability to provide relief in cases where the common law defendant no longer exists.

Consider the transaction costs of Superfund. Since 1980, the EPA has consumed over $4.4 billion in management costs. Multiplying the three-year total of $410 million for insurance-related litigation costs by seven to obtain an approximation of expenditures through 1995 and adding the result to $4.4 billion brings the total to $7.2 billion, just for transaction costs and not counting the costs imposed on potentially responsible parties.

How do these numbers compare with tort actions under common law? There is no separate estimate of such costs for suits involving hazardous wastes, but consider this: In 1985 the direct cost of litigating all non-auto tort claims in the United States came to a total of $11.0 billion for all parties, which was for 405,000 cases.[21] Just the insurers' and EPA's Superfund transaction costs could amount to $7.2 billion for handling some 1,400 sites.

Applying insurance concepts to Superfund allows us to identify crucial weaknesses that relate to regulatory risk reduction. When the major actors are considered, we find the following:

- Superfund lacks the benefit of a price system that links benefits provided to cost. Communities can externalize cleanup costs.
- Communities are subsidized to assist them in dealing with Superfund procedures. These incentives increase demand for the cleanup of low risk sites.
- The fees paid by chemical, petroleum, and other corporations have no relationship to risk-producing activities. Funds collected today are used to deal with past actions taken by different parties.
- The rule of strict, joint and several liability applied to potentially responsible parties and their insurers causes transaction costs to rise significantly. Administrative costs soar. Little is accomplished by way of risk reduction.

3. Just How Much Risk is Reduced?

Risk Reduction in the Normal Course of Doing Business

Risk reduction ordinarily begins when risks are somehow identified. Experience and learning cause people to assess outcomes, avoid hazards, and to require additional rewards when knowingly assuming risk. Risk premiums are a common feature of wages, prices, and yields on investment. Before the fact, estimates affect contracts. Knowledge obtained after the fact causes future contracts to change.

When a large number of people in a community feel threatened by an apparent hazard, common-law courts can issue injunctions so long as the prospect of imminent harm is demonstrated. After the fact, losses are the purview of the law of torts, which also reduces risky behavior. In the ordinary course of doing business, firms and organizations that engage in activities that pose threats to innocent parties logically take into account the losses that will be imposed on them when they do not exercise due care. Competition and the hope for profits demand such behavior. In cases of accidents, negligence, or simple actions that cause damages, the damaging party can be sued, made to pay damages, and enjoined so that future risks are reduced.

Common law cannot address the risks posed by abandoned waste sites, which result when chemical-using plants are closed by bankrupt firms and waste haulers disappear in the night. But community action can and does address those problems. Ordinary people of common sense do not allow themselves and their children to be exposed to costly risks. Communities band together and organize actions either to wall off the risk, move away from it, or clean the site. In short, public works projects funded by those who are most sensitive to costs and benefits can address the problem of abandoned sites.

Risk Reduction under Superfund

Under Superfund, risk reduction takes a different route. The program is designed to avoid future harms. Risk is identified prospectively. An abandoned waste site may enter the risk identification process and candidacy for cleanup when someone calls a public official and expresses concern about a landfill or any other location where wastes have been stored or disposed. Somehow, a complaint is registered. In any number of ways, the EPA or its designated state agency receives notice of a potential problem site. Early in the process, a search begins to identify the potentially responsible parties, who may become liable for cleanup costs.

The level of risk posed by a candidate site is identified by the EPA through a structured investigation known as the Preliminary Assessment/Site Inspection. How many and how often real people are exposed to real risks is not a part of the analysis. The risk assessment process is based on an assumption that human beings will be exposed.

The analysis uses an EPA ranking system that generates a risk index.[22] The various components in the index relate to such things as risk to human populations and migration of hazardous materials into groundwater. Any site that has an index above the EPA's critical number is then placed on the National Priority List, which means that the site must eventually be cleaned up. However, a costly analysis first must be made to determine just how to proceed with cleanup. If the site is a critical hazard, which means there is danger of imminent harm, the EPA can take immediate mitigating action.

A Sample of Superfund in South Carolina

To illustrate just how risk is assessed and reduced, consider a part of the South Carolina experience.[23] In September 1990, the EPA had 423 candidate sites listed for South Carolina. In some cases, preliminary assessments were yet to be done. In other instances, sites had progressed through the process. At that time, there were twenty-three Superfund sites in the state. In September 1996, the number had risen to twenty-six. Some had just been designated; others had almost been completely cleaned. Work had been underway on some for more than ten years. Complete documentation of the sites and risk assessments are provided when a site enters the National Priority List. Verbatim excerpts from

EPA documents are provided here to describe two of the sites that offer a relatively complete history.

The South Carolina Recycling and Disposal, Inc., Site

This site is located on state highway 48, about seven miles south of Columbia, South Carolina. The property covers about seven acres, of which two were used for waste storage. About 7,200 drums of toxic, flammable, and reactive wastes were on the site at the time of listing, as well as numerous smaller containers. Two small ponds at the northern end of the site are remnants of lime slurry disposal ponds, used by an acetylene manufacturer that once occupied the property. The storage area is partially fenced. Air, groundwater, and surface water are contaminated. It is ranked the top priority site in South Carolina.[24]

According to the EPA and state environmental officials, the site near Columbia, also called the Bluff Road site, is South Carolina's top priority site for cleanup. In 1991, it ranked 83rd of over 1,200 sites nationwide on the EPA's National Priority List. According to EPA records, this site came to their attention in November 1979. The preliminary assessment was completed on February 1, 1980, and the site investigation was completed on November 1, 1984. The remediation plan was completed on October 19, 1989, almost five years after the site investigation was concluded. The feasibility study was completed in March of 1990. In sum, the trip from first notice to completion of plans for action took more than ten years.

As of September 1996, the 7,200 drums of wastes had been removed from the site by the potentially responsible parties, as well as some additional laboratory waste. The full cleanup has not been completed, but tests for contamination of neighborhood water supply are all negative. In short, the site was studied, researched, litigated, and discussed for over fifteen years without receiving the permanent cleaning as mandated by Superfund.

Consider the history of the site. During the early 1970s, the site, which includes two lagoons, was used by an acetylene manufacturer. One lagoon contains lime and the other lagoon contains lime sludge, covered with about six inches of water. The lagoons seem to hold the only remnant of wastes left by the acetylene manufacturer.

In 1975, the site became the property of Columbia Organic Chemicals. A subsidiary of Columbia Organic Chemicals, South Carolina

Recycling and Disposal, Inc., used the site to store, recycle, and dispose of chemical wastes. The operation was closed in 1982 after tests of soil and groundwater conditions were conducted by the EPA and the S.C. DHEC.

In 1982 and 1983, partial cleanup of the site was performed to limit further contamination. Clean soil and gravel replaced the excavated soil. The lagoons were left undisturbed. A remedial investigation of the site was initiated in 1984, under the direction of the state environmental agency. The report was never completed, and there were some problems with the accuracy of the tests that were conducted. For these reasons, another contractor began work on a second investigation, which was completed in October 1989.

The new contractor was employed by a subset of the potentially responsible parties, known as the Bluff Road Group. As of September 1996, only the members of the Bluff Road Group have spent funds on cleanup, study, or planning, with regard to the Superfund process at Bluff Road. However, other parties are being pursued to contribute to the cost of cleanup and design at the site.

The potential scope of the enforcement liability and settlement process can be seen by the fact that there are over 288 potentially responsible parties involved in settlement. Some of the alleged polluters are agencies dedicated to improving the environment and public health and welfare. A number of hospitals, as well as the state medical university, are being held liable for pollution at the site. In addition, the U.S. Center for Disease Control, which investigates the health effects of Superfund sites, is on the list. The EPA is ranked 20th in amount of pollution disposed at the site. The state environmental protection agency is listed as a responsible party for pollution at the site. The two agencies that are responsible for enforcing Superfund are targets of their own enforcement.

The Risk and Remediation Analysis

To assess risk at the Columbia site, twenty-three monitoring wells were drilled, twenty-nine soil borings were collected, and thirty-nine surface soil samples were studied, along with numerous surface water, lagoon water, and sediment samples. According to field investigations, soil at the most contaminated part of the site is penetrated by organic chemicals to a depth of three to eleven feet. There is no contamination of the deep-water aquifer that flows beneath the site. No significant contamination was found in the surface water drainage system or in the air samples. A shallow water aquifer, which is not used as a source of drinking water, showed limited contamination coming from the contaminated soil and the lagoons. Regardless of the pollution, according to the studies, the water in the area is not considered a good source of potable water. A creek that flows near the site shows no contamination.

Current and future use scenarios were developed in the studies to determine the health risks posed by the site. Once the investigation began, the site was fenced and access was restricted. There have been no reports or observances of trespassing. Furthermore, since there is no runoff contamination to the creek that flows on the property, there is no exposure to people by this path. As mentioned, no one draws drinking water from the shallow water aquifer, and regulatory restraints could be employed to ensure that the aquifer is not used in the future.

Under the current use scenario reported in the remediation analysis, the only possible way for a person to become exposed to carcinogens at the site would be for hunters to consume wildlife that contained concentrated levels of residual carcinogens ingested at the site. To achieve the risk trigger point, a deer hunter was assumed to have consumed venison of site-fed deer twice a week for a seventy-year lifetime. The analysis assumed that all of the deer eaten by this person came from wildlife grazing at the site. The assumption is far-fetched, since the disposal site encompasses only two acres. Further, the two acres are basically cleared of vegetation and therefore provide no food for deer or other wildlife.[25] Even with these extreme assumptions, the estimated levels of public health or environmental risk associated with the chemicals at the site are negligible.

Following the development of the current use scenario, the contractor's analysis developed a future use scenario as required by the EPA that contains certain EPA-required assumptions. As the contractor's report to the EPA put it:

> It should be stressed that these scenarios are hypothetical in nature with little likelihood of occurring . . . .They have been developed and evaluated at the agency's request to complete the intent of the agency's guidance documents for the conduct of an RI at a CERCLA site . . . .For the hypothetical future use scenarios, there appear to be concentrations of site-related chemicals in the shallow aquifer that may result in unacceptable levels of exposure only if all the health protective assumptions of the scenario are realized.[26]

Despite effective fencing, the isolated location, and the absence of detectable trespassing, site remediation plans are required to assume that children will use the area as a shortcut and as a playground. The EPA also requires that the area be considered a common area for adult trespassers. Even under these assumptions, the pollutants at the site "do not present a potential risk to the health of potential adult or child trespassers on the site."[27]

The official analysis of the site also assumed that the contaminated shallow aquifer would discharge into the creek on the seven-acre plot. Under this assumption, there was still no significant increase in the threat to either wildlife or humans. To meet the mandates of the EPA, the analysts included in its future use scenario the development of the area into a residential neighborhood in which residents took their drinking water directly from the contaminated portion of the shallow water aquifer. Under this assumption, there was a significant increase in the incremental cancer risk faced by those drinking from the shallow aquifer. The contractor explained the situation this way:

> In weighing acceptable residential exposures to potentially carcinogenic compounds, an acceptable level of risk must be determined. Cancer is a significant cause of death in the United States, with a background incidence of about three in ten (280,000 cases in a population of 1,000,000) (American Cancer Society 1988). Approximately 80 percent of these cases result in death directly attributable to the disease. Incremental lifetime cancer risk (also referred to as excess cancer risk) is defined as the estimated increased risk that occurs over an assumed average life span of seventy years (EPA, 1986) as the result of exposure to a specific known carcinogen. Thus, an incremental lifetime cancer risk of one in a million (one x 10-6) may be interpreted as an increase in the baseline cancer incidence from 280,000 per million population to 280,001 per million population.

Based on the scientific evidence and the regulatory precedence of the acceptable risk ranges set for exposures to carcinogens in drinking water and at Superfund site cleanups, rigid adherence to an incremental lifetime cancer risk of one in a million (one x 10-6) was shown to be clearly unwarranted in the exposure scenario developed in the current risk assessment. A more reasonable approach for a residential exposure scenario that maintains a level of health protection compatible with the rural environment surrounding the SCRDI site would be to use an incremental lifetime cancer risk of one in ten thousand (one x 10-4). This implies that if 10,000 people were to be located in the zone of impact of the SCRDI site, one additional cancer occurrence might be predicted.[28]

The risks are calculated based on individuals being exposed to selected carcinogens for a seventy-year lifetime.[29] It is also assumed that all 10,000 people would be exposed to the maximum contaminate concentrations found on-site. For instance, all 10,000 people would have to drink from the most contaminated well at the site every day for a seventy-year lifetime to incur the risk of seeing one additional case of cancer during this seventy-year period. The analysis assumes implicitly that if 10,000 people lived directly on the two-acre site for a seventy-year lifetime, and they obtained their drinking water from the most contaminated shallow water aquifer, ingested the most contaminated soil, and inhaled the most contaminated dust, one additional case of cancer above the background levels could be predicted.[30] This is one example of Superfund risk reduction.[31]

The Estimated Expenditures

It is estimated that the cost of cleaning the Columbia site is $7 million, which is well below the $25 million national average for Superfund site cleanup.[32] However, the estimates do not include legal fees incurred during litigation. Litigation costs, which can continue long after cleanup has begun, often exceed cleanup costs by several-fold. In this case, certain unidentified, potentially responsible parties paid costs of $3.7 million for the remedial investigations and feasibility study.[33] These potentially responsible parties are now in the process of suing several hundred other polluters at the site for their share of the cleanup and site investigation costs.

The Independent Nail Site

The Independent Nail Company manufactures metallic screws on a site in Beaufort County, South Carolina. The company bought the site from Blake and Johnson, a manufacturer of screws and fasteners, in 1980. There is a one-acre lagoon on the site into which Blake and Johnson placed waste water containing cyanide, chromium, cadmium, lead, nickel, zinc, copper, and iron. When Independent Nail bought the site, it asked Blake and Johnson to investigate the quality of groundwater. In response, Blake and Johnson installed three monitoring wells of intermediate depth. Analyses revealed that groundwater had been impacted by the lagoon. Further tests conducted by the state showed that the contaminated groundwater had moved outside of the area covered by the monitoring wells. Surface water had been locally contaminated by the lagoon, but is not used as a source of drinking water.[34]

The Independent Nail site, located three miles northwest of the town of Beaufort,[35] is immediately across from the bases operated by the U.S. Air Force and Marine Corps. From 1969 until 1980, the site was used by the firm of Blake and Johnson. During that time, approximately 33,000 gallons of wastewater were pumped daily into unlined lagoons. Independent Nail Company purchased the property in the summer of 1980. From all reports, Independent Nail never contributed pollution to the site.

The documented pollution history of the site began in 1975 when waste from one of the lagoons leaked into a drainage ditch. State health officials then initiated groundwater testing.[36] The 1975 tests detected levels of lead, mercury, chromium, and iron that exceeded drinking water standards. Subsequent 1980 tests found evidence of chromium and lead. The Independent Nail site was added to the national priority list in September 1983. Further 1985 tests revealed absolutely no contamination or elevated levels of metals or volatile compounds at the site. Soil tests indicated surface contamination in the lagoon and in two areas about 500 feet outside the lagoon.

The site risk assessment made the assumption that on-site workers would be exposed to carcinogens by eating and breathing certain daily doses over their entire employment lifetimes. Even with that assumption, the report indicated "that cumulative exposure. . .would not pose health risks."[37] An even more extreme assumption was required to cross the risk threshold. In effect, the same person would have to ingest contaminated soil 250 times a year over a seventy-year lifetime or that same person would have to inhale contaminated air 250 times a year for seventy years.[38]

Contamination at the Independent Nail site was confined to surface soil; there was no evidence of contaminated groundwater. The recommended remediation called for treatment of the soil and replanting at an estimated cost of $2 million. In 1988, some thirteen years after the first soil tests, cleanup activity was completed. While two anecdotes cannot be accepted as full documentation of all Superfund risk analyses, these two are representative of what is found when Superfund is reviewed in South Carolina. Other sites have other problems, and in some cases, the risks may be higher. In any case, the problem should not be trivialized. However, the evidence for the national program, as reported by the EPA, is consistent with these two stories. Indeed, when the EPA assessed its own mandate, the agency gave Superfund risk reduction a low priority.[39] The message communicated is a simple one: If Superfund is a risk reducing activity, it is a very costly one.

4. Superfund as a Risk Increaser

Considering Superfund's output, it takes an optimistic soul to argue that this regulatory approach for reducing risks from hazardous waste is superior to common law rules. Too little has been done to mitigate sites; the cost incurred has been enormous. While the EPA has engaged in numerous emergency actions that go beyond the Superfund count, the question of net benefits still hangs in the balance.

We cannot know with any degree of certainty if Superfund has reduced major risks to human populations generated by hazardous waste, but there is little doubt that the program has imposed risks of a different type. Consider this: When regulation befalls major sectors of the economy, costs are imposed immediately in two ways. Investors estimate the effects of pending regulations on the future cash flows of affected companies. The shares of those firms are bid down to adjust for losses that were not previously perceived. Along with these actions, investors worldwide assess the economic environment established by major regulations and reassign risk to affected sectors. There can be unexpected gains or losses from the first effect and unexpected changes in risk from the second.

In an effort to estimate these effects in conjunction with the 1980 Superfund act, Dalton, Riggs, and Yandle (1996) assembled a portfolio of stocks of chemical and petroleum firms.[40] A chronology of the passage of the Superfund law was then developed, which started when Love Canal was declared a national emergency in early August 1978. Then, the date of each major legislative proposal was included all the way to the first week in December 1980, when the final law passed by both houses of Congress. Estimates of unexpected losses or gains and changes in risk were estimated for the chemical/petroleum portfolio using the modified capital assets pricing model and the seven event windows mentioned above. Overall the paper finds that the passage of Superfund, which was for the presumed purpose of reducing risks from harmful hazardous wastes, increased the riskiness of investments in the chemical and petroleum industry. In addition, the law imposed significant financial losses to the owners of firms in the two industries. And what can we predict from these effects? Two basic industries that provide support to the growth of the U.S. economy appear to have become riskier and less profitable. Logic tells us that industries so affected will grow less rapidly, and those who invest will require a higher rate of return, a return that compensates them for additional risk. At the margin, future incomes are lower and basic investments are riskier.

5. Conclusion

This chapter has reviewed Superfund as a legislative effort to reduce the risk of hazardous wastes. At the outset, the statutory effort was briefly contrasted with common law rules that prevailed before the 1980 law was passed. The alleged weaknesses of common law were mentioned. Clearly, common law rules are imperfect and costly.

The basis for Superfund was discussed through the use of an insurance analogy. Of course, Superfund was not intended to be insurance, even though the fund has some surety characteristics. But using the insurance straw man helps to identify costly Superfund characteristics. Then, the question of risk was addressed directly by illustrating how risk was assessed for two Superfund sites in South Carolina. The sometimes trivial level of risk to be reduced was set off against the record of enormous costs imposed by the Superfund process.

Finally, another kind of risk was introduced, a risk imposed by the legislative effort to reduce risk. Evidence presented indicates that two major U.S. industries were affected adversely by the Superfund law. The chemical and petroleum industries sustained losses and faced newly increased risks as the Superfund law was built.

What can we say about Superfund? The program is basically about finding revenues, no matter how, for the purpose of addressing emotional concerns about risks that can be quite small. The program is also about emergency cleanups of truly hazardous waste sites. The dynamics of the program assure excess demand for scarce cleanup resources. Excess demand for political action gives power to politicians. The costly litigation induced by Superfund's rule of liability decreases the supply of cleanup activity. Subsidies provided to Superfund communities increase demand for action. Excess demand grows.

The legislative blueprint had the effect of increasing the risk of two major U.S. industries, the same industries called on to fund the cleanup activity. Lower profits and higher risk yield fewer assets to be used for environmental control.

So what can we conclude from all this? If ordinary people wish to guard themselves from harm that may result from hazardous waste, the rules of common law seem to offer a positive prospect. But what if the owner of the hazardous waste site is gone? What if there is no one to sue? In those cases, the concerned community must organize and take action to clean up the waste or protect themselves from it. A public works project seems to be the logical outcome.

Reaching this conclusion leaves a crucial unanswered question on the table: Why has Superfund continued for more than a decade? The weaknesses and failings of the legislation to achieve cleanup goals are widely recognized. The extraordinary cost imposed on recognizably innocent parties has been publicized many times. Hearings have been held, and Superfund was reauthorized with its basic flaws intact.

The mystery that remains takes on the following elements. If regulation is provided to give appropriable rents to a few highly organized interest groups, who are they? Answering by naming the environmental bar and environmentalists is not enough. Those groups surround and support all environmental regulation. If regulation is designed, at least partly, to serve the public interest, how can Superfund survive when so little has been done?

At this point, the mystery of Superfund remains unsolved. With reauthorization once again on the congressional calendar, we may discover the sources of the political wealth generated by a statute that appears to raise rather than reduce risk. In the process, Superfund will likely be reauthorized in some modified form, perhaps removing the strict, joint and several strict liability rule that has caused so much mischief.[41] But there is an alternate scenario to consider. It could be that the Superfund saga will come to a halt. If the program was more about emotion and cheap federal dollars than fact, then given fiscal stress, the next Superfund may be sharply modified.

There are reasons to think this could happen. In October 1994, the Fifth U.S. Circuit Court of Appeals in New Orleans ruled against Superfund's strict, joint and several liability feature where there is a logical basis to apportion damages.[42] The case involved damages of $1.8 million to be paid by one of three firms charged with contaminating an Odessa, Texas, water system. The appeals court remanded the case to the lower court for retrial.

More to the point of community action and the common sense of ordinary people, Michigan legislation passed in 1991 has reduced the costs of cleaning hazardous wastes from locations that could become the sites of new industrial plants. Quoted in the *New York Times*, Russell Harding, Deputy Director of the Michigan Department of Natural Resources, said, "We can't get our old industrial sites redeveloped. We're losing jobs and tax base in cities. And if the laws aren't adjusted, our green spaces will become gray spaces. It's not what a lot of people want to hear, especially people in the environmental community, but we just have to get real on this stuff."[43]

Then there is the story about the forty-six inhabitants of Triumph, Idaho, who, in 1993, fought the EPA tooth and nail to stop the agency's effort from destroying their community, a former mining site, which had been named a Superfund site.[44] The citizens of Triumph are happy in their town, do not agree with the EPA's assessment of risks, and want to keep their land and their homes as they are.

At some point, real science and common sense tend to prevail in all of man's efforts to reduce risk. That point arrives rather quickly when huge amounts of scarce resources are diverted from reducing risks that are real and costly to churning away in a process that reduces little risk at all.

Finally, in spite of the apparent attractiveness of statute-based remedies for hazardous waste problems, some suggest that common law environmental protection is resurging. How could this be? And is the common law that seems to be finding its way through statute law cracks and crevices really the same as the old common law?

The evidence for resurgence is seen primarily in actions involving hazardous waste cleanup under the Superfund program.[45] Oddly enough, one commentary on resurgence begins with this criticism of statute law remedies: "Federal statutory causes of action, the cost recovery tools of the 1980s, have become burdensome and costly," critical words previously reserved for common law.[46]

The problems that seem to be addressed more effectively by common law first relate to recovery of damages, a recourse not available under statute law where all penalties are paid to the government or to court-designated environmental programs. In other words, environmental rights protected for ordinary people at common law are not recognized in statute law. A field of leaking drums may do serious damage to the property of adjacent landowners, but the affected landowners cannot recover. As mentioned earlier, there is a stigma effect that can impose severe property-value losses on entire neighborhoods. After all, Superfund cleanup efforts generally take more that ten years for the typical site. Land values fall throughout the area; real estate goes on sale. In some cases, decay sets in that leads to permanent destruction of a community. None of these consequential losses are recoverable under statute law.

The next problem relates to exemptions. As a result of special interest struggles, damages from petroleum spills are exempted by Superfund.[47] A water supply contaminated by leaking gasoline tanks, one of the most common environmental problems encountered by landowners, may eventually be cleaned by Superfund action, but the individuals who bore the cost of finding another source of drinking water or moving to secure the same, cannot recover the related damages. Nuisance and trespass suits at common law provide an avenue for relief.

But how has common law suddenly overcome the alleged handicaps that made it so unworkable that statute law had to come to the rescue? After all, critics said that common law was ineffective in dealing with environmental harms involving multiple culprits. Cause and effect, not mere speculation, is required by common law rule. And what about the problem of many receivers of harm, none of whom is damaged sufficiently to have an incentive to bring an action?

Advances in science address the first problem. Computer modeling and the development of schemes for matching waste with its source make it more feasible to identify defendants. And greater expectations that public prosecutors will pursue public nuisance cases bring a more effective response from attorneys general when a few citizens complain. For these and the previously mentioned reasons, members of the environmental bar now recommend that common law pleadings be combined with complaints of statute law violations when citizens bring suit against a hazardous waste polluter.

Perhaps, we have come full circle and, in doing so, have rediscovered the merits of common law. But we have paid dearly while traversing the circle. Having learned more about risks and high-cost efforts to reduce them, we may be at a point where common sense will again prevail. However, before becoming too optimistic about the future prospects of Superfund, we should remember that statutes that allow us to spend other people's money always tend to be more attractive than rules that require us to bear the cost of our own behavior.

Notes

1. Richard Epstein offers the logic of pending catastrophe as the basis for regulating environmental use, which he sees as a precondition for contracting. See Richard A. Epstein, "Regulation and Contracting Environmental Law," *West Virginia Law Review* 93 (Summer 1991): 859-891. For a brief discussion of Love Canal and related incidents, see Peter Huber, "Environmental Hazards and Liability Law," Robert E. Litan and Clifford Winston, eds. *Liability Perspectives and Policy*, Washington, DC: Brookings Institution, (1988): 128-153. For more on Love Canal, see Bruce Yandle, *Common Sense and Common Law for the Environment*, Lanham, MD: Rowman & Littlefield Publishers, 1997, 78-80.
2. On this, see Katherine N. Probst, Don Fullerton, Robert E. Litan, and Paul M. Portney, *Footing the Bill for Superfund Cleanup*, Washington, DC: The Brookings Institution, 1995. Also see U.S. General Accounting Office, "Superfund: Current Progress and Issues Needing Further Attention," GAO/T-RCED-92-56 (June 11, 1992): 1.
3. This is the basic premise that underlies Robert Higgs' extensive examination of the growth of federal programs. (See Robert W. Higgs, *Crisis and Leviathan*, San Francisco: Pacific Research Institute for Public Policy, 1989.)
4. See Roger E. Meiners, "Elements of Property Rights: The Common Law Alternative," Bruce Yandle, ed. *Land Rights: The 1990s Property Rights Rebellion*, Lanham, MD: Rowman & Littlefield Publishers, (1995): ch. 8, and Roger E. Meiners and Bruce Yandle, "Common Law Environmentalism," *Public Choice* 94, 1998, pp. 49-66.
5. For an interesting discussion of environmental common law and a comparison of its features with statute law, see Donald Dewees, "Tort Law and the Deterrence of Environmental Pollution," T.H. Tietenberg, ed., *Innovative Environmental Policy*, Aldershot, England: Edward Elgar Publishing Limited, 1992: 139-164.
6. See Bruce Yandle, "Can Superfund's Fatal Flaws be Fixed?" *National Environmental Enforcement Journal* 4 (October 1989): 3-10.
7. This risk-producing trait is not uncommon for regulation. Indeed, it is almost a law that reducing risk in one dimension of life will always increase risk for another dimension. Of course, the important question relates to the gain in risk reduction. On the general issue of risky risk reduction, see John D. Graham and Jonathan B. Wiener, eds. *Risk Versus Risk: Tradeoffs in Protecting Health and the Environment*, Cambridge, MA: Harvard University Press, 1995.
8. On this, see Bruce Yandle, "Taxation, Political Action, and Superfund," *Cato Journal* 8 (Winter 1989): 751-764.

9. See • HYPERLINK http://www.epa.gov/superfund/whatissf/mgmtrpt.htm ••http://www.epa.gov/superfund/whatissf/mgmtrpt.htm, June 2, 1999.
10. The 1995 levels are reported in Richard L. Stroup, "Superfund: The Shortcut that Failed," Issue No. PS-5, Bozeman, MT: PERC (May 1996): 2. Also see "How to Rescue Superfund: Bringing Common Sense to the Process," *Backgrounder*, Washington, DC: The Heritage Foundation, July 1995. Also see, Richard L. Hembra, "EPA's Superfund TAG Program," Washington, DC: U.S. General Accounting Office, GAO/T-RCED-93-1, 1993, p. 2, and Katherine N. Probst, Don Fullerton, Robert E. Litan, et al., *Footing the Bill for Superfund Cleanups*, Washington, DC: The Brookings Institution, (1995): ch. 2.
11. See abstract of GAO Report, "Superfund: Extent of Nation's Potential Hazardous Waste Problem Still Unknown" (Dec. 17, 1987), in U.S. General Accounting Office, Environmental Protection: Bibliography of GAO Documents, January 1985-August 1988, (1988): 228.
12. The fund was set at $1.6 billion in the 1980 legislation. The 1986 amendments added another $8.5 billion, and in 1990, an additional $5.1 billion was added. In early debates leading to the passage of the 1980 Superfund law, some thought $200 million would do the job. Others called for $4 billion, and an EPA assistant administrator argued that $40 billion would be required. From all indications, the EPA official had a better crystal ball. (For discussion, see Brett Dalton, David Riggs, and Bruce Yandle, 'The Political Production of Superfund: Some Financial Market Results," Eastern Economic Journal 22 (Winter 1996): 75-87.)
13. Recognizing the difference between risk and uncertainty is crucial to the analysis that follows. Since uncertainty by definition cannot be estimated a priori and therefore is not associated with a stable probability density function, it is impossible for an insurer to establish risk-based prices. Establishing controls that limit uncertainty-generating behaviors is a major insurer task. For discussion of uncertainty and risk in a regulation context, see Robert Kneuper and Bruce Yandle, "Auto Insurers and the Air Bag," Journal of Risk and Insurance 61, No. 1 (March 1994): 107-116.
14. It should be noted that Superfund activity can impose heavy costs on the community. Indeed, the announcement of an official site designation has been shown to be associated with a significant decline in property values in the vicinity of the site. In fact, efforts to mitigate sites seem to impose even more losses, contrary to what one might expect from an effective cleanup program. (See Katherine A. Kiel, "Measuring the Impact of the Discovery and Cleaning of Identified Hazardous Waste Sites on House Values," Land Economics 71(4) November 1995: 428-435.)
15. U.S. General Accounting Office, "EPA's Superfund TAG Program: Grants Benefit Citizens But Administrative Barriers Remain," Washington, DC: GAO, GAO\T-RCED-93-1 (1993): 3.
16. The data reflected here are from "Superfund Liability: Who are the PRPs?" Superfund Issues Forum, American International Group, Inc. (June 1991).

17. Council of Environmental Quality and Office of Management and Budget, Improving Federal Facilities Cleanup, Washington, DC (October 1995). See also Bruce Benson's chapter on "Toxic Torts by Government" in this volume.

18. See Jan Paul Acton and Lloyd S. Dixon, *Superfund and Transaction Costs: The Experience of Insurers and Very Large Industrial Firms*, Santa Monica, CA: RAND, 1992, p. 10.

19. According to the U.S. General Accounting Office, the administrative costs of Superfund had absorbed 48 percent ($4.4 billion) of the $9.1 billion total Superfund appropriations through fiscal year 1991. See U.S. General Accounting Office, "Superfund: Current Progress and Issues Needing Further Attention," Washington, DC: GAO, GAO/T-RCED-92-56, June 11, 1992: 7.

20. U.S. General Accounting Office, "Hazardous Waste: Issues Surrounding the Insurance Availability," GAO/RCED-88-2, Washington, DC: U.S. GAO, October 1987.

21. These data are taken from Committee for Economic Development, *Who Should Be Liable?* Washington, DC: CED (1989): 51-53.

22. EPA uses a number of 28.5 as the threshold index for entering its National Priority List. The number, which obviously has no meaning outside the analysis and does not give a direct reading of risk, is interesting for one reason; it emerged early on as the value that generated the 400 sites required by Congress. A higher or lower number would have created either too few or too many sites. (For discussion, see James Bovard, CATO Policy Analysis No. 89, Washington, DC: CATO Institute, August 14, 1987).

23. This material is based on Brett A. Dalton, "Superfund: The South Carolina Experience," Roger E. Meiners and Bruce Yandle, eds. *Taking the Environment Seriously*, Lanham, MD: Rowman & Littlefield Publishers, Inc., 1993: 103-139.

24. This description was provided by Mr. H. Kirk Lucius, Freedom of Information Coordinator, Atlanta Regional Office, U.S. Environmental Protection Agency.

25. In the Columbia area, the most productive deer land yields fifty deer per square mile. If these two contaminated acres were wooded, covered with lush vegetation, and accessible to deer, they could sustain 1/7th of a deer. Hence, the most radical assumptions of bioaccumulation make the health hazard an extremely remote and unlikely possibility. Bioaccumulation normally results in pollutant accumulation in the internal organs and fatty tissue of a deer, both of which are rarely eaten by hunters. The current use scenario does not presume deer would be fenced out of the area. Even assuming the worst case scenario, the lifetime incremental increase in risk of cancer was still not significant enough to call for cleanup action.

26. Remedial Investigation Report, SCRDI-Bluff Road Site, Knoxville, TN: IT Corporation (October 1989): 6-3.

27. Remedial Investigation Report, SCRDI-Bluff Road Site, Knoxville, TN: IT Corporation (October 1989): 6-4.

28. Ibid.
29. According to the EPA's National Contingency Plan, which sets forth the guidelines and mandates to be followed when conducting Superfund site assessments, an increase in the incremental incidence of cancer of one in one million may be unacceptable. The range of unacceptable increases may be as high as one in 10,000. When performing a Superfund assessment, the investigator must go to extreme measures to justify the risk level that is selected by EPA (Conversation with James Dee, IT Corporation, Knoxville, TN).
30. This interpretation is based on conversation with James Dee, IT Corporation, Knoxville, TN, September 5, 1991.
31. After completing the site analysis, the contractor developed a response to the pollution at the Bluff Road site, which included alternatives for remedial activity, their corresponding costs, and their advantages and disadvantages. The response analysis indicated that site contamination was limited to the soil at the site ranging in depth from 3 to 11 feet. In addition, the shallow aquifer, ranging in depth from 10-15 feet below the surface, is contaminated. The contractor estimated about 263 million gallons of water to be cleaned, about 28,000 to 45,000 yards of contaminated soil to be removed or treated. According to EPA mandates, the cleanup methods must provide permanent solutions and comply with all relevant statutory preferences for treatment.
32. The data for the national average were provided by the American International Group, Washington, DC, in correspondence from Jan M. Edelstein, March 29, 1991.
33. This information was provided by the law firm of Edwards & Angell, New York, with the permission of the unnamed clients. Legal costs are not included.
34. This description was provided by Mr. H. Kirk Lucius, Freedom of Information Coordinator, Atlanta Regional Office, U.S. Environmental Protection Agency.
35. The discussion here is drawn from U.S. EPA, First Operable Unit, Remedial Investigation Report for the Independent Nail Company, Beaufort, SC, June 8, 1987, on file with the South Carolina Department of Health and Environmental Control, Columbia, SC.
36. This discussion is based on First Operable Unit, Remedial Investigation Report for the Independent Nail Company Site, Beaufort, SC, June 8, 1987, on file with the South Carolina Department of Health and Environmental Control.
37. Ibid., p. 35.
38. Ibid.
39. U.S. EPA, "Unfinished Business: A Comparative Assessment of Environmental Problems: Overview Report," Washington, DC: U.S. EPA, February 1987. For a stronger opinion on the same topic, see Paul R. Portney, "Reforming Environmental Regulation: Three Modest Proposals," Issues in Science and Technology 4 (Winter 1988): 74-81.

40. See Brett Dalton, David Riggs, and Bruce Yandle, "The Political Production of Superfund: Some Financial Market Results," *Eastern Economic Journal* 22 (Winter 1996): 75-87.
41. For example, see "How to Rescue Superfund: Bringing Common Sense to the Process," *Backgrounder,* Washington, DC: Heritage Foundation (July 31, 1995). For earlier comments, see John H. Cushman, Jr., "Administration Plans Revision to Ease Toxic Cleanup Criteria," *New York Times,* January 31, 1994, A1; Timothy Noah, "Clinton Today Begins an Attempt to Clean Up Big Toxic-Waste Problem: Superfund Law Itself," *Wall Street Journal,* February 3, 1994, A16; and James Lis and Melinda Warren*, Reforming Superfund,* St. Louis, MO: Center for the Study of American Business (February 1994). But also see, "Superfund Status Quo: Why the Reauthorization Bills Won't Fix Superfund's Fatal Flaws," Issue Bulletin, Washington, DC: Heritage Foundation (October 3, 1994).
42. See Margaret A. Jacobs, "Paying for the Cleanup," *Wall Street Journal,* October 5, 1994, B5. The case was Environmental Protection Agency vs. Sequa Corp., Fifth U.S. Circuit Court of Appeals, New Orleans, 91-8080.
43. Keith Schneider, "Rules Easing the Urban Toxic Cleanups," *New York Times,* September 20, 1993, A8.
44. Tony Horwitz, "For These Residents, EPA Cleanup Ruling Means Paradise Lost," *Wall Street Journal,* September 21, 1993, A1-10.
45. See Randall G. Vickery and Robert M. Baratta, Jr., "Back to the Legal Future," Environmental Law (June 10, 1996): C1, C3; Eric E. Nelson and Curt R. Fransen, "Playing with a Full Deck: State Use of Common Law Theories to Complement Relief Available Through CERCLA," Idaho Law Review 25(1988/1989): 493-519; and Tom Kuhnle, "The Rebirth of Common Law Actions for Addressing Hazardous Waste Contamination," Stanford Environmental Law Journal 15 (January 1996): 187-229.
46. Randall G. Vickery and Robert M. Baratta, Jr., op. cit., p. C3.
47. A number of states have established funds for clearing away abandoned fuel tanks, but still do not provide compensation to damaged parties. For example, see John Charles Ormond, Jr., "Cleaning Up Contaminated Sites," Business & Economic Review 40 (January, February, March 1994): 28-29.

References

Acton, Jan Pau, and Lloyd S. Dixon. 1992. *Superfund and Transaction Costs: The Experience of Insurers and Very Large Industrial Firms.* Santa Monica: RAND.

American International Group. 1991. "Superfund Liability: Who are the PRPs?" Superfund Issues Forum. Washington: American International Group, Inc. (June).

Bovard, James. 1987. CATO Policy Analysis No. 89. Washington: CATO Institute (August 14).

Committee for Economic Development. 1989. *Who Should Be Liable?* Washington: Committee for Economic Development.

Council on Environmental Quality and Office of Management and Budget. 1995. *Improving Federal Facilities Cleanup.* Report of the Federal Facilities Policy Group, Washington, DC (October).

Cushman, John H., Jr. 1994. "Administration Plans Revision to Ease Toxic Cleanup Criteria*," New York Times* (Jan. 31): A1.

Dalton, Brett A. 1995. "Superfund: The South Carolina Experience." Roger E. Meiners and Bruce Yandle, eds. *Taking the Environment Seriously.* Lanham, MD: Rowman & Littlefield Publishers, 103-139.

Dalton, Brett, David Riggs, and Bruce Yandle. 1996. "The Political Production of Superfund: Some Financial Market Results." *Eastern Economic Journal* 22 (Winter): 75-87.

Dewees, Donald. "Tort Law and the Deterrence of Environmental Pollution," T.H. Tietenberg, ed. *Innovative Environmental Policy.* Aderhold, England: Edgar Elgar Publishing Limited, 1992, 139-164.

Epstein, Richard A. 1991. "Regulation—and contract—in Environmental Law." *West Virginia Law Review* 93 (Summer): 859-891.

Graham, John D., and Jonathan B. Wiener, eds. 1995. *Risk Versus Risk: Tradeoffs in Protecting Health and the Environment.* Cambridge, MA: Harvard University Press.

Hembra, Richard L. 1993. "EPA's Superfund TAG Program." Washington, DC: U.S. General Accounting Office, GAO/T-RCED-93-1.

Higgs, Robert W. 1989. *Crisis and Leviathan.* San Francisco: Pacific Research Institute for Public Policy.

Horwitz, Tony. 1993. "For These Residents, EPA Cleanup Ruling Means Paradise Lost." *Wall Street Journal* (September 21): A1, A10.

"How To Rescue Superfund: Bringing Common Sense to the Process." 1995. *Backgrounder.* Washington, DC: Heritage Foundation (July 31).

Huber, Peter. 1988. "Environmental Hazards and Liability Law," Robert E. Litan and Clifford Winston, eds. *Liability Perspectives and Policy.* Washington, DC: Brookings Institution, 128-153.

Jacobs, Margaret A. 1994. "Paying for the Cleanup." *Wall Street Journal* (October 5): B5. The case was Environmental Protection Agency vs. Sequa Corp., Fifth U.S. Circuit Court of Appeals, New Orleans, 91-8080.

Kiel, Katherine A. 1995. "Measuring the Impact of the Discovery and Cleaning of Identified Hazardous Waste Sites on House Values." *Land Economics* 71(4) (November): 428-435.

Kneuper, Robert and Bruce Yandle. 1994. "Auto Insurers and the Air Bag," *Journal of Risk and Insurance* 61, No. 1 (March): 107-116.

Kuhnle, Tom. 1996. "The Rebirth of Common Law Actions for Addressing Hazardous Waste Contamination." *Stanford Environmental Law Journal* 15 (January): 187-229.

Lis, James, and Melinda Warren. 1994. *Reforming Superfund.* St. Louis, MO: Center for the Study of American Business (February).

Meiners, Roger E. 1995. "Elements of Property Rights: The Common Law Alternative," Bruce Yandle, ed., *Land Rights: The 1990s Property Rights Rebellion*. Lanham, MD: Rowman & Littlefield Publishers.

Miners, Roger E., and Bruce Yandle. 1998. "Common Law Environmentalism." *Public Choice* 94: pp. 49-66.

Nelson, Eric E., and Curt R. Fransen. 1988/1989. "Playing with a Full Deck: State Use of Common Law Theories to Complement Relief Available Through CERCLA." *Idaho Law Review* 25: 493-519.

Noah, Timothy. 1994. "Clinton Today Begins an Attempt to Clean Up Big Toxic-Waste Problem: Superfund Law Itself." *Wall Street Journal* (Feb. 3): A16.

Ormond, John Charles, Jr. 1994. "Cleaning Up Contaminated Sites." *Business & Economic Review* 40 (January, February, March):28-29.

Portney, Paul R. 1988. "Reforming Environmental Regulation: Three Modest Proposals." *Issues in Science and Technology* 4 (Winter): 74-81.

Probst, Katherine N., Don Fullerton, Robert E. Litan, and Paul R. Portney. 1995. *Footing the Bill for Superfund Cleanup*. Washington, DC: The Brookings Institution.

Schneider, Keith. 1993. "Rules Easing the Urban Toxic Cleanups." *New York Times* (September 20): A8.

Schwert, G. William. 1981. "Using Financial Data to Measure Effects of Regulation." *Journal of Law and Economics* 24 (April): 121-158.

Stroup, Richard L. 1996. "Superfund: The Shortcut that Failed." Issue Number PS-5, Bozeman, MT: PERC (May).

"Superfund Status Quo: Why the Reauthorization Bills Won't Fix Superfund's Fatal Flaws." *Issue Bulletin*. Washington, DC: Heritage Foundation (Oct. 3, 1994).

U.S. Environmental Protection Agency. 1987. "Unfinished Business: A Comparative Assessment of Environmental Problems: Overview Report." Washington, DC: U.S. EPA (February).

U.S. General Accounting Office. 1987. *Hazardous Waste: Issues Surround Insurance Availability*. GAO/RCED-88-2, Washington, DC: U.S. GAO (October).

_____. 1989. "Abstract," of "Superfund: Extent of Nation's Potential Hazardous Waste Problem Still Unknown." Environmental Protection Bibliography of GAO Documents, January 1985-August 1988. Washington, DC: U.S. GAO.

_____.1992. "Superfund: Current Progress and Issues Needing Further Attention." GAO/T-RCED-92-56. Washington, DC: U.S. GAO (June 11).

_____.1993. "EPA's Superfund TAG Program: Grants Benefit Citizens But Administrative Barriers Remain." Washington, DC: GAO, GAO\T-RCED-93-1.

Vickery, Randall G. and Robert M. Baratta, Jr. 1996. "Back to the Legal Future," *Environmental Law* (June 10): C1, C3.

Yandle, Bruce. 1989. "Taxation, Political Action, and Superfund," *Cato Journal* 8 (Winter): 751- 764.

_____.1989. "Can Superfund's Fatal Flaws be Fixed?" *National Environmental Enforcement Journal* 8(Winter): 751-764.

_____. 1997. *Common Sense and Common Law for the Environment*. Lanham, MD: Rowman & Littlefield Publishers.

2

Air Toxics Policy: Liabilities from Thin Air

Richard L. Stroup[1]

In 1990, Congress passed the Clean Air Act Amendments. Title III
of the Act required the Environmental Protection Agency to regulate
emissions of toxic or hazardous pollutants. This controversial Title was
the culmination of pressure by environmentalists to impose more stringent controls on airborne emissions of 189 potentially toxic airborne
substances in addition to the "criteria" pollutants such as sulfur dioxide, carbon monoxide, and ozone. About 250 major industrial emitters
of the newly listed substances, including chemical plants and oil refineries, would have to install "maximum achievable control technology"
(MACT).[1]

There are two major problems with Title III. The first is that the
available evidence indicates that the costs greatly outweigh the benefits. In fact, the evidence suggests that while the costs of added controls are large, there may be little or no benefit from the added regulation. The strongest claim to a benefit from Title III is that it may reduce
the risk of getting cancer. But the risk from these pollutants is uncertain, and now appears to be smaller than risks that we accept every day
—risks caused by others without the consent of the person at risk, just
as any risks from air toxics would be for people who lived near the
toxic emissions. In contrast, the costs of the regulation are substantial
and are directly paid by a relatively small number of industrial firms.

The concentration of large costs raises a second major problem: that of fairness. A few firms are being made liable for very costly control measures even though there is little evidence that they have done anything wrong or that the emissions to be controlled at such great cost are actually imposing (or would impose) significant harm on anyone or on the environment. To impose such costs to avoid unconscionable harm on even one person might be perfectly just. But no such harms or risks are demonstrable.

The following section of this chapter will discuss what is known about the health risks from air toxics. This information will help to make clear the nature and extent of the problems in the 1990 Act. The next section will explain why, even with highly competent and dedicated individuals guiding the process, the incentives facing voters, politicians, and bureaucrats bring about these results. Finally, some recommendations will be offered on how these incentives could be changed, so that potential harms from air toxics might be controlled in a more positive way.

What Do We Know About the Risks of Air Toxics and the Benefits of Reducing Them?

Pollution control is special as a policy issue only because of the knowledge problem. If a particular pollutant is released at a specified time and place, who is (or would be) harmed, and by how much? If we could easily learn the answer to that question for any given (or contemplated) release, a simple solution would be available: Forbid those that would be truly harmful, just as we stop personal assaults, and for the same reason—imposing such harm is wrong. In fact, common law has for many decades offered this sort of protection. A plaintiff who can show that he suffers real and substantial harm (or is about to be harmed) by an emitter of air toxics can generally get court-ordered relief in the form of damage payments from the polluter and/or an injunction to stop the emissions.

With full knowledge of this sort, potential polluters know that harmful emissions will not be tolerated and will find other times and places for their releases, making them less harmful and thus permissible. Or they will simply find other, nonpolluting ways to achieve their goals just as would-be perpetrators of assault refrain from using assault to get what they want when a court, knowing who did what to whom, would discipline them.

The problem, of course, both in the case of polluters and muggers, is in showing who has done what to whom. Has the accused polluter or mugger actually caused harm? Often the knowledge about what effects pollution is having, or will have, is difficult to obtain. The common law will not punish defendants for alleged harms if plaintiffs cannot show by weight of the evidence that the defendants caused those harms. So instead, possible victims of pollution turn to regulation precisely because the information available to judges and juries about pollution and its effects is so sparse. Unfortunately, turning to regulators does not provide the information that is needed for rational pollution control: how harmful is the pollution in various situations?

Information is not the only requirement for proper control. There is also the question of what constitutes harm. Even the clear presence of some degree of unwanted activity should not automatically trigger enforcement activity by government. For example, if I run to get on a train and accidentally knock you down, but then apologize and help you up, my "assault" will probably go unpunished so long as there is no intent to harm and no damage. Some tolerance is expected, in any society, when incidental damage is below some threshold, or *de minimis* level.

How do we decide what is a *de minimis* outcome or air pollution risk? One revealing way to think about this question is to ask whether the public accepts risks in other circumstances that are comparable to the risk from air toxics. Goldstein et al. offer data on an involuntary risk that is comparable to harm from airborne toxins: the risk of death due to an airborne object—a falling airplane—faced by people on the ground. Their statistics show that for a seventy-year lifetime, the risk of death from that source is 4.2 per million. Today, in other words, an average American citizen who otherwise would live to age seventy is exposed to a 4.2 per million chance of being killed by a falling airplane over her lifetime.[2] This danger is very small, and appears to be acceptable to U.S. citizens. How do we know that? One indicator is that there is little pressure to decrease the risk by sharply upgrading regulatory requirements for airplane maintenance and pilot training or by, say, banning pleasure flights. Arguably most existing regulations were put in place to protect passengers and crew, rather than groundlings. Perhaps somewhat greater risks of air crashes would be tolerated if only the groundlings' safety were at stake, rather than the safety of other, additional people at the same time. A second indicator of the risk's acceptance is that people living near airports seem not to take any special precautions.

How does the 4.2 per million risk of death compare to the equally involuntary risks to neighboring citizens from the air toxics emissions regulated under Title III of the Clean Air Act Amendments? Title III has two general requirements for the 189 listed air pollutants. The First ignores risk entirely, requiring polluters to use Maximum Achievable Control Technology (MACT),regardless of actual risks from the specific emissions to be controlled. The MACT requirement is expected to bring any risk to the most exposed resident of the area down to less than one per million—one-quarter the risk of death from a falling airplane. The second requirement is that if the residual risk exceeds that level, in the judgment of the EPS, the EPA Administrator must promulgate standards to further reduce the remaining risks.[4] Most pollution standards are driven either by technology requirements, such as MACT, or by a requirement to keep the risk of harm below stated risk levels. Title III imposes both constraints on emission levels for the listed air toxics. But as we will see below, the EPA calculates expected risks from any facility by using procedures that are expected to far overstate the risks. Let us look more closely at what is known about the risk of cancer from airborne toxic emissions, and how that information is utilized by the EPA.

The EPA has estimated that the leading hazardous air pollutants (about 100 in total) may cause between 1,700 and 2,700 additional cancer cases per year.[3] But outside analysts disagree. Paul Portney of Resources for the Future says that the EPA's risk assessment methodology is "designed to overestimate risk far more often than it underestimates it," and he estimates that 500 cases is "probably a generous estimate of the reduced cancer risk associated with the emissions controls."[4] In fact, he estimates that benefits may be as low as zero, because it is possible that no cancers will actually be prevented. His "most likely" estimate is that the benefits will be worth $1 billion,[5] still far less than the cost of the control technologies required under Title III of the Act, projected to run from $6 to $10 billion per year.

John Graham, of the Harvard School of Public Health, has also been skeptical. He testified before a congressional committee in 1989 that the daily inhaled levels of air toxics are several orders of magnitude (several multiples of ten) smaller than levels studied by scientists. "There is no direct evidence that outdoor exposure to toxic air pollutants is responsible for a significant fraction of disease or mortality.... EPA's widely quoted cancer risk estimates should be treated with caution because they are based on some questionable assumptions—for example, that humans are as sensitive as the most sensitive tested animal species, and that any human exposure to a carcinogen, no matter how small, results in some increase in cancer risk."[6] These assumptions and many others are made by the EPA in order to make their risk assessments conservative—that is, more likely to overstate than to understate the true risk. Often the overstatement is severe.[7]

As an illustration of the EPA's conservative assumptions and their effects, consider the EPA estimates of the risk of cancer caused by airborne emissions from coke ovens, which make coke from coal for use as a fuel in steel-making. The risk assessment process has been carefully critiqued by Frederick Reuter in an industry-sponsored consultant's report[8] to the National Research Council Committee on Risk Assessment of Hazardous Air Pollutants.

The EPA cites studies of steel workers inside coke-making plants as having revealed, as summarized by Reuter, "coherent patterns of excess relative risks of death due to respiratory system cancer, relating directly to both the level and the duration of exposure to coke oven emissions..."[9] The EPA used this to justify costly control measures to protect people who lived nearby. Using primarily data from the EPA, Reuter pointed out that workers in coke plants who were not working close to the ovens, and thus had much lower exposure to the emissions, showed "no coherent patterns of excess relative cancer risks."[10] In fact, the studies utilized by the EPA showed that among lower-exposure workers at coke plants, those with less exposure actually had a greater level of cancer mortality. Apart from the high-exposure workers located at the coke ovens, the studies showed no apparent added respiratory cancer risk from working in coke plants, even though the exposure to emissions there was much larger than in the surrounding neighborhoods.

Given the absence of evidence that coke oven emissions cause off-property cancer risks, how could the EPA justify its demand for expensive controls to reduce further these very low neighborhood exposures? The agency's risk assessments show that for maximally exposed individuals outside the plant, a risk estimate can be constructed that will suggest the possibility of an unacceptable risk.

In the absence of good data on human exposures outside the plant to air toxics emissions, on the carcinogenicity of calculated exposures, and on the other components needed to assess the resulting risk, the EPA fills in data gaps with a number of assumptions, each tending to overstate the health risks. The resulting upwardly biased numbers are typically multiplied together, resulting in a compounding effect that is likely to further overstate the risk. Reuter's report provides details on the product of the effects of EPA assumptions, each of which is constructed to bias the risk assessments upward, in order to avoid possible understatement of the risks. (Use of unbiased estimators, if the EPA had done so, would have resulted in estimates equally likely to understate or overstate the true risk.) The biased estimators that the EPA actually used can be hundreds, thousands, or tens of thousands of times higher than the unbiased or "maximum likelihood" estimators.

Reuter uses EPA data and risk assessments at the Clairton, Pennsylvania, coke-making plants of USX Co. to show the degree of overstatement when several conservative assumptions are used in place of actual data on the various components of EPA risk assessments. EPA models used to assess the cancer risk from coke oven emissions to the hypothetical "Maximally Exposed Individual" (MEI) supposedly living at the edge of the plant's property,[11] yield a strongly biased risk assessment. Reuter's study uses primarily EPA data and some more sophisticated EPA models to challenge the validity of those risk assessment findings. Table 2.1, reproduced from Reuter[12] shows, for each of eight components of the risk assessment, a minimum and a maximum magnitude adjustment required to change the EPA's biased component estimator to a "most likely" estimate of each component. Reuter estimates that together, these components must be changed by enough to reduce the estimated cancer risk by about five orders of magnitude.

The EPA had calculated that without additional controls cancer deaths caused by the Clairton facility would amount to 2.83 per hundred people, indicating that a person so exposed would have a risk of dying from cancer that is increased by more than one chance in forty. Instead, Reuter calculates from much the same data base that the best estimate of increased risk of death from cancer is not one in forty persons, but about one in 10,000,000 persons.

As calculated by Reuter, the risk to a person on the ground of death from an airplane falling from the sky is roughly 250,000 times greater than the risk of dying from cancer from maximum exposure to coke oven emissions. If Reuter is correct, the Clairton risk might best be viewed as *de minimis*, or well below the threshold level of risk that Americans find tolerable, since the far greater risk from the falling airplane is easily accepted by regulators and by the public. In trying to be so protective by distorting specific risks, the EPA may well distort resource use so much as to reduce incomes sufficiently to cause more, rather than fewer health problems. The recent literature on this general topic supports that hypothesis.[13]

If the Clairton controls are representative, then Title III appears to cause a massive waste of resources, since the expenditures represent

Table 2.1
Estimated Reductions in Maximum Individual Risk
As a Proportion of Baseline Risk*

Revision	Revised Risk Estimate as a Proportion of Baseline Risk Estimate	
	Minimum	Maximum
1. Maximum likelihood estimate of unit risk factor	0.064	0.064
2. Nonlinear dose-response relationship evaluated at predicted ambient pollutant concentration	0.11	0.11
3. Up-to-date emissions data	0.25	0.40
4. Detailed representation of multiple emissions release points	0.25	0.75
5. Site-specific air dispersion modeling	0.10	0.10
6. Concentration estimates for actual residence locations	0.50	0.50
7. Expected duration of residence	0.28	0.43
8. Spatial activity patterns	0.40	0.70
Cumulative reduction	2.5×10^{-6}	3.2×10^{-5}

*Associated with individual revisions to EPA risk assessments for coke-making facilities at Clairton,Pennsylvania. Source: Reuter (1991)

real resources devoted to actions that in many cases are probably useless. It seems quite likely that at many facilities, congressional assumptions that health risks are present that justify the imposition of MACT standards are false. Similarly, it seems likely that EPA's risk assessments exaggerate estimated risks so much that their findings will often be invalid. Even if the risks are real, a far simpler, less disruptive and less costly solution could be adopted. For example, the polluter could be allowed to try to purchase the land on which exposures otherwise would occur and to see that no one spends enough time there to be harmed. Title III is, after all, intended to protect residents from air toxics.

The fact that EPA (and other agency) risk-reduction or health-based directives are not well-founded is well known and understood by the professionals involved. The extreme cost of the tiny additions to personal safety associated with some environmental rules has also been well known since before the CAAA were passed in 1990.[14] Yet air toxics policy is one of many similar outcomes of a regulatory process that has been going on for more than twenty years. The following section provides some insights on how that has happened.

How We Got There and Why: The History of Title III

The Clean Air Act Amendments of 1990 were not written on a blank slate.[15] The 1970 Clean Air Act had previously instructed the Environmental Protection Agency to regulate hazardous air pollutants. The EPA came up with regulations for seven chemicals over the next nineteen years. And, as of 1989, four of those were subject to court-ordered reconsideration. One major problem for the EPA was the fact that over all those years—and even now—there has been little or no good evidence of health problems caused by most air toxics at the low outdoor concentrations most citizens experience. (The exposure of workers to these pollutants at much higher concentrations is another matter, regulated by another agency under different laws.) However, a health-conscious public, worried by Rachel Carson's 1960 book, *Silent Spring* and other environmentalist writings, dramatized stories in the press, and organized environmental groups that fed public fears, pushed a sensitized Congress toward ever more stringent environmental controls. EPA activities themselves influenced this process.

The Role of the EPA

Prior to the 1990 Amendments, the Clean Air Act required the EPA to promulgate air toxic standards that would protect health with an "ample margin of safety." Cancer was a particular concern, and the EPA used a linear dose-response model for carcinogens. In this model, the risk of cancer causation is proportional to the person's exposure to the chemical. The model allows no possibility of a threshold level of exposure below which illness is not caused. So, for any air pollutant that could at some level of dosage (however high) cause cancer in any animal, and would thus be considered by the EPA to be a carcinogen in humans, no dose could be considered to be completely safe. In assuming that any scientific study showing that a pollutant could cause cancer in any animal was a carcinogen, and by using the linear model despite the objections of many outside professionals, the EPA had put itself in a box. With no direct evidence of harm to human health, or of a threshold level beyond which harm occurs for most toxic substances at the very low exposure levels that were likely outside, how could the EPA rationally set a standard? If the agency tried to do so using its no-threshold risk assessment models, no level would be completely safe; so how could the EPA allow any emissions at all?

The change to a technological standard was one answer. The technology could be required of all emitters of specified substances, without a need to show any particular harm or imposition of risk from the emission. Title III of the 1990 Act told the EPA to require each emitter to use the "maximum achievable control technology." The EPA administrator was required to report to the Congress within six years on the risk to public health remaining, or likely to remain, from regulated substances after MACT standards are applied.

To err on the side of caution in regard to potentially toxic substances, and spurred by the support of environmental groups and their congressional supporters, the EPA has interpreted the limited available evidence for these substances in a such a way as to exaggerate risks (as noted above), feeding the fears of citizens and encouraging demands for more stringent controls. The end result has often been to expend large amounts of resources to achieve marginal gains in risk reduction.[18]

Quite apart from the skewed estimates of risk, but intensifying their effect, is the added problem caused by the misleading way that the EPA often reports its risk estimates in the summaries that are all that most members of the public (and most EPA decisionmakers) read. That is true throughout the agency programs, and risks from air toxics are no exception. George Gray and John Graham of the Harvard Center for Risk Analysis reviewed the EPA's *Cancer Risk from Outdoor Exposure to Air Toxics*[16] and concluded: "In our opinion, the 'Executive Summary' of the EPA's report—and especially the summary table—has misled journalists, policy makers, and the American people about what is known about the carcinogenic effects of certain air pollutants."[17] They provide quotes from several important journalists, from William Reilly (EPA Administrator at that time) and from a prominent environmental leader, indicating clearly that all of them had misinterpreted the report in ways that reflected the report's summary, rather than its more careful and accurate main narrative, which had supporting data and calculations. Due in part to the presentation of information by the EPA, many members of the public were in no mood to put up with carcinogens and other "toxics" in their air. The public, too, played a part moving us to the 1990 Amendments.

The Role of the Public

Concern by the public about air pollutants, and especially about those that may cause cancer, is easy to understand, and it is not simply the creation of exaggerated risks emanating from EPA risk assessments and public statements. Cancer is a very large problem; about one American in four will die of cancer. And it is well known that in sufficient doses, certain toxic chemicals emitted into the air can cause cancer. Is there a connection?

In at least one area of the nation, where several of the toxic air pollutants are emitted due to a concentration of petrochemical plants, the death rate from cancer is abnormally high. This is the area of the lower Mississippi River called "cancer alley," much covered by the press. For decades, Louisiana's Gulf coast has been a center of oil and chemical plants. Many people assume that the chemical plants are causing cancer along the Gulf Coast.

But available evidence indicates that assumption is wrong. In 1992, the Louisiana Cancer and Lung Trust Fund Board reported on its study of cancer incidence and mortality. The incidence rate of virtually all cancers was "at or below the national average," reported Joel Nitzkin, Director of Special Projects for the Louisiana Office of Public Health.[18] (The exception was lung cancer in males, but Nitzkin attributed more than 90 percent of these cancers to cigarette smoking.)

Yes, death rates from cancer were high. But the reason, Nitzkin explained, was not that chemical plants were inducing more cancer, but that people with cancer were not getting timely, adequate medical care. That is, shutting down the plants would reduce pollution but not cancer. In fact, if shutting down the chemical plants reduced residents' income, it might reduce their ability to get adequate medical assistance, thus increasing cancer deaths.

This incident illustrates how easy it is for the public (and even experts) to misread information about risk. When misreading is combined with exaggerated risk estimates and a political and bureaucratic system ready for action on the issue, the result is a series of major government programs that are unlikely to improve public health or welfare.

Information is an important determinant of the effectiveness of public policy, but incentives are another important factor. Much of the environmental policy in the U.S. is driven by the fact that voters see gains from reducing risk through the political process. The same can be said of support for police forces and the military—bureaus that reduce risk from crime or from foreign invaders. There is a major difference, however.

Voters know that as taxpayers, reducing risk will be costly to them. They also know that giving police the power of arrest entails some risk of false arrest for the voter, too. So they limit the power given the police, both by insisting on a budget and by limiting police powers. An alleged mugger or rapist is assumed to be innocent until the police and the prosecutor can prove him guilty beyond a reasonable doubt. Each citizen knows that he, or a relative or friend, could one day stand accused, even if they had in fact done nothing wrong. So each is wary of tightening the rules, for fear that the cost of doing so might be personalized. Similarly, in the case of regulating airplanes for added safety, most of us want to be able to fly at a reasonable cost, so we are reluctant to demand costly regulation of airplane safety unless we are confident that the benefits make those costs worthwhile.

In the case of the EPA and air toxics, however, voters may believe that they can benefit from risk reduction, however slight, while giant, faceless corporations pay for those benefits. Voters seldom see the potential for the EPA to bother them, since they seldom are emitters of toxic chemicals. We should not be surprised that they may support giving strong police powers to the EPA while demanding little proof when the EPA (or the Congress) alleges harm or risk from emitters of pollutants. It is especially interesting to note that public utilities got special breaks under Title III—no MACT standards were imposed on them until the EPA could show that existing risks justify the standards. Proof is required before utilities (whose rate are regulated so that by law increased costs are passed on to all customers) must meet MACT standards.[19]

Voters who see little cost to themselves will tend to support heavy expenditures to reduce risks, however small, even if evidence of harm is flimsy. From their point of view, they are simply "making the polluter pay." Seeing little chance of "false arrest" or unjust punishment for themselves, they may see no need to demand that the validity of the alleged harms and risks from emitters of chemicals are seriously questioned before large costs are imposed on those accused of the harms. As in the case of "Cancer Alley," support for environmental programs may be based on conditions which are simply misinterpreted. Whatever its basis, public sentiment on the issue is one important factor—but not the only one—that must be taken into account by the politicians who make the laws and have much influence over rule making.

The Role of Politicians

Candidates for office and elected politicians face pressure from voters to provide, among other things, protection from environmental harms and risks, real and suspected, especially protection at the expense of others. Politicians can deliver that by passing legislation to establish environmental policies and programs such as the CAAA of 1990, and its Title III addressing air toxics. But they also face concerted pressure in the opposite direction from those who will be liable for the cost of reducing emissions. On many issues at the federal level, the concentrated interests of affected firms (or unions or other organized entities) can have influence greater than their numbers as voters, partly because as organized interest groups, they have a strong lobbying incentive. In this situation, the economic theory of public choice predicts that a politician who wants to maintain a coalition of support to stay in office will tend to support the policy. The special favors granted over several decades to the airline industry and the trucking industry at the expense of consumers provide examples.[20]

The 1970 Clean Air Act, and the 1977 and 1990 amendments to it, imposed clean-up costs on well-established industry groups and offered the promise of cleaner air to the public. They can be viewed as a political response to the large increase in popular support for environmental legislation that began to develop strongly in the late 1960s.[21] The powerful, though ephemeral, political force exerted by unorganized citizens can be thought of as popular pressure. These acts featured distributed benefits and concentrated costs, suggesting that the political power of organized interest groups had been diminished. While public choice offers little explanation for these apparent reversals in the degree to which special interests influence policy, recent political and legal scholarship suggest that dynamic changes in the amount of political power held by unorganized citizens play an important role.[22]

During periods of heightened public awareness and activity on a specific issue such as clean air, politicians tend to be less responsive to interest group pressures and more responsive to public opinion. The loss of popular support on such a timely, salient issue that would result from opposing such legislation may have been judged so large that any loss of campaign contributions from the concentrated interests at that time became a secondary issue.

In fact, far from reducing such contributions, passage of the Clean Air legislation in the form that it took, both in 1970 and 1990, almost surely increased the need for those concentrated interests to increase their contributions. The legislation meant that those very interests needed more than ever to increase their access to politicians and their influence with them, for the Clean Air Act and its amendments largely delegated the lawmaking process to the EPA. The politicking on how the rules would be written, which would determine which firms, which industries, and which regions would be most affected and so on, had just begun when the legislation passed. The acts that were passed contained strong wording that declared that the EPA must take decisive actions within specified time periods, and that the people would then be breathing air clean enough to be protective of their health.

David Schoenbrod, formerly an attorney with the Natural Resources Defense Council and now a law professor, uses the Clean Air Act as a primary example in making his case that delegating the writing of the rules to the EPA gave politicians the best of both worlds: They promised tough rules and clean air, to be delivered within specified time limits, but they could avoid making the hard choices of which pollutants would be subject to which standards, and by whom.[23] Congress had taken the credit; the EPA would take the heat on the tough decisions, and meanwhile would have to pay attention to legislators who were pressing them to do what helped the businesses and other constituents, who had to constantly lobby the politicians as the rule-making process inched along. And when one of a legislator's strongest constituents was about to be harmed by a new rule, the legislator was likely to demand that the EPA inflict no such harm. Having claimed credit for protecting the environment, the politician could now claim credit for protecting the specific polluters from harm under the Act.

The various versions of the Clean Air Act had not come close to delivering on their promises. The EPA simply could not do all that was demanded, though some benefits in the form of cleaner air were produced, in exchange for the imposition of substantial costs.[24] The acts were, Schoenbrod and others argue, masterpieces of political craftsmanship.

Politicians have had a large role in regulating air toxics through the passage of legislation, although a smaller role than they sometimes claim. Lobbying by organized groups, from industry to environmental groups, has helped to shape the stands taken by various politicians. Elected officials who help shape environmental legislation are also influenced by popular pressure, which in turn is strongly influenced by press coverage.

The Role of the Press

What determines the press coverage of environmental issues such as air toxics? For the media, a health threat is always a good news story. The editor seeking readers and the reporter competing for space recognize that what is frightening catches readers' attention. Facts that reduce such fears do not make for an exciting story, unless the facts support a scandal on the part of those claiming crisis, or unless one scientist is flatly contradicting others.[28] The claim of crisis, however, can bring press coverage that stirs strong action.

The recent history of Uniroyal's Alar provides a case study in how a group, the Natural Resources Defense Council (NRDC) in this case, can use the press and populist pressure to further its goals.[25] Alar is a plant growth regulator that offers many benefits to apple growers. In 1980, the EPA began to review the risks and benefits of Alar to determine if it should be reapproved for use on food crops. Over several years, the agency was bombarded with pressure from both Uniroyal and NRDC, the former supporting approval and the latter opposed. Although each side had minor victories, neither source of pressure was overwhelming, and by 1989 the EPA had still not made a binding decision.

In early 1989, NRDC hired a public relations firm to dramatize the potential danger from Alar to sway public opinion strongly in favor of eliminating its use. Utilizing a new scientific study (which subsequently was widely criticized), along with a famous actress as spokesperson to gain media attention, NRDC brought the chemical to the attention of the press and the public in a very dramatic way, enhancing the entertainment value of the coverage it was cultivating. NRDC asserted that the danger to pre-school children from eating apples treated with Alar was intolerable. Alar was referred to in the television news media as the most potent cancer-causing agent in the food supply. In March and April of 1989, stories about Alar were featured almost daily in newspapers across the country.

As a result of this coverage, it appears that many citizens formed the opinion that Alar was harmful. Sales of red delicious apples, a variety of apple on which Alar was used, dropped dramatically during the period of heavy publicity. Some school districts banned apples from children's lunches. Some citizens were sufficiently frightened to write to congresssional representatives, who responded by considering a bill that would bypass the normal EPA pesticide procedures to ban Alar. The manufacturer took Alar off the market. The Alar issue, after nine years in a gridlock caused by competing interest groups, had been quickly resolved when popular pressure aligned with one of the interest groups. Donations to NRDC rose dramatically.

It is instructive to note that the EPA had earlier cited evidence (which was also subsequently revealed to be faulty) that Alar was at least as dangerous as later claimed by NRDC. This information was released to the press in a matter-of-fact format, without the link to pre-school children or the aid of celebrities. Without the emotional connection later presented by NRDC, the story did not receive follow-up attention from the press. Hence little populist pressure was formed. No change in the availability of Alar resulted from these earlier EPA claims. The role of the press, even when its members have no specific agenda beyond gaining readers and viewers, can be very large in determining what voters believe and thus the pressures facing politicians and regulators.

Improving Incentives, Reducing Regulatory Failures

It is one thing to recognize problems and their causes. It is another to reduce those problems. Reducing the problems in programs like the EPA's air toxics requires changing the incentives that brought those problems on. The EPA, the public, politicians, and the press are all parts of the problem, due to the incentives they face. Dealing with them is necessary if we are to improve the control of air toxics. This problem seems to be relatively new in the environmental area. How were pollution control policies controlled in the past? Perhaps we can learn from our history. To step back for a look takes us back to the common law.

The Move from Common Law to Regulation: Progress?

Beginning around 1970, when the environmental movement became a major political force, regulation largely supplanted the common law in the U.S. as a way to protect individuals against many classes of invasion and injury by others, including pollution. However, the common law was the traditional method of dealing with such invasion and harm, and despite its limitations, it has much to recommend it.[26] In addition, its standards can help us think more carefully about the goals and tools of regulation.

Historically in both the U.S. and Great Britain, courts required polluters to compensate those they damaged. The courts made these decisions in light of relevant precedents and context. Judges were usually local or state judges, rather than federal, and juries were the "peers" of the plaintiffs and defendants. Rules for acceptable levels of pollution were determined not by distant regulators, but by what was desirable by local community standards. At some times and in some places, a "balancing test" was used, weighing pollution harms against such factors as economic productivity.[27] Often, however, such balancing was left to the market: the court established the rights of the disputing parties, and then the parties could trade those rights for consideration, if both parties were willing. If a plant was determined to be harming a downwind party, and thus polluting illegally, the plant owner could try to buy either the permission or the property of the downwind party, who then had the right to trade or not. Alternatively, if the judge or jury gave the polluter the right to continue, anyone offended could attempt to make a trade with the polluter that would convince him to reduce or eliminate the pollution.

The common law had two major benefits. First, whoever was judged to have the right in a pollution case—the emitter or the receptor—the end result was decided by a local judge and therefore was ruled to be just by local standards. In addition, any inefficiency of the judgment was very unlikely to last long. Once the responsible party was determined, trades were possible to avoid inefficient outcomes. In common law, justice is decided first, and the "cost-benefit" comparisons are after that, with any changes unanimously agreed upon by the trading parties themselves.

Another benefit of the common law approach is that its basic rules, and the concept of a fair trial, are in the interest of each person who recognizes that he might be a plaintiff or a defendant. To bend the rules case-by-case for one side or the other is unwise, for one could be on either side next. Balanced and consistent rules and procedures are the result.

The body of common law was flexible. It evolved in response to technical change and other factors. Gradually, implicit rules developed that identified acceptable dangers and reasonable conduct. Awards to plaintiffs gave potential polluters an incentive to foresee and avoid damages that could result in costly liability awards. With advances in science, common law changed also, allowing for procedural and technical remedies that made it easier for judges and juries to pinpoint causes of environmental harms, to provide for appropriate redress, and to enjoin imminent dangers. The threat of costly judgments against polluters provided the incentive to foresee and avoid damages. Businesses were free to innovate technologies as long as their conduct caused no actual harm. The changes, on balance, brought an ever-safer nation.

When prospective risks were shown to present a "dangerous probability" of harm, judges could provide anticipatory injunctive relief by enjoining the activity. Thus, prior restraint of dangerous behavior was provided when clearly warranted. Freedom to carry on productive and beneficial activities was the rule, but with the accompanying responsibility not to injure others.

Of course, the common law was imperfect. Environmentalists correctly pointed out that such a system could fail to provide justice in cases where it was difficult for the court to know just what damages would result from a given emission or stream of a pollutant over time. Some harms, such as cancer, may manifest themselves in the future. Physical and medical science are incomplete in these areas. When the sources of pollutants are many and widely dispersed, it is difficult to pinpoint the culprit or assess the proportional harm caused by each source. This problem is serious in the case of air pollution that covers a wide area, such as smog in the Los Angeles area or acid rain that may occur over broad areas hundreds of miles from its source. And juries, of course, are not ideally qualified to assess science and health risk issues. Transactions costs are another problem in common law. Trials are costly, with the need for lawyers and experts.

So, direct government regulation—politically controlled, bureaucratic regulation—began to augment and replace the imperfect results of common law protection of rights. In the case of air pollution in the U.S., local and state governments took the lead, although technological advance and economic pressures were already helping to clean industrial pollution.[28]

Unfortunately, moving pollution control to the public sector did not eliminate the lack of knowledge that limits the effectiveness of the common law process. If courts cannot get the necessary information, then how can the voters, or Congress, or public servants? Voters, for example, are the same individuals who would comprise juries, but when they vote, and have their impact on regulation, they have not sat through a trial with its careful balance of presentations from both sides. Instead they rely for information mainly on a press that is not allowed by its consumers to present great amounts of detail and balance: entertainment in the form of villains and heroes, and of dramatic statements about risk, sells. Do political and bureaucratic decisions really utilize superior information? Certainly, regulators have access to legions of Ph.Ds, but they are constrained by what voters tolerate and demand politically. Politicians, installed by voters, have often ordered the EPA to avoid comparing benefits to costs, for example. Such limitations, combined with the narrowly focused goals of each office within the EPA, reduces the value of whatever information regulators possess.

While the transaction costs under common law can be large, they are large, too, under regulation—perhaps more so. Lobbying for legislation and for rule making is costly, and lawsuits are plentiful in regulation as well. Are transactions costs really smaller, in a regulatory regime? The common law itself, evolving as it had been prior to the rise of agency regulation, might have better promise in air toxics cases, especially where a single polluter is causing the alleged problem, and only a few people, who live (or who might hypothetically be assumed to live) nearby, are affected.

Because the common law cannot handle all problems, and because political forces have reduced its use in the current era, we turn our attention now to some changes that may help citizens to control regulatory projects such as the air toxics program, to bring its benefits more into line with its costs.

Controlling the EPA

Since EPA, like other agencies, is subject to—in the words of Justice Breyer—"tunnel vision," it is important to have checks on the behavior of its decision makers. Breyer suggests that a professional, government-wide cadre of professional risk assessors and risk management experts, housed in a central bureau akin to today's Office of Management and Budget be established. It would review line agencies' proposed rules and regulations, and propose a risk reduction agenda. It would not have the tunnel vision of similar professionals housed in line agencies, because their task would be more broad. They would seek alternative ways, for example, to reduce risk. Such a cadre of professionals, however, would not escape the influence or the vicissitudes of politics.[29]

A second form of control is legal review. That might come from the courts, or from a specialized administrative court within the agency. Are policies set by Congress and by presidential appointees being followed by the bureaucrats? Often they are not. As one career EPA lawyer put it, in his call for legal review of the EPA decisions, here in the context of risk assessments, "From personal experience, I know that my [line decision maker] colleagues will continue to ignore risk-assessment guidelines, and high-level managers will not require subordinates to comply. Unless the law imposes legal obligations, EPA's behavior will not change."[30] For review to be effective, the law and its purpose must be specified in advance, which is something we have seen that politicians who delegate are loath to do. The convenience of delegation is precisely to be able to state grand and ultimate goals, without getting into the specifics, where all the political pain resides. Still, as we will see below, there are ways to deal with that as well.

Informing the Public

How can the public, especially as voters, be given incentives to become more informed about risks and realistic alternatives for reducing various risks, so that they are better able to control politicians and the EPA? What would help make them better customers for serious press reports rather than relying mainly on the more sensationalized (and thus perhaps more intriguing) but often less accurate accounts of environmental problems and policy alternatives?

One way to sharpen voter interest in the facts is to bring the policy decisions as close to home as possible. Dan Benjamin, in his chapter in this volume, shows that risks which affect people most directly are the ones they are most likely to know about. That is part of the genius of the property rights, or common law approach: once rights are established, the individual decision maker controls her own destiny to a much larger degree. How much should be sacrificed to avoid a certain type of risk? One whose decision is decisive for herself will be much more likely to ponder that question seriously, and to seek solid information on it, than one who is simply one voter, and who thus will be statistically unlikely to have any effect on what actually happens at the federal level. The person who lives near a source of possibly dangerous toxics has a strong incentive to become knowledgeable about that risk, to learn whether legal action against it might be warranted. An investor has a strong incentive to avoid investing in a firm that is polluting in a way that will generate legal liability.

Decisions at the local level lie between private decision and national decision on the continuum of individual incentives to be informed. The issues at a specific pollution source, for example, are more directly relevant to the local individual, and both the costs and the benefits of better policy are more concentrated on fewer voters. Devolution of government decisions to the lowest level possible will help involve voters in the information-seeking and decision process.

Controlling Politicians

Schoenbrod points out that the ability to effectively delegate law-making powers to regulatory agencies has a pernicious effect on holding politicians accountable for the costs of their decisions. The politician can take the credit for declaring that the air will be made clean and safe, but the bureau, under delegation, can be stuck with the tough decisions about how the goals might be achieved, and who will be forced to pay. Schoenbrod calls for the Supreme Court to force an end to delegation by enforcing the delegation doctrine of the Constitution.[31] Others, such as Judge Douglas Ginsberg, think that the court cannot do the job, but think that the people and some leadership in Congress could force Congress to approve major rules under a new law, before those rules are allowed to take effect.[32] That would surely reduce the incentive to delegate decisions and should help hold politicians accountable. The problem, of course, is that Congress has little incentive to limit its freedom in such a way. It would take a strong act of political entrepreneurship, probably with much favorable press coverage, to bring pressure on Members to act against their own interest in such a way. Such limits were the purpose of the Constitution which, Schoenbrod argues, can indeed support such limits. At this point, however, neither the Supreme Court action that Schoenbrod wants, nor the entrepreneurial political effort needed by the Ginsberg plan seems to be forthcoming.

Controlling the Press

Since the public gets so much of its information on issues from the press, and since the press is instrumental in getting across whatever point the public might absorb, its role is critical to any improvement. Unfortunately in some respects, but thankfully in other ways, in a free nation only the public can effectively discipline the press. If the public, perhaps through a more localized governmental treatment of most environmental problems, can demand more of the press and be willing to spend the time and effort to absorb it, the highly competitive press in this country will almost surely deliver.

Concluding Remarks

A wide array of commentators has noted the problems inherent in the Clean Air Act, its various amendments, and in Title III of the 1990 Amendments. The sources of these problems can be found without resort to asserting stupidity among decision makers at any level, corruption, or lack of dedication among EPA officials. The problems stem from a lack of proper incentives throughout the public sector, including the lack of incentives for voters to become better informed. The move from common law to regulation has not been without the introduction of new problems in the case of air toxics. Institutional changes like those suggested can help to bring incentives more in line to help produce a regulatory regime that is less unjust and less wasteful of limited resources.

Notes

1. Bradley K. Townsend and Jane S. Shaw helped in the research and writing of this chapter, and PERC provided support for our work.
2. See "Clean Air Act Rewritten, Tightened," *Congressional Quarterly 1990 Almanac*, p. 275.
3. Bernard D. Goldstein, Michele Demak, Mary Northridge, and Daniel Wartenberg, "Risk to Groundlings of Death Due to Airplane Accidents: A Risk Communication Tool," *Risk Analysis*, 12, 3 (September 1992), pp. 339-41.
4. A concise overview of Title III can be found in Stephanie B. Oshita and Christian Seigneur, "Risk and Technology in Air Toxics Control: California and the Clean Air Act Amendments," *Air and Waste*, 43 (May 1993), pp. 723-28.
5. The air toxics regulations do not address all these supposed cases. According to EPA publication *Cancer Risk From Outdoor Exposure to Air Toxics*, (USEPA Office of Air Quality Planning and Standards, EPA-450/

89, 1989) only about 20 percent of the risk associated with toxic air pollutants comes from the major sources that the EPA is required to regulate. The rest come from mobile sources such as automobiles and minor sources like backyard barbecues.

6. Paul R. Portney, "Policy Watch: Economics and the Clean Air Act," *Journal of Economic Perspectives,* 4 (Fall 1990), pp. 173-81. The quote is from p. 178.

7. Ibid., pp. 178-79.

8. "Air Toxics: How Serious A Threat," *Consumer's Research* (March 1990), p. 21.

9. For example, the "maximally exposed individual" methodology used by EPA can be expected to estimate the 99.99th percentile exposure to the most exposed person, according to N. C. Hawkins (1984), "Conservatism in maximally exposed individual (MEI)predictive exposure assessments: A first cut analysis," *Regulatory Toxicological Pharmacology* 14 1984, pp. 107-17.

10. Frederick H. Rueter, "Supplemental Comments on EPA's Risk Assessment Methods, with Examples from EPA's Assessments of Cancer Risks Associated with Coke Oven Emissions," report presented to NRC Committee on Risk Assessment of Hazardous Air Pollutants, November 14, 1991.

11. Ibid., p. 4.

12. Ibid.

13. U.S. E.P.A., *Coke Oven Emissions from Wet-Coal Charged By-Product Coke Oven Batteries—Background Information for Proposed Standards,* EPA-450/3-85-028a, Research Triangle Park, North Carolina, Office of Air Quality Planning and Standards, April 1987. (Cited in Reuter.)

14. Reuter, p. 29.

15. Tammy O. Tengs, Miriam Adams, Joseph Pliskin, Dana Fafran, Joanna Siegel, Milton Weinstein, and John Graham, "Five Hundred Life-Saving Interventions and Their Cost-Effectiveness," *Risk Analysis,* 15, 3 (1995), pp. 369-90. See also, Albert L. Nicholas and Richard J. Zeckhauser, "The Perils of Prudence," *Regulatory Toxicology and Pharmacology,* 8 (1988), pp. 61-75 for a more thorough discussion of the logic behind this issue.

16. See, for example, John Morrall, "A Review of the Record," *Regulation,* 10 (November/December 1986), p. 25. More recently, see Tengs, Tammy Tengs, et al.

17. For some historical background, see Robert W. Hahn, "United States Environmental Policy: Past, Present and Future," *Natural Resources Journal,* 34, 1 (Winter 1994), pp. 305-48.

18. Stephen Breyer, *Breaking the Vicious Circle* (Cambridge, MA: Harvard University Press, 1993), pp. 10-19.

19. Ibid.

20. George M. Gray and John D. Graham, "Risk Assessment and Clean Air Policy," *Journal of Policy Analysis and Management,* 10, 2 1991, pp. 286-95. The quote may be found on p. 286.

21. Joel L. Nitzkin, "Cancer in Louisiana: A Public Health Perspective," *Journal of the Louisiana Medical Society*, April 1992, p. 162. Op cit. 4,
22. See Oshita and Seigneur, p. 724.
23. See Thomas Moore, "Rail and Truck Reform: The Record So Far," *Regulation*, November/December 1988, pp. 57-62; and A. Kahn, "Surprises of Airline Deregulation," *American Economic Review*, 78, 2 (1988), pp. 316-22 for example.
24. See D. Elliott, B. Ackerman, and J. Millian, "Toward a Theory of Statutory Evolution: The Federation of Environmental Law," *Journal of Law, Economics, and Organization*, 1, 2 (1985), pp. 313-40.
25. See J. Pope, "Republican Moments: The Role of Direct Popular Power in the American Constitutional Order," *University of Pennsylvania Law Review*, 139, 2 (1990), pp. 287-368; D. Farber, "Politics and Procedure in Environmental Law," *Journal of Law, Economics, and Organization*, 8, 1 (1992), pp. 59-81; and M. Levine and J. Forrence, "Regulatory Capture, Public Interest, and the Public Agenda: Toward a Synthesis," *Journal of Law, Economics, and Organization*, Vol. 6 (1990), pp. 167-98.
26. See David Schoenbrod, *Power Without Responsibility: How Congress Abuses People Through Delegation* (New Haven, CT: Yale University Press, 1993).
27. See Howard Latin, "Overview and Critique: Regulatory Failure, Administrative Incentives, and the New Clean Air Act," *Environmental Law*, 21 (Summer 1991), pp. 1647-720, for a strong critique of the Clean Air Act and amendments. Latin blames bureaucratic incentives, or a lack of them, for the lack of proper results. Unlike Schoenbrod, Latin takes the Acts as the will of Congress, rather than as a cynical ploy by politicians trying to "have it both ways" by intentionally passing legislation that it knows will fall far short of its stated demands and objectives.
28. See Stanley Rothman and S. Robert Lichter, "Elite Ideology and Risk Perception in Nuclear Energy Policy," *American Political Science Review*, 8, 2 (June 1987), pp. 387-404. This was the lead article in the leading political science journal. It was not subject to substantial attack by comments in subsequent issues. Their results seem not to have been refuted.
29. This account was drawn from Andrew Yates, "Risk Assessment, Risk Communication, and the Media: The Case of Alar," Working Paper 93-7 (1993), Political Economy Research Center, Bozeman, MT.
30. See the Meiners and Brown chapter in this volume. For a history of common law in regard to environment, as it evolved in the U.S. (becoming weaker in most jurisdictions to accommodate industrial development, then stronger again against polluters in the 1960s), see H. Marlow Green, "Common Law, Property Rights and the Environment: Analysis of Historical Developments and a Model for the Future," PERC Working Paper 95-12 (1995), Bozeman MT, PERC. For an excellent detailed account of the usefulness of common law in defending citizens and the environment against pollution, primarily in the Canadian context, see Elizabeth

Brubaker, *Property Rights in Defense of Nature* (London: Earthscan Publications, 1995).

31. For an overview of these aspects of common law in the United States, and their development, see H. Marlow Green.

32. See Indur M. Goklany, "Richer is Cleaner: Long-Term Trends in Global Air Quality," *The True State of the Planet* (New York: The Free Press, 1995), pp. 347-48.

33. For example, the Congress's Office of Technology Assessment, which had been staffed almost entirely under the direction of a Democratic-controlled Congress since its inception in 1972, was criticized by Republicans and abolished by a Republican Congress in 1985.

34. Alan Carpien, letter to the editor, *Washington Times*, June 18, 1995, p. B5.

35. Schoenbrod, chapter 11.

36. Douglas Ginsberg, "Delegation Running Riot", *Regulation*, (Summer 1995), pp. 83-87.

3

Toxic Torts by Government

Bruce L. Benson

The explosion in the number of toxic tort suits filed against private firms suggests that the toxic tort issue is a problem of "market failure" (Trauberman 1983, 3-7). That is, toxic torts apparently arise because private firms create significant negative externalities in the form of health costs for their workers, customers, and neighbors. These costs may be inflicted intentionally, through avoidable negligence, or under strict liability, simply by doing certain things regardless of care. Thus, toxic tort litigation is presumed to involve an appropriate use of the courts to correct for market failure.

The facts are often quite different than these presumptions. There is considerable debate within the scientific community as to whether many of the alleged harms are valid. But evaluation of this debate is left to the experts in order to focus on another issue: the fact that even if alleged harms are real, a substantial portion of the "blame" for the toxic tort explosion should be placed on government. The rapid growth in toxic tort litigation is largely a reflection of government failure, rather than market failure.

There are several dimensions to this government failure, just as there are many dimensions to the toxic tort problem. To the degree that allegations about toxic harms are valid, the costs of toxic torts includes tragic consequences for human health. Additional costs include investments in avoiding future toxic exposure through greater care in the production and use of toxic products, clean-up efforts for toxic dumps and spills, the production and dissemination of information about toxic risks, and dispute resolution that arises from allegations of toxic exposure. Government failures have raised all of these costs.

This chapter focuses on two of the government failures that contribute to the toxic tort dilemma. First, section I explains that as a major demander of toxic products and a major producer of toxic waste, the government is often the cause of toxic exposures. When government immunity from tort liability applies, however, government decision-makers have little incentive to disseminate information to potential victims of potential toxic harms. Assuming that exposure to toxic substances generates harm, the costs that arise from government use of toxic substances tend to be higher than the costs from harms arising from activities subject to tort liability. Second, government immunity can be extended to firms that contract with the government. As explained in section II, firms with government provided immunity also have weaker incentives to protect affected parties.[1] Concluding comments appear in section III.

I. Government as the Proximate Cause and
Last Clear Chance Avoider of Toxic Harms

Government agencies may be the largest single source of demand for potentially toxic substances. They use large quantities of allegedly toxic products in ways that put people at risk, often without warning the parties at risk or even acknowledging the risk. Yet government is often immune from tort liability (Federal Tort Claim Act, 1946; *Feres v. U.S.*, 340 U.S. 135 [1950]). Even when government is not immune, the decision makers responsible for exposing others to toxic harms are almost always immune. Immunity means that government decision-makers have relatively weak incentives to warn others about the hazards that government activities create, thereby increasing likely exposure to potential toxic substances. When general government immunity applies, those claiming toxic harms have little recourse but to sue the companies that supplied the products to the government.[2]

An Example: Agent Orange[3]

Agricultural chemical research had produced a number of synthetic compounds to regulate and suppress plant growth as early as the 1930s, but much of the research on herbicides had been undertaken by the U.S. Army. During World War II, the Army performed defoliant research at Fort Detrick, Maryland, where a number of defoliant compounds, including 2,4,5-T, were formulated (Schuck 1986, 16). The compounds developed by the Army were regarded as more effective, easier to apply, and safer than existing weed killers, so after the war they were made available to the private sector. The Army continued to test many herbicides in the laboratory and in the field. President Kennedy approved Operation Hades (later called Operation Ranch Hand) in which various mixtures (called Agents Blue, White, Purple, Green, and Pink) of 2,4,5-T, another powerful herbicide, 2,4-D, and other chemicals were sprayed on the jungles of Vietnam by low-flying aircraft. As U.S. involvement in Vietnam increased, so did the defoliation efforts. Agent Orange, consisting of equal parts of 2,4,5-T and 2,4-D, was introduced in 1965. Several chemical companies were compelled to provide the Army with Agent Orange under the Defense Production Act (Glasser 1986, 514).[4]

Agent Orange was very effective and largely displaced Agents Pink and Green, although White and Blue continued in use. The application of Agent Orange intensified after 1965, reaching a peak of 3.25 million gallons in 1969. Its use ended in 1970, but by then 11.2 million gallons had been sprayed over about 10 percent of South Vietnam's land area.

Potential dangers of herbicide toxicity in general and of Agent Orange in particular had been known by Army officials for some time. Monsanto, one of the largest producers of Agent Orange, informed army officials that 2,4,5-T was a toxic substance as early as 1952. A 1963 Army review of toxicity studies of 2,4,5-T concluded that there was an increased risk of chloracne (a severe but often treatable skin condition) and respiratory irritations, and that the risk was heightened when the chemical was applied in high concentrations by inexperienced personnel. (Note that there is little or no evidence to support many of the other alleged harms of Agent Orange; see Franklin [1994, 3-4], for instance.) The Army knew as much, and probably more, about the potential dangers of the herbicides as any company that manufactured them. The Joint Chiefs of Staff were also informed of potential health dangers of herbicides by the President's Science Advisory Committee in 1963. Hence, knowledge of potential dangers existed outside the Army. President Johnson's Science Advisory Committee apparently discussed the potential toxicity of 2,4,5-T in meetings between April and June of 1965.

The National Cancer Institute contracted with Bionetic Research Laboratories in 1965 to study the potential toxicity of a number of herbicides and pesticides, including both 2,4-D and 2,4,5-T. A preliminary report on that research was filed with the National Cancer Institute in 1966, indicating that 2,4,5-T caused birth defects in mice and rats exposed to relatively high levels, and that 2,4-D was also potentially teratogenic. However, these findings were not made public until 1969 when a scientist and opponent of the defoliation program leaked them to Ralph Nader, and also convinced President Nixon's science advisor to convene a meeting of scientists to discuss it. Before this, the Army had denied (perhaps correctly [Franklin 1994, 3-4]) that any serious danger existed. But while some officials insisted that the exposure levels in the experiments were much higher than what military personnel in Vietnam faced, the Army suspended all military use of 2,4,5-T, including Agent Orange, on April 15, 1970. They never reauthorized its use.

In light of such evidence, Judge Pratt, in reaching summary judgment on the chemical companies' government contractor defense in the early stages of the Agent Orange mass tort litigation, concluded "that the government and the military possessed rather extensive knowledge tending to show that its use of Agent Orange in Vietnam created significant, though undetermined, risks to our military personnel" (Schuck 1986, 99).

Similarly, Judge Weinstein, who took over the Agent Orange case and ultimately forced a settlement, found that the government had a central role in the Agent Orange case and that liability could not be fairly determined without the government's active participation. He noted that it was "not unlikely [that] the information necessary for an informed decision on Agent Orange was readily available to the government while it was ordering and using [it]" (Schuck 1986, 182). Since the government probably knew even more than the chemical companies about potential adverse effects of Agent Orange (Schuck 1986, 251), Judge Weinstein is reported to have said that litigating Agent Orange without the government being involved was "Like playing Hamlet without the Prince of Denmark" (Schuck 1986, 58). He believed the best way to settle the case was for the Veterans' Administration and Congress to be brought in, but "we cannot do that because of limited jurisdiction. The intelligent way to handle it would be if there is any liability [on the part of the chemical companies, for]... the VA to take over the whole thing, then to just have the manufacturers make a lump sum donation paying the cost of the damages, if any, attributable to Agent Orange" (Schuck 1986, 115).

Despite evidence of substantial knowledge by government officials of potential health hazards of Agent Orange, the government denied virtually all liability. As Schuck (1986, 24) notes, "If the VA had one principle firmly embedded in its bureaucratic mind, it was that any problems suffered by veterans exposed to Agent Orange ... were not, with the exception of chloracne, caused by the herbicide." Similarly, as Judge Weinstein worked to force a settlement in the Agent Orange case, and the defendants demanded that the government contribute to that settlement, the government's lead attorney on the case responded: "The United States declines to attend or participate in settlement negotiations or court settlement of this case because any settlement that calls for contribution by the United States is not warranted. This is the United States' firm position, and we anticipate no change whatever in any aspect of it" (Schuck 1986, 148). Judge Weinstein described the government's view as "benign detachment," characterizing it as "cruel" to veterans.

There are substantial questions regarding the scientific evidence of causality with respect to many of Agent Orange's alleged health effects (Franklin 1994, 3-4). Indeed, Judge Weinstein viewed the causal evidence to be insufficient to warrant a judgment for the plaintiffs in the Agent Orange cases. Thus, it may be that government officials were correct in denying any liability, but if this is the case then the producers of the product should also have been free of liability. However, the producers were manipulated by Judge Weinstein into a $180 million settlement. The defendants in the Agent Orange cases also incurred large legal fees, estimated to be in the $100 million range (Schuck 1986, 5). The defendants in the case brought action against the federal government to recover their litigation expenses and settlement costs. They were not successful, however, as the Supreme Court affirmed lower court rulings against them in 1996 (*Hercules Inc., et al. v. United States*, Yock, J., 25 Cl. Ct. 616; 26 Cl. Ct. 17; *aff'd*, 24 F. 3d 188; *aff'd*, 116 S.Ct. 981 [1996]).

Without doubt, however, the government must be considered the responsible party for any harm that can legitimately be attributed to Agent Orange. After all, (1) government agents had developed and tested herbicides since World War II and developed one of the two herbicides used in Agent Orange; (2) the government contracted for the production of the product, drawing upon the Defense Production Act to compel the firms to supply it; and (3) the government had been told of potential health harms by numerous sources, including Agent Orange producers, and still directed the application of the herbicide by inexperienced personnel who often were not warned of potential harms and not required to take protective precautions (Franklin 1994, 3-4).

Assuming alleged harms are valid, because the government is immune from prosecution for such harms, particularly in the case of military activity, officials who had knowledge of such harms had little incentive to warn or train personnel to mitigate the danger. Of course, the defoliants may have saved many U.S. military personnel from harm or death in combat. Thus, subjecting some people to the alleged dangers of Agent Orange might have been justified on some cost-benefit grounds; but if some of the alleged harms are valid, then failure to train personnel in safety procedures and require their use made the costs of using it substantially higher than they had to be.

To the degree that Agent Orange is a producer of toxic harms, the case appears to be one of gross negligence on the part of government officials, but immunity for such officials can apply even when potential harms are intentionally inflicted.[5] For instance, in *U.S., et al. v. James B. Stanley*, (483 U.S. 669, 107 S. Ct. 3045 [1987]) the Supreme Court ruled against a former serviceman who filed a suit against military officers and civilian employees of the government, seeking to recover for harms arising as a result of a secret and intentional Army experiment. Master Sergeant Stanley had volunteered for a chemical weapons testing program in 1958 but was secretly administered LSD as part of an experiment to test its effects on human subjects. He was not informed of what he had been administered until 1975, when he received a letter from the Army asking for his cooperation in a study of the long-term effects of LSD on "volunteers who participated" in the 1958 tests.

Stanley allegedly suffered severe personality changes as a result of the LSD, including violent behavior which he was unable to recall, and this in turn led to his discharge in 1969 and the end of his marriage a year later. An administrative claim for compensation was denied, so he filed suit alleging negligence in the administration, supervision, and subsequent monitoring of the LSD testing program. However, the government denied liability, so the claim was amended to include a claim against individual officers and civilians involved in the experiment for violations of his constitutional rights. The 1987 Supreme Court ruling denied these claims as well.

What about other toxic tort problems? Consider just a few examples. The Department of Defense (DOD) apparently knew of potential risks associated with experimental drugs and vaccines before the Gulf War, but administered them to troops anyway, with no warning of potential effects and no monitoring afterwards (Ritter 1994). The DOD has also identified 10,439 suspected hazardous waste sites on active military installations that require cleanup or additional investigation. Over 100 of these DOD facilities are on the Environmental Protection Agency's "Superfund" National Priorities List of the worst contaminated sites in America (Calhoun 1994).

Similarly, "Decades of environmental, health, and safety abuses at U.S. Department of Energy nuclear weapons laboratory, production, and test facilitates have left an estimated 4,500 contamination sites covering tens of thousands of acres of land," and nine of these facilities are on the Superfund priorities list (Center for Defense Information 1994, 1). These nuclear weapons facilities have produced more than 99 percent of all high-level radioactive waste in America, no suitable method exists to permanently dispose of this waste, and several facilities are so contaminated that they probably will simply be sealed off from public access as "national sacrifice zones."[6]

Furthermore, while industry is recognized as the major producer of chlorofluorocarbons (CFCs such as Freon) which allegedly deplete the stratospheric ozone layer and pose hazards to human health as well as the world environment, 1989 estimates suggest that the armed services and their contracted weapons producers were responsible for about 37 percent of the nation's Freon emissions (Siegel 1992, 1). The DOD was still requiring the use of CFCs in specifications for weapons production and maintenance at the time, and it used CFCs directly for various cleaning and maintenance purposes. (The Pentagon used 5.75 million pounds of Freon in 1989 as a solvent to clean electric equipment and metal parts, for instance.) Similarly, every launch of the Air Force Titan 4 expendable launch vehicle and of NASA's space shuttle does as much to deplete the ozone as the total annual release of CFCs from most individual industrial plants during 1990 (Siegel 1990). And the list goes on and on.

Clearly, the government is a primary cause of many alleged toxic problems, and, as with Agent Orange, the government generally refuses to accept liability. For instance, as Chen (1984, 26) notes, "Even though the vast majority of asbestos victims worked in government shipyards, the United States Government continues to reject any suggestion that it bears a moral, if not a legal, obligation to the victims."[7] Millions of tons of asbestos were used to insulate ships built in naval shipyards, and the way it was applied exposed many thousands of workers to dangerous levels of asbestos, as compared to the relatively safe way that asbestos was used in building construction.

II. Extending Immunity: The Government Contractor Defense

When the victims of a toxic harm cannot sue the most likely responsible party, the government, they sue a more remote party. Traditionally, more remote parties were not likely to be liable under tort action, but as explained in chapter 5 of this book, tort law has changed dramatically over the last three decades. For instance, in a leading case, the 5th Circuit Court of Appeals held that asbestos producers had a duty to warn anyone likely to come in contact with asbestos that it posed a health risk, and therefore, that all asbestos suppliers were liable for damages to a worker who had worked in a government shipyard *(Borel v. Fiberboard Paper Products Corp.*, 473 F.2d 1076 [5th Cir. 1973], cert. denied, 419 U.S. 839 [1974]). An explosion of asbestos suits followed, as discussed in some detail in chapter 5, but they were filed against manufacturers as third parties rather than as employers (Chen 1984, 26).

The immunity that the government grants itself has been extended in some situations to private firms in the form of the "government contractor" defense.[8] Therefore, this defense can shield companies supplying toxic products to the government. In *Boyle v. United Technologies Corp.*, 487 U.S. 500, 108 S.Ct. (1988) at 2518, the Supreme Court considered this defense to resolve conflicting interpretations that had been developing, and stated that "Liability for design defects in military equipment cannot be imposed, pursuant to state law, when: (1) the United States approved reasonably precise specifications; (2) the equipment conformed to those specifications; and (3) the supplier warned the United States about the dangers in the use of the equipment that were known to the supplier but not to the United States."

Prior to *Boyle*, the government contractor defense had been raised for Agent Orange, and the judges involved indicated that the defense may have been valid, but pre-trial settlement precluded resolution. The defense was also raised in some asbestos suits, for instance, but with only minimal success prior to *Boyle*.[9] When the courts have rejected the defense, they generally relied on lines of reasoning that have apparently been altered by *Boyle* (Seifert 1989, 202-4). There are strong arguments for application of the defense in asbestos cases (Seifert 1989, 205-6; Slawotsky 1996, 938-41), and the defense has been successful in some recent asbestos cases.[10]

Boyle left open the issue of whether or not the product must have been manufactured for the military under a government contract (Seifert 1989, 204; Hensinger 1990, 373).[11] However, in *Carley v. Wheeled Coach* (991 F.2d 1117 (3d Cir.), *cert.denied*, 114 S. Ct. 191 (1993) the government contractor defense was expanded to encompass nonmilitary equipment (Buckhold and Goekjian 1994; Deibert 1994; Loy 1995; Moore 1994; Slawotsky 1996; and Townsend 1994).[12] As the defense expands, government suppliers, like government decision-makers, have weaker incentives to inform others about the hazards they face, as long as they inform the government, further increasing the potential for exposure to toxins used by government. Thus, the costs of health damages and/or deaths may rise, although perhaps the costs associated with toxic tort litigation—discussed in chapter 5—would fall if injured parties had no one to sue.

The potential implications of an expanded government contractor defense are illustrated by the class action suit against NLO Inc., the owners of a uranium processing plant in Fernald, Ohio. The plant was charged with emitting thousands of pounds of uranium dust into the air, discharging waste into a nearby river, and storing materials from the World War II Manhattan Atomic Project in leaking concrete silos. NLO was under contract with the Department of Energy (DOE). DOE officials revealed that the government knew of the problems for decades, and that the problems actually "stemmed from the government's capital investment and policy decisions" (Hensinger 1990, 376).[13] These officials also admitted that it was a government decision "not to spend money to stop the pollution or tell the public about it" (Sherman 1988, 3).

This admission by DOE officials was an attempt to protect NLO from liability under the government contractor defense, apparently to "encourage participation in the risky nuclear industry and to allow for the failure to implement health and safety measures" (Hensinger 1990, 376). Whether this tactic might work cannot be determined, however, since the NLO lawsuit was settled on July 26, 1994 (see Chesley et al. 1995, 587-90) for details).

Several legal scholars have been critical of the expansions in the scope of the government contractor defense (e.g., Deibert 1994; Loy 1995; Townsend 1994), but not for the incentive reasons suggested here.

Others have been more supportive (e.g., Buckhold and Goekjian 1994; Slawotsky 1996), and *given government immunity,* this support seems well founded. The real problem is that the government and its employees cannot be held accountable when they are the proximate cause of an alleged harm, and under these circumstances justice does not appear to be served by shifting liability onto a third party who acted under government orders and/or supervision.

III. Conclusions

When government is a direct and proximate causal agent for toxic tort harms, government officials have weak incentives to warn potential victims, either because of general government immunity, or because of effective immunity for government officials when damages are paid by taxpayers. Government is a major user of allegedly toxic products, ranging from Agent Orange to asbestos in shipyards and schools, and a major producer of toxic waste. Indeed, it would appear that the United States government is directly responsible for more potential toxic harms than any other single entity in the   country.

But the government can cause toxic harms in other ways as well. For example, legislative mandates, such as the Defense Procurement Act which was used to force production of Agent Orange, can directly lead to toxic exposure. Furthermore, the biggest remaining danger from asbestos may arise because of government mandates. A 1983 EPA survey found that only 2,600 of the 126,000 schools nationwide had asbestos problems (Chen 1984, 30). Nonetheless, in the Asbestos Hazard Emergency Response Act of 1986, Congress mandated that all schools start removing asbestos despite the fact that the EPA, the American Medical Association, and other health groups have advised against general removal because it will unnecessarily put many people at risk (Bailey 1992, 2).

Government created law can also lead to greater toxic exposure by influencing the incentives of important actors in the toxic substances arena. Government immunity may be extended to various producers of toxic substances through the government contractor defense, for instance. Such liability rules provides incentives for firms producing and using a potentially toxic product to be unconcerned about health problems.

Notes

1. Other types of government failure have toxic tort implications. For instance, government efforts to determine liability for toxic tort problems can be far more costly than the problems themselves, as explained in chapter 5 of this volume (Benson). In addition, laws can lead to greater exposures to toxic substances in at least two ways: (a) liability rules established by statute or court precedent can provide incentives for firms producing and using a potentially toxic product to be unconcerned about health problems (e.g., the government contractor defense), and (b) the government may mandate actions that put people at significant toxic risk. Examples of these and other costs arising from government actions are discussed in the context of the primary issues examined in sections I and II and chapter 5, but only tangentially. Additional examples are mentioned in the concluding comments in chapter 5 in order to suggest the potential scope of the role of government failure in toxic torts.
2. This statement should be qualified. In some cases the supplier may also not be liable. Prior to the 1960s, the suppliers of toxic substances to government and non-government entities alike had good reason to believe that they were not liable for harms resulting from government use, as explained in section II.
3. Much of this subsection draws from Schuck (1986).
4. Under this statute, performance of a procurement contract is compelled, and failure to do so involves potential criminal penalties of up to a year in prison or $10,000 in fines. Of course, the chemical companies did not object to being compelled, and apparently enjoyed substantial profit from the sale of Agent Orange.
5. Intentional government experiments on uninformed subjects have been commonplace. A large number of radiation experiments on human subjects were conducted between 1945 and 1970, for instance, some of which exposed large civilian populations ("65 Nuke Fallout Hits L.A.," *Oakland Tribune,* August 26, 1994).
6. See note 12 for more information on the extent of government involvement with hazardous waste sites.
7. Asbestos related torts are discussed in chapter 5. Also see Brodeur (1985, 250-53).
8. Key cases include *Feres v. United States* 340 U.S. 135 (1950); *In re "Agent Orange" Product Liability Litigation,* 534 F. Supp. 1045 (E.D.N.Y. 1982); *Mckay v. Rockwell International Corp.,* 704 F.2d 444 (9th Cir. 1983); *In re "Agent Orange" Product Liability Litigation,* 818 F.2d 187 (2d Cir. 1987); *Boyle v. United Technology Corp.* 108 S. Ct. 2510 (1988); and *Carley v. Wheeled Coach* 991 F.2d 1117 (3d Cir.), *cert.denied,* 114 S. Ct. 191 (1993). Additional cases are cited below and for discussion, see for example, Buckhold and Goekjian (1994); Deibert (1994); Loy (1995); Hensinger (1990); Glasser (1986); Moore (1994); Scadron (1988-89); Seifert (1989); Slawotsky (1996); Townsend (1994).

9. See for example, *Teff v. A, C, & S, Inc.* No. C-80-924M (W.D. Wash. Sept. 15, 1982); *Teff v. A, C, & S, Inc.* No. C-84-154M (W.D. Wash. Oct. 12, 1984); *In re: Related Asbestos Cases*, 543 F.Supp. 1142 (N.D. Cal. 1982); *Chapin v. Johns-Manville Sales Corp.* (No. S79- 072 (S.D. Miss Jan. 27, 1982); *Hammond v. North American Asbestos Corp.*, 97 Ill. 2d 195 (1983); *Nobriga v. Johns-Manville Sales Corp.* No. 55624 (Haw. Cir. Ct., May 24, 1982); *Plas v. Raymark Industries, Inc.*, No. C-78-946 (N.D. Ohio May 3, 1983); *Hansen v. Johns- Manville Products Corp.*, 734 F.2d 1036 (5th Cir. 1984), *cert. denied* 470 U.S. 1051; *In re: Maine Asbestos Litigation*, 575 F. Supp. 1375 (D.C. Me. 1983); and *McCrae v. Pittsburgh Corning Corp.*, 97 F.R.D. 490 (E.D. Pa. 1983). See Seifert (1989, 202-4) for discussion, but see notes 9, 10, and 11, and related discussion for post-*Boyle* cases.
10. See *In re Brooklyn Navy Yard Asbestos Litig.*, 971 F. 2d 831 (2d Cir. 1992) and *Niemann v. McDonnell Douglas Corp.*, 721 F. Supp. 1019 (S.D. Ill. 1989) in which the defense was applied to asbestos products for military purposes. But see Scadron (1988-89) for an analysis of asbestos cases in the context of *Boyle*, which concludes that the defense is not likely to be successful, and see *In re Hawaii Federal Asbestos Cases*, 960 F.2d 806 (9th Cir. 1992) for an apparently contradictory ruling. The uncertainty about the applicability of this defense simply adds to the uncertainty about many aspects of tort law discussed in chapter 5.
11. Other issues were left unresolved as well. For instance, *Boyle* considered a claim of defective design, but not claims of defective manufacturing or of failure to warn. Subsequent decisions have extended the *Boyle* doctrine to include failure to warn, however. See *Nicholson v. United Technologies*, 697 F. Supp. 598 (D. Conn. 1988), and see Hensinger (1990, 372) for discussion. Then, in *Harduvel v. General Dynamics Corp.* (878 F.2d at 1317 (11th Cir. 1989), *cert. denied*, 494 U.S. 1030 (1990)), a failure to warn case, the court found that the characterization of the claim was irrelevant in determining the applicability of the defense. Also see *Bailey v. McDonnell Douglas Corp.* 989 F.2d 794, 801 (5th Cir. 1993), *reb'g denied* 995 F.2d 225 (5th Cir. 1993), and for discussion see Slawotsky (1996, 941-45). Similarly, *Boyle* dealt with a procurement contract but the defense has since been applied for service contracts as well (*Guillory v. Ree's Contract Serv., Inc.*, 872 F. Supp. 344, 346 (S.D. Miss. 1994; *Richland-Lexington Airport Dist. v. Atlas Properties, Inc.*, 854 F. Supp. 400, 422 (D.S.C. 1994); *Lamb v. Martin Marietta Energy Sys., Inc.*, 835 F. Supp. 959, 966 & n.7 (W.D. Ky. 1993)). The issue of whether or not the defense can be raised when a product meets government specifications under government contracts but has also been supplied to nongovernment entities apparently is yet to be clarified, although some courts have differentiated between ordinary military equipment that has civilian applications and equipment which only has military uses (e.g., see *Bentslin v. Hughes Aircraft Co.*, 833 F. Supp. 1486, 1490 (C.C. Cal. 1993), finding that the defense is stronger when the equipment does not have a civilian use.

12. Also see *Wisner v. Unisys Corp.*, 917 F. Supp. 1501, 1509 (D. Kan. 1996); *Stone v. FWD Corp.*, 822 F. Supp. 1211 (D. Md. 1993); and *Johnson v. Grumman Corp.*, 806 F. Supp. 212 (W.D. Wis. 1992).

13. The federal government generates, transports, stores, and disposes of huge amounts of hazardous waste. A 1990 Congressional Budget Office (CBO) report noted that about 2,300 current federally-owned facilities are involved in hazardous waste activities, that more than 7,100 sites formerly owned by the federal government may develop hazardous waste problems, that an uncounted number of sites which had been in private hands were seized by the federal government, and that the federal government had contributed contaminants to many privately owned hazardous waste sites, such as the site owned by NLO, Inc. The extent of government liability was unclear according to the CBO, depending on how broadly Congress ultimately defines the government's liability (CBO 1990, 6). At that time, however, the Federal Agency Hazardous Waste Docket identified 1,099 federal facilities that were not in full compliance with standards for hazardous waste management, some of which had serious hazardous waste contamination problems. The Docket was incomplete, omitting all formerly owned sites, private sites where the government had contributed to the contamination (e.g., the NLO site), and federal facilities with only "small quantities" of hazardous waste. Furthermore, the Docket omitted facilities for which federal agencies had not yet reported their hazardous waste activities to the EPA. As another indicator of the federal government's contribution to the hazardous waste problem, of the 1,219 facilities on or proposed for EPA's National Priorities List of hazardous waste sites that pose the greatest risk as of November 1989, 114 were federally owned. It is not clear how many of the remainder were federally owned and then transferred to private hands (but some were [CBO 1990, 17-18]) or were privately owned but recipients of government generated wastes (although at least eighty of the 1,099 sites fall into this category, and the identification of contributors was far from complete [CBO 1990, 18-20]). Even if the Congress ultimately admits liability for cleanup of the hazardous waste sites that are federal government owned, created, or sponsored under contract (e.g., NLO, Inc.), the CBO noted that the Congress may not actually provide sufficient funds to meet the requirements that they set. Thus, they may admit "fault" and continue to generate toxic harms without cleaning them up, thereby creating toxic tort issues such as those involved in the NLO case.

References

Bailey, Glenn W. 1992. "Asbestos Litigation Monster Rewards Plaintiffs' Lawyers While Devouring Jobs and Economic Growth." *Legal Backgrounder* 24 (August 28): 1-4.

Benson, Bruce. "Rent Seeking on the Legal Frontier." Chapter 5 of this volume.

Brodeur, Paul. 1985. *Outrageous Misconduct*. New York: Pantheon Books.

Buckhold, William C., and Lisa D. Goekjian. "The Government Contractor's Defense to Product Liability Claims." 1994. *Commercial Law Journal* 99 (Spring): 64-92.

Calhoun, Martin. 1994. "The Military and the Environment." *The Defense Monitor* 23, 9: 1-7.

Center for Defense Information. 1994. "Nuclear Threat at Home: The Cold War's Lethal Leftovers." *The Defense Monitor* 23, 2: 1-8.

Chen, Edwin. 1984. "Asbestos Litigation is a Growth Industry." *Atlantic* 254 (July): 24-32.

Chesley, Stanley M., Louise M. Roselle, and Paul M. De Marco. 1995."The Plaintiff's Perspective: the Need for Class Certification in Radiation Cases." *Gonzaga Law Review* 30: 573-98.

Congressional Budget Office. May 1990. *Federal Liabilities Under Hazardous Waste Laws*. Washington, DC: Congress of the United States, Congressional Budget Office.

Deibert, Thomas S. 1994. "Product Liability - The Third Circuit Extends Scope of Immunity Provided by Federal Government Contractor Defense." *Temple Law Review* 67 (Winter): 1421-42.

Franklin, Jon. 1994. "Poisons of the Mind." *Activities* 14: 1-6.

Glasser, Jonathan. 1986. "The Government Contractor Defense: Is Sovereign Immunity a Necessary Perquisite?" *Brooklyn Law Review* 52: 495-531.

Hensinger, Mary C. 1990. "*Agent Orange* and *Boyle*: Leading the Way in Mass Tort Action." *Journal of Contemporary Health Law and Policy* 6 (Spring): 359-77.

Loy, Steven Brian. 1995. "The Government Contractor Defense: Is it a Weapon Only for the Military?" *Kentucky Law Journal* 83 (Winter): 505-28.

Moore, Kelly A. 1994. "The Third Circuit Expands the Government Contractor Defense to Include Nonmilitary Contractors." *Washington University Law Quarterly* 72 (Fall): 1435-47.

Neely, Richard. 1982. *Why Courts Don't Work*. New York: McGraw-Hill.

Ritter, John. 1994. "Report: Military Knew Risks of Gulf War Drugs." *USA Today* (December 8).

Scadron, Michael S. 1988-89. "The New Government Contractor Defense: Will it Insulate Asbestos Manufacturers form Liability for the Harm Caused by Their Insulation Products?" *Idaho Law Review* 25: 375-98.

Schuck, Peter H. 1986. *Agent Orange on Trial: Mass Toxic Disasters in the Courts*. Cambridge, MA: Belknap Press.

Seifert, Timothy D. 1989. "*Boyle v. United Technologies Corp.*: The Government Contractor Defense." *St. Louis University Public Law Review* 8: 189-206.

Sherman, Rorie. 1988. "Atom-Plant Disclosure: Knowledge as a Sword." *National Law Journal* (November 28): pp. 3, 22.

Siegel, Lenny. 1990. *No Free Launch: The Toxic Impact of America's Space Programs*. Sabattus, ME: Military Toxics Project.

————. 1992. *Operation Ozone Shield: The Pentagon's War on the Stratosphere*. Sabattus, ME: Military Toxics Project.

Slawotsky, Joel. 1996. "The Expansion of the Government Contractor Defense." *Tort & Insurance Law Journal* 31 (Summer): 929-45.

Townsend, Jake Thomas. 1994. "Ambulance Chasers Beware: The Questionable Expansion of the Government Contractor Defense." *Minnesota Law Review* 78 (June): 1545-73.

Trauberman, Jeffery. 1983. *Statutory Reform of "Toxic Torts": Relieving Legal, Scientific, and Economic Burdens on the Chemical Victim*. Washington, DC: Environmental Law Institute.

4

Common Law Approaches to Pollution and Toxic Tort Litigation

Jo-Christy Brown and Roger E. Meiners

Environmental law is now dominated by statutes and regulations. Less studied in recent years are the traditional common law theories that have been used to address legal issues related to environmental pollution and toxic tort litigation. The common law doctrines most frequently relied upon in environmental litigation, primarily because they create a framework that addresses and regulates land use and environmental assaults, are: (1) nuisance, (2) trespass, and (3) strict liability for abnormally dangerous activities.[1]

The following discussion of these three theories highlights points of interest and import related to each theory. Because the nuisance theory, as interpreted by American courts, has been most widely applicable to issues related to environmental litigation, it is discussed most extensively. A brief discussion of environmental issues related to water law, with its basis in riparian common law, is included. Water pollution has been the subject of the most litigation under the common law, so it will be discussed in some detail. At various points some of the major differences between statutory standards and those established in common law are contrasted. We finish with a review of how the law applied in general to instances of water and air pollution and how the enactment of environmental statutes have been interpreted by various jurisdictions to permeate some of the common law remedies for pollution.

I. Nuisance

The *nuisance* cause of action provides the backbone of common law environmental (pollution) litigation. The term "nuisance" is derived from the French word for "harm" (*nuire*).[2] While sometimes difficult to apply, the concept of nuisance is generally commonsensical. As Justice Sutherland said, in a famous bit of dictum, "Nuisance may be merely a right thing in a wrong place like a pig in the parlor instead of the barnyard."[3] Causes of action for nuisance claims can be either public or private, but nuisances may be combined public and private nuisances.[4]

A. Public Nuisance

Public nuisance is an act or omission which causes inconvenience or damage to the public health or public order, or an act which constitutes an obstruction of public rights.[5] Normally, only public officers (attorney generals or district attorneys) have standing to sue to abate public nuisances. However, individuals who show they suffer harms distinct from those suffered by the general public may also be granted standing to sue to abate a public nuisance.

Public nuisances are not limited to protecting interests in land. Claims of nuisance may arise from conducting certain businesses. For example, the Court in *Ballenger v. City of Grand Saline,*[6] heard a suit in which residents living near a "chicken house" complained of offensive odors. *Ballenger* involved the City seeking a permanent injunction to abate the operation of a wholesale chicken business. There, the City alleged various nuisances, including noise and odor. The Court of Appeals upheld the lower court's decision that the operation was "in fact and in law" a public nuisance, and that the City was entitled to permanently abate the activity.

The *Restatement, Second, Torts* ' SEC.821(C)(2) states that: "In order to enjoin or abate the public nuisance, one must have standing to sue as a representative of the general public, as a member of a class in a class action, or as a citizen in a citizen's action."[7] Unlike in past decades, when standing by parties other than public officials was rare, the common law theory of public nuisance can now be used effectively to address a wide range of environmental ills.[8] In actuality, any person who is injured can sue for equitable and monetary damages; courts will merely refer to the nuisance resulting in money damages as a "private nuisance."

B. *Private Nuisance*

Nuisance, it has been said, "is a good word to beg a question with."[9] *Private nuisance* is defined as a *substantial and unreasonable interference* with the use and enjoyment of an interest in land.[10] The interference may be intentional or reckless. In pollution law, the typical case involves a defendant which is operating in a way that is alleged to be offensive or harmful to the plaintiffs. The legal issue posed in such cases is whether the "act" of the defendant is an intrusion that is "sufficiently noxious" to give rise to a finding of nuisance.

While private nuisance claims have been traditionally limited to protecting the interest to "land," modern courts have been more generous, holding that the action protects all pleasures, comforts, etc., normally associated with occupancy of land. Current case law also allows for consequential damages to the "possessor of land interest" to allow recovery for injuries to his own health[11] and for loss of services of family members.[12]

Cases based on common law private nuisance claims generally discuss the "weighing process," of various factors involved. Generally speaking, the actor's conduct will be condemned as "unreasonable" if the gravity of the harm outweighs the utility of the conduct.[13] This weighing process is used by courts in both public and private nuisance actions. In discussing what factors illuminate the "gravity of harm," courts have considered:

1. Extent and character of the injury alleged;
2. Social value of the use invaded;
3. Burden of avoiding harm by the harmed party.

In discussing what factors illuminate the "utility of conduct," courts have considered:

1. Social value of the conduct;
2. Suitability to the locality to the activity;
3. Impracticality of avoiding the invasion.

In modern American case law, it is common for one or even several of these factors to emerge as determinative.[14] As a result, cases can be found to support virtually any and every proposition.

An interesting dilemma is created under today's statutory environmental law by the fact that some actions, which might have in the past been considered nuisances, are now permitted under various environmental regulations. For example, acid mine drainage that damaged marine life was held immune from a public nuisance claim because the discharges were administratively sanctioned under state law.[15] In some cases, what had been a common law nuisance was allowed to become respectable business.

C. Factors Analysis and Weighing Process

The following is an overview of the "factors" used by American courts to "weigh" whether or not a certain activity constitutes a nuisance.

Extent of Harm. The primary factor which influences courts hearing common law nuisance claims concerning pollution is the "extent of harm" being done. Remember that, under common law nuisance theory, liability must be premised on "substantial harm."[16] The determination of what is "substantial" harm is intended to be objective, and thus should deny relief to the particularly vulnerable plaintiff. Most courts concur that the harm must be "real and appreciable" to constitute a nuisance. However, the standard of substantial harm is not a difficult standard to meet.[17]

This requirement is particularly useful in the environmental context because it results in focusing limited resources at eliminating real (i.e., "substantial") problems. In this regard, environmental decisions based upon the common law nuisance principle may be more economically sound than action mandated by regulatory control, inasmuch as mandatory statutory compliance can (and often does) deplete limited funds by forcing industry to address tiny details, thus leaving nothing to address subsequently discovered larger and more serious environmental problems. Inflexible statutory compliance, with no regard to the "extent of harm," may also create an unreasonable economic burden on industry.[18]

When comparing statutory environmental law to common law theories, the common law "extent of harm" factor equates to the statutorily directed air or water quality "standards." Unlike common law requirements, the statutory standards are often based on extra-sensitive individuals. For example, the Clean Air Act of 1970 establishes standards specifically designed to protect the "old and infirm."[19]

However, a major issue in successfully litigating pollution cases is that they often deal with risks (i.e., events or harm that will not surface until the future), so issues of extent of harm (i.e., "substantial" vs. "no harm") are difficult for plaintiffs to prove under the requirements of common law. In addition, under nuisance law, the burden of proving the "extent of harm" falls upon the plaintiff, thereby creating a greater burden on the plaintiff than, for example, merely demonstrating statutory noncompliance.

Also of import in evaluating the extent of harm is whether the tortious conduct is "continuous and persistent" or "inflicts long lasting effects." Courts have generally held that, while a plaintiff may be expected to endure a short-term or slight annoyance, long-term or more serious and/or permanent harm may be remedied by a nuisance suit.[20] While nuisance law may excuse a single or brief failure of pollution control technology, which may result, for example, in an isolated invasion by chemicals or a one-time sewage overflow,[21] such acts may be actionable if a court, "weighing" the factors, determines that the extent of harm is sufficient to merit remedy.[22]

To establish harm at common law, when the harm is not obvious but arises from an increase in the risk of cancer due to exposure to certain chemicals, the risk is supposed to be one recognized by competent science. The Federal Rules of Evidence were codified in 1975. They state that scientific evidence is admissible if the expert qualifies as such and the testimony "will assist the trier of fact to understand the evidence or to determine a fact in issue."[23] Confusion about the standard for expert testimony, which arose because Congress chose to codify what had been a common law standard, was only partly clarified by the Supreme Court in 1993 in *Daubert v. Merrell Dow Pharmaceuticals.*[24] That case held that under the Rules, courts are to determine the validity of scientific evidence offered at trial that helps the trier of fact to understand the evidence. Some critics assert that the courts have little expertise to know what constitutes "scientific validity," so that, given the complexity of some cases, judges may not be able to consistently serve as effective gatekeepers to keep dubious expert testimony out of court.[25] Whether popular fears that have no basis in "generally accepted" scientific evidence will be allowed to help determine cases is now a major issue in toxic tort litigation and one reason why industry has been looking to Congress for statutory protection from litigation that may have been spawned by Congress.

Utility of the Offending Activity. .A second factor evaluated by courts in deciding common law nuisance pollution cases is, "How useful is the activity that is causing the nuisance?" Traditional nuisance cases assign liability to actors who are inspired solely by the desire to cause harm.[26] Likewise, activities with negligible social value (e.g., houses of prostitution, illegal gambling parlors, etc.), which result in nuisances, are easily dispatched by American courts. But in modern pollution cases, it is generally the defendant's goal to produce a useful product or provide a useful service, while employing many workers and investing millions of dollars. This combination of factors sorely tests the "weighing process" applied by courts in nuisance law.[27]

One related issue, which has posed problems for courts in pollution nuisance cases, is whether the judicial weighing process should consider broader social questions related to the welfare of individuals not before the court (i.e., the workers who will lose their jobs if the activity is enjoined by the court). Clearly, statutory approaches to environmental torts do not consider this issue. Likewise, traditional nuisance law prefers the narrower view, which does not take account of such issues. However, this narrow view is not universally embraced and, in fact, was ignored in one well known modern pollution case, *Boomer v. Atlantic Cement Co.*[28]

In *Boomer,* the claimants were landowners who sought an injunction to restrain the operator of a cement plant from emitting dust, noise, and vibration resulting from blasting activities. The Court noted the monetary investment in the plant (i.e., $45,000,000) and the fact that the plant beneficially employed 300 people in its discussion of "balancing factors," which might justify a less onerous remedy than the permanent injunction of the operation sought by the plaintiffs.

The Court also stated that the risk of having to pay permanent damages to the injured landowners should be sufficient to spur the operator to research and develop improved technical controls for minimizing or preventing the nuisance. Additionally, the Court emphasized that this civil "common law" action did not foreclose a regulatory action by a public agency, seeking proper relief.

The Court in *Boomer,* in support of its holding, noted that other jurisdictions have ruled that the granting of permanent damages is appropriate where "the loss recoverable would obviously be small compared with the cost of removal of the nuisance."[29] The Court held "...[I]t seems fair to both sides to grant permanent damages to plaintiffs which will terminate this private litigation."[30] Thus, it is clear that, in determining the utility of the activity involved, the pivotal question is "What goes on the scale in weighing the utility of the offending activity?"

One leading (but older) environmental case on "balancing factors" of the utility of the activity involved is *Madison v. Ducktown Sulfur, Copper & Iron Co.*,[31] in which the Court awarded damages but not injunction. There, the Court decided not to "destroy property worth $2,000,000 and wreck two great mining and manufacturing enterprises" merely to protect several small tracts worth an aggregate of less than $1,000.

The opposing view is expressed in *Hulbert v. California Portland Cement Co.*,[32] where the Court stated, "If the smaller interest must yield to the larger, all small property rights ... would sooner or later be absorbed by the large, more powerful, few..." That is, so far as possible, rights that have been involuntarily stripped do not have price tags for those who wish to have their rights restored rather than be compensated for the injury suffered.

In general, the value of defendant's enterprise should not be an issue in determining whether a nuisance exists. The courts have been split on whether that should be a valid issue in determining what remedy should be designed. That environmental regulations can take the view that relative costs are relevant may be seen in *Portland Cement Ass'n. v. Ruckelshaus*,[33] in which the Court held that an administrator had adequately taken into account information that concluded that the costs of air pollution control equipment could be passed on with no substantial competitive effects on the business.

The use of the balancing doctrine, for remedy purposes, is often politically popular and justified as economically efficient. This approach has dominated certain regulated industries, such as nuclear reactors.[34] A more mundane excellent example of this approach is seen in *Harrison v. Indiana Auto Shredders Co.*[35] There, a district court, deciding a nuisance claim based on allegations of noise, dust, and vibration pollution, permanently enjoined the operation of a company which was in the business of shredding junk automobiles. In addition, the court had also awarded compensatory and punitive damages, which totaled over $530,000.[36]

The Court of Appeals stated, "This case presents a very difficult question of how to balance the legitimate demands of an urban neighborhood for clean air and a comfortable environment against the utility and economic enterprise of a beneficial, but polluting, industry."[37] Seeking justification for its decision from a leading scholar, the Court opined, "Nuisance has always been a difficult area for the courts; the conflict of precedents and the confusing theoretical foundations of nuisance led Prosser to tag the area a legal garbage can.'"[38]

In discussing the "weight" of these factors, the Court of Appeals noted: ..."[T]he problem of zoning is a local one, governed by local law ... the appropriate local authority has zoned the property specifically for shredder use; and appellant has been issued a permit to so use the property."[39] Ultimately, the Court of Appeals overturned the injunction and held that the operator should have sufficient time to correct the offensive operations. In addition, the compensatory and punitive damages were overturned.

Technological Controls. A third weighing factor, considered by courts, is whether the offending activity can be abated by use of advanced technological controls. In this regard, nuisance law recognizes and encourages innovative technology to avoid "shut down" as the remedy to the nuisance. As early as 1907, judicial decisions associated with the copper and lead smelting industry exerted pressure on industry to develop technology to solve environmental problems.[40] In fact, some courts anticipated and even ordered technological control, regardless of the availability of such technological fixes.[41]

The decision in *Bove v. Donner-Hanna Coke Corp.* is an interesting example of one common law approach to complaints related to air pollution from the burning of coke. There, the Court held that the resulting pollution was only a "petty annoyance" to nearby residents and that the industry involved was "indispensable to progress."[42] Later, in 1955, the company remained immune to complaints based in common law, but on the basis that the plant was properly operating emission control equipment, and, thus, any emissions were "unavoidable."[43] Decisions such as these evidence the flexibility given to courts in deciding common law issues related to pollution, by use of the various "weighing factors."

Like common law, statutory and regulatory environmental law demands use of the most technologically advanced control available to industry. However, at common law, in the area of ordering technological controls, the assignment of burdens can be decisive. Arguably, the burden could be placed on the plaintiff to demonstrate that controls being used are not adequate, or not sufficiently advanced. However, another approach might be to place the burden on the defendant to demonstrate that all operational controls have been exhausted and, therefore, any remaining intrusion should be viewed as reasonable. Because investigation, studies, and expert testimony in this area require experience and scientific sophistication, the assignment regarding the burden of carrying this point is significant.

Location of Activity. A fourth weighing factor is the consideration by courts of the locale of the nuisance producing activity. A nuisance is often just the right activity in the wrong place.[44] A "nuisance per se" is defined as an activity which constitutes a nuisance anywhere it is done and regardless of how it is done (e.g., houses of prostitution, illegal stills). A "nuisance in fact" is only a nuisance because of where it's done. But even if done in right place, some activities may result in damage to neighbors.[45]

Priority of Use. Often, a factor in a claim of nuisance is, "Who was there first?" Case law has traditionally held that it is sound economic policy to treat a new owner who finds his property impinged upon by his neighbor's pollution activity of long duration (i.e., one who has "moved to the problem") differently from an owner who was situated on his land before the pollution generating activity commenced. The reason for this, of course, is that a new owner who comes to the nuisance has had the opportunity to bargain for a reduced value or price for the already impaired property, inasmuch as, by the exercise of reasonable diligence, he should have been aware of the pollution or nuisance. Additionally, past judicial decisions, based in common law, have not been sympathetic to purchasers who buy property next door to a problem, when there is a hint that the purchaser may have bought the property for the purpose of "blackmailing" the polluter into providing damages.[46]

Defense to Nuisance Claims. Because the vast majority of cases sounding in common law nuisance are premised on *intentional conduct* by the defendant, a claim of contributory negligence, which is not a defense to an intentional tort,[47] will not defeat a plaintiff's nuisance claim.[48] Other potentially more promising defenses are statute of limitations, estoppel, or laches.

Damages Available to Nuisance Claims. Today, courts hearing allegations of common law nuisance, connected with environmental issues, use remedies that fall into four basic categories: (1) damages; (2) land use accommodations; (3) technological accommodations; and (4) operational controls. Thus, remedy under common law is potentially more flexible than that afforded by a regulatory framework. An example of an imaginatively crafted remedy to an alleged environmental common law nuisance is demonstrated by Learned Hand's decision in *Smith v. Staso Milling Co.,*[49] where the owner of a summer residence alleged air, noise, and water pollution from a nearby mill. There, Judge Hand designed a remedy to deal separately with each alleged nuisance. Specifically, with regard to the water pollution, Judge Hand held that the injury was so substantial and deliberate that the defendant was flatly required to avoid further injury whether by installation of sufficient technological controls or by ceasing operation. As to the air pollution, the defendant was not required to close its operation, but rather was required to operate under the "best available technical remedy" to the air pollution. Regarding noise pollution, the defendant was required only to discontinue blasting operations at night, so as to avoid unreasonable noise disturbing the adjacent residents. Monetary damages were also awarded.

Two oft-cited modern cases illustrate how common law courts attempt to craft remedies in pollution situations. In *Boomer v. Atlantic Cement Co.,*[50] plaintiff-homeowners sued a neighboring cement plant that caused damage through dirt, smoke, and vibration. Rather than grant the permanent injunction against operation of the plant that the homeowners requested, the New York high court ordered the plant to pay permanent damages to plaintiffs. The court noted that it did not want to destroy an economically valuable operation, the cement plant, but that it should have economic incentives to reduce the damage inflicted on neighbors. If the plant could reduce the offensive results from its operation, its damage payments would be reduced. This case illustrates how courts attempt to weigh costs and benefits of the various parties involved.

Another approach was taken by the Arizona high court in *Spur Industries, Inc. v. Del E. Webb Development Co.*[51] As a land developer was expanding outside of Phoenix, it ran into the ugly fact of a cattle feed lot, which produces strong odors and flies. Even though the developer came later than the feed lot, the court recognized that as populations expand, land uses change. An injunction against the feed lot was ordered. It was given time to move to a new location; the cost of the move would be paid by the land developer.

The Arizona court joined the courts in some other states that do not go with the traditional rule of "moving to the nuisance." Under that rule, it would have been up to the developer to strike a deal with the feed lot operator to make it worth his while to move. Common-law rules differ across states, allowing us to observe the differences that evolve from competitive rules of law, rather than a national standard determined by strong political interests in the federal legislature.

Generally, the rules on damages concerning nuisance are straight forward. If the nuisance is permanent, the measure of damages is the depreciation in the market value of the real estate caused by reason of the nuisance.[52] Where the nuisance is temporary, and abatement can be accomplished, the plaintiff usually recovers for the depreciation in the value of the property during the period in which the nuisance existed, plus any special damages which have resulted.[53] Special damages recoverable in nuisance cases include direct damage to property, including crop losses, injury to livestock or domestic animals, plants, clothing, or homes. Likewise, owners of riparian land can recover economic losses associated with a decline in fishing, and/or destruction of swimming facilities.[54] In addition, nuisance plaintiffs occasionally recover reasonable costs to defend against further invasion. This might include costs associated with sound-proofing, landscaping, etc.[55]

Most often, courts in designing equitable relief combine technological and operational limitations, such as requirements to reduce production, install emission control devices, limit sulfur content of ore, control emissions released, control biproducts captured, compile and make available complete operation records, pay for experts or medical committees, provide medical examinations to persons who are allegedly affected, and/or establish funds to protect those who have been injured in the event that future injuries develop. Few cases based on common law claims have dealt with remediation of past pollution. Thus, while these cases may "stop future" pollution, they have not, traditionally, addressed the cleanup of the environment, which is the target of statutory litigation.

Some opponents of the use of common law claims in environmental litigation allege that the highly technical nature of pollution actions does not lend itself to interpretation by inexperienced judges. However, court appointed experts and masters have been relied upon in the past, and could be even more so in the future, to help formulate and monitor relief in complex environmental cases.[56]

II. Trespass

Trespass, occasionally invoked in environmental pollution cases, is a common law theory closely related to "nuisance." The distinction being that, where there is a direct and immediate physical invasion of the property (e.g., poison killing vegetation on property), the cause of action lay in trespass. If, however, the invasion was indirect (e.g., smoke and/or fumes), the cause of action lay in nuisance. Thus, in trespass cases, defendants are usually alleged to have conducted some activity which has resulted in an encroachment by "something" upon the plaintiff's exclusive right to possession of land. However, in recent times, the line between trespass and nuisance has become less clear. Now, the basic distinction is that trespass is defined as an *intentional invasion* of a person's exclusive possession of property, whereas a nuisance is defined as a *substantial and unreasonable interference* with the use and enjoyment of property.

In environmental litigation, however, both concepts of nuisance and trespass are often discussed without distinct differentiation between the elements of recovery, inasmuch as many types of conduct interfere both with a plaintiff's interest in possession and use as well as the enjoyment of his property. However, the theory of trespass may be of an advantage to plaintiffs, because the tort of trespass is completed upon the slightest tangible invasion of the plaintiff's property, whereas a cause of action for nuisance requires the proof that the use and enjoyment of the property has been substantially and unreasonably interfered with.

Two potential barriers to the use of trespass to address environmental pollution are that trespass requires a direct invasion of an object onto the plaintiff's land. Historically, some courts have concluded that when pollution is carried onto land by an intervening force such as wind, the contact was not direct. Other courts have held that the definition of "object" requires something larger and more substantial than smoke, dust, gas, or fumes.[57] However, other courts have held that gaseous and particulate fluorides from aluminum smelters constituted a trespass.[58]

The remedies available under claims for trespass and nuisance are very similar. However, one important difference between the two is that in many jurisdictions, the statute of limitations for nuisance is shorter than that for trespass.

III. Strict Liability for Abnormally Dangerous Activities

Another potential common law basis for environmental torts can be found in the theory of strict liability for abnormally dangerous conditions and activities. To succeed, a plaintiff need only prove that an activity qualifies as "abnormally dangerous" and was in fact the cause of his harm. However, the questions of whether or not an activity is "abnormally dangerous" is a question of law for the court.

The theory of strict liability for abnormally dangerous conditions is generally credited to *Rylands v. Fletcher.*[59] In *Rylands,* a mill owner in a coal mining country constructed a water reservoir upon his property. The water burst through the filled-up shafts of abandoned mines and flowed into adjoining mines causing damage. Justice Blackburn stated, "We think that the true rule of law is that the person who for his own purposes brings on his land and collects and keeps there anything likely to do mischief if it escapes, must keep it at his peril, and if he does not do so, is *prima facie* answerable for all the damage which is the natural consequence of its escape."[60]

American jurisprudence has followed this example but has placed the emphasis on the type of risk created and the location of the particular activity.[61] American judicial decisions have found that the following activities constituted abnormally dangerous conduct: (1) water collected in quantity in a dangerous place; (2) explosives or inflammable liquids stored in a quantity in the midst of a city; (3) blasting; (4) pile driving; (5) crop dusting; (6) fumigation of buildings with cyanide gas; (7) operation of oil refineries in densely-settled residential areas; and (8) operation of factories which emit smoke, dust, or noxious gases in the midst of town.

However, decisions of what may constitute abnormally dangerous conduct may vary from jurisdiction to jurisdiction. For example, the Court in *Turner v. Big Lake Oil Co.* refused to impose strict liability for the escape of salt water from ponds used by defendants in their oil operation, inasmuch as the Court concluded that the construction of such basins was necessary to the oil business.[62] Likewise, the Court in *Fritz v. E.I. DuPont de Nemours & Co.* declined to hold DuPont strictly liable for injury caused by chlorine fumes, inasmuch as chemical manufacturing is an accepted and recognized industry in Delaware and the use of chlorine gas is recognized in that industry.[63] Thus, the location of the activity may be decisive in such cases.

Nonetheless, tort of strict liability for abnormally dangerous activities is of increasing significance in environmental litigation, as today's courts generally reason that, quite simply, persons engaged in dangerous activity should bear the costs of damages caused thereby.[64]

IV. Water Related Common Law

Generally speaking, there are two basic doctrines for allocating water rights among water users. They are the common law of riparian rights and the basically statutory doctrine of prior appropriation. While there is some overlap, riparian rights generally control in the eastern states where water supplies have historically been no problem, while the doctrine of prior appropriation is generally followed in the western states, where water is scarce.

A. *Riparian Rights*

Common law riparian rights are basically concerned with the rights of riparian water owners with respect to their use of water in water bodies, such as lakes and/or streams. Riparian owners are people or entities that possess riparian land—that is, land which includes a part of a bed of a water course or lake or which borders upon a public water course or lake.

Riparian law is subdivided into two fundamentally different theories. First is the "natural flow theory" which was adopted from English courts by some American jurisdictions. This theory holds that all riparian owners are entitled to have the stream flow past their land as it would normally do in its natural state. The second riparian common law theory is the "reasonable use theory," under which each riparian water owner is protected merely from any "unreasonable uses" of water by other riparian owners, which causes harm to his own use of the water.

However, the *Restatement, Second, Torts* reflects a decision by its drafters (under the guidance of water authority Frank J. Trelease) to separate water pollution cases from the common law of riparian rights. This is seen in ' 849 which (Tent. Draft No. 17, 1971) which reads:

[xx]' 849. In General—Pollution

(1) An interference with the use of water caused by an act of conduct that is not itself a use of water but that affects the quality or quantity of the water may subject the actor to liability if the act or conduct (a) constitutes a nuisance, (b) constitutes a trespass, or (c) is negligent, reckless or abnormally dangerous with respect to the use.

(2) The pollution of water by a riparian proprietor that creates a nuisance by causing harm to another person's interest in land or water is not the exercise of a riparian right.

Thus, most disputes regarding water pollution between riparian owners are not dealt with under the theory of riparian rights, but are dealt with under the doctrines of nuisance, trespass, or strict liability for abnormally dangerous activities. For example, an industry with riparian rights which withdrew water from a stream and returned it in a polluted condition, in litigating the issue of whether or not such use deprived a downstream riparian owner of his water supply would base its action on riparian rights. However, to the extent that the cause of action was based on pollution of the water, the case would be analyzed under traditional common law theories of nuisance, trespass, or strict liability. Thus, the distinction is between "water quality" versus "water quantity" controversies.[65]

This *Restatement* analysis is based on common sense and is most often followed in all jurisdictions. Courts consistently treat water pollution cases as common law nuisance cases. However, some jurisdictions have invented a new tort of "pollution" which is for all intents and purposes indistinguishable from nuisance.[66] In fact, riparian rights pollution cases consider the basic nuisance issues such as (1) who was there first; (2) the state-of- the-art technology; and (3) extent of harm.[67]

However, in some cases, a claim under a riparian rights theory is more favorable to plaintiffs than other common law actions. For example, in jurisdictions which take seriously the natural flow preference, relief can be granted even when there has been no actual damage (e.g., where downstream landowner does not actually use the water).[68] In this way, the natural flow theory represents absolutism in common law water pollution control, in that it requires "clean water in, clean water out."

In addition, riparian law reflects the same remedial compromises used in nuisance cases—damages, land use accommodations, technological and operational adjustments. Likewise, an emphasis similar to that found in other common law actions is placed on the extent of hurt, and the character of the uses on the water course, as well as state-of-the-art technology to prevent pollution of the water.

B. Prior Appropriation

Under the theory of prior appropriation, the rule is that the one who first appropriates the water and puts it to its use acquires the right to continue to use the water against all claimants that follow—i.e., "first in time, first in right." This water rights doctrine clearly favors development and consumption of water as opposed to preservation of water quality, even if the consumption is blatantly wasteful. (Conversely, the natural flow theory limits every use which might adversely effect downstream quality or quantity.)

It is unclear whether jurisdictions which use the appropriation theory will shift the requirements of that theory for the protection of the environment. Many of the prior appropriation pollution cases that have been decided have been based on conflicts between junior upstream miners and senior downstream farmers. Those cases have indicated that the general rule is that, while a proper use of a stream for mining purposes necessarily contaminates it to some extent, such contamination or deterioration of the quality of the water cannot be carried to such a degree as to inflict substantial injury upon another user of the waters of the stream.[69] Thus, the standard seems to be one of "reasonable use" which is, of course, the standard in nuisance law. However, there is still the troublesome concept in appropriation riparian law which dictates that when a senior water appropriator pollutes the water to the detriment of junior appropriators, there may be no remedy, since the senior had the right to destroy the junior's rights, anyway, by taking all of the water for its own use. Thus, in appropriation water law jurisdictions, pollution of water is often addressed under other common law theories, such as nuisance.

V. Application of Common Law Doctrines to Pollution Cases

While not necessarily predictive of future judicial interpretation and application of the common law doctrines, a review of past decisions provides an overview of the wide variety of approaches used in applying common law theories to toxic tort claims. Remedies available to plaintiffs complaining of pollution can range from an order enjoining the defendant's operation to the imposition of monetary damages, sometimes including exemplary damages. The primary factor that will determine what remedy the court will provide appears to be simply "What relief does the plaintiff seek?"

While historically parties claiming "public nuisance" (i.e., others similarly situated suffered the same harm) were barred from seeking monetary damages and, thus, could only seek injunctive relief, such is no longer the case. Courts today tend to hold that any injured person is able to seek both equitable relief and monetary damages.[70] Nonetheless, when examining older cases, it should be remembered that plaintiffs remedial options were proscribed by the nature of the alleged polluting activity and, thus, courts occasionally strained to reach the outcome they "desired."

A. Case Law Related to Air Pollution

Frequently, courts deciding older air pollution cases attempted to craft monetary remedies that were arrived at through application of the various "balancing factors" discussed above at section I.C. For example, in a 1911 decision, the court in *Bliss v. Washoe Copper Co.*[71] heard a case in which citizens of Idaho sought to obtain an injunction to permanently close a large smelting facility. *In dicta*, the court noted that nine-tenths of the inhabitants of the city would be compelled to move in order to obtain a livelihood if the smelter's operations were halted.[72] Ultimately, the Bliss court denied the injunctive relief sought and remitted the case to the lower court for a determination of monetary damages.[73] Similarly, the court in *Mountain Copper Co. v. U.S.*[74] noted that where a plaintiff seeks injunctive relief to halt a lawful enterprise which is causing pollution, the court will balance the comparative injury to the plaintiff versus the potential loss to the defendant and, if necessary, will provide monetary damages rather than equitable injunctive relief.

Occasionally, courts have appeared to issue injunctions but have, in reality, crafted decisions that provide monetary damages to the plaintiffs. One such example is *Boomer v. Atlantic Cement Co.*[75] There, the court reviewed various balancing facts and, ultimately, granted an injunction which prohibited the operation of a cement plant. However, the court "conditioned" the injunction on payment by the defendant, *and acceptance by the landowners*, of permanent damages in compensation for servitude on the land. The court apparently believed this solution would force all parties to take into effect the balancing factors that would motivate both sides to settle their differences.

Other courts, in applying balancing factors, have provided plaintiffs with monetary damages beyond those necessary to compensate for actual damages. The court in *McElwain v. Georgia Pacific Corp.*[76] imposed punitive damages upon an owner for exposing neighbors to "noxious and toxic fumes, gases, smoke, and particles" from its paper mill. The court deemed exemplary damages appropriate because the defendant had not "done everything possible" to eliminate the problem.[77] Indeed, numerous other decisions, applying the balancing factors, have avoided the injunctive relief requested by plaintiffs in favor of awarding monetary damages.[78]

But in spite of the use of balancing factors by some courts, others have taken the approach that injunctive relief is the appropriate solution to pollution disputes.[79] An early example of this approach is seen in the decision of *Sullivan v. Jones & Lughlin Steel Co.,*[80] where the court enjoined the operation of furnaces that emitted smoke and dust on neighboring land. There, the court stated, "There can be no balancing of conveniences when such balancing involves the preservation of an established right...which will be extinguished if the relief is not granted against one who would destroy it in artificially using his own land." But even in this approach, courts have occasionally attempted to avoid totally enjoining a defendant's operation. For example, the court in *Georgia v. Tennessee Copper Co.*[81] enjoined the operation of a copper smelter, but ordered the injunction stayed if the operator performed four tasks ordered to improve conditions related to the facility's operations (e.g., reduced emissions and kept operation records available for review by a local doctor).

Since regulations have come to dominate air pollution law, not many common law cases have been litigated in recent years, but *Bradley v. American Smelting and Refining Co.*[82] gives an indication of how courts would be likely to handle such matters. The Bradleys, who lived four miles from a copper refinery run by American Smelting (ASARCO), sued ASARCO for damages in trespass and nuisance caused by the deposit on their property of airborne particles of heavy metals from ASARCO's smelter. The smelter had operated since 1905; it was regulated by state and federal air pollution laws and was in compliance with all regulations. The gases that passed over and landed on the Bradley's land could not be detected by humans without scientific instruments. The Washington supreme court held that ASARCO had the requisite intent to commit intentional trespass even though no harm was intended. The company had to know that it emitted particles from its facility. The deposit of microscopic particulates gave rise to an action for trespass as well as a claim of nuisance. However, for either cause of action to be successful, there must be proof of actual and substantial damages. The case was dismissed for lack of evidence of damage to plaintiffs or their property from the pollution.

B. Cases Applying Common Law Relief for Water Pollution

Frequently, courts have confused and intertwined common law doctrines of nuisance (private and public) and trespass with landowner and riparian rights. The result is a strange and inconsistent variety of precedential decisions, which seem to be particularly confusing in the area of water law.

Some courts have attempted to apply balancing factors to riparian rights cases, and on this basis have awarded monetary damages for pollution to a variety of water bodies including creeks,[83] "natural water holes"[84] and water wells.[85] In discussing various "balancing" issues and ultimately awarding monetary damages, the court in *Branch v. Western Petroleum* noted "We know of no acceptable rule of jurisprudence which permits those engaged in...important enterprises to injure with impunity those who are engaged in enterprises of lesser economic significance.[86]

As in the area of air pollution, some courts have determined that water pollution is best remedied by injunction.[87] An early example of this philosophy is seen in the decision in *Indianapolis Water Co. v. American Strawboard Co.*[88] There, the court heard a complaint against a manufacturer that discharged its waste into a river used by a water company to provide water to a city. The court stated, "[N]o plain, adequate, and complete remedy exists at law, and injunction will lie to restrain such discharge."[89] Likewise, the court in *Sandusky Portland Cement Co. v. Dixon Pure Ice Co.*[90] enjoined the use of streamwater by an upstream cement company upon finding that such use resulted in an increase of the temperature of the river (near the plaintiff's ice plant), which the court determined was sufficient to retard the formation of an ice field used by plaintiff to harvest and market commercial ice.

These examples demonstrate the diversity of approaches taken, as well as the wide variety of outcomes achieved by various courts, in attempting to apply common law doctrines and remedies to both air and water pollution.

VI. The Impact of Environmental Statutes on Common Law Remedies

In addition to the variety of common law doctrines and remedies imposed by various jurisdictions, courts have been forced, in recent years, to grapple with the effect of numerous federal and state environmental statutes on the rights of persons exposed to environmental pollution. In this regard, some courts have attempted to couch the issue of whether environmental regulations preempt common law claims in terms of "exhaustion of administrative remedies."

One example of this approach is found in *Ellison v. Rayonier, Inc.*[91] In *Ellison,* tideland owners brought an action for "wrongful pollution" against the operators of a cellulose mill. There, the court noted that because the defendant's plant was (assumed to be) operated pursuant to an agency permit and had not discharged effluent in violation of that permit, it could not be found to have caused a nuisance or trespass. The court referred to this as the "doctrine of primary jurisdiction" and stated that, where an administrative agency is vested with the authority (under a regulatory scheme) to set standards for industrial activity, courts will suspend their jurisdiction to entertain common law actions pending administrative decisions on matters within the agency's jurisdiction.[92]

Similarly, the court in *People v. New Penn Mines, Inc.*[93] noted that the (State of California's) Dickey Water Pollution Act had granted the state water regulatory agency with exclusive jurisdictional authority over cases involving water pollution, when raised by the state. (Note, however, that the *New Penn* court also mentioned that other California decisions have held to the contrary on exactly the same issue.[94]) Thus, with clearly divergent opinions, even within a single jurisdiction, the uncertainty of the effect of statutory law on common law claims remains.

Some courts have, apparently, spoken more definitively on the issue of the impact of state regulatory schemes on state common law claims. For example, in *Atlas Chemical Industries, Inc. v. Anderson,*[95] an appeals court held that the fact that a defendant had been issued a permit to discharge effluent into a stream did not constitute a defense to pollution claims raised by the plaintiff.[96] The *Atlas* court noted, however, that a defendant may present evidence of permit compliance and the use of modern pollution control devices in an attempt to mitigate damages.[97]

Another facet of this issue is what impact federal regulatory schemes have on common law, both federal and state. As early as 1981, courts held that federal acts preempted federal common law with respect to water discharge effluent standards.[98] These decisions held that federal courts lacked the authority to impose more stringent effluent limitation standards than those imposed by the federal agencies responsible for determining appropriate standards. Moreover, in 1981 the court in *Middlesex County Sewerage Authority v. Nat'l. Sea Clammers Assn.*[99] held that the federal common law of nuisance, as related to ocean pollution, had been completely preempted by two federal acts (i.e., Federal Water Pollution Control Act of 1972 and the Marine Protection, Research and Sanctuaries Act of 1972).[100]

Additionally, several decisions in the mid-1980's clarified that federal regulation preempts federal common law related to pollution of *interstate* waters.[101] Likewise, courts have held that federal acts may preempt the affected state's common law in the case of interstate discharges.[102] However, most state and federal environmental acts contain "savings clauses" which preserve a plaintiff's right to bring state common law actions for nuisance and trespass claims intrastate.[103] For example, the Federal Clean Water Act contains a savings clause: "Nothing in this section shall restrict any right which any person (or class of persons) may have under any statute or common law to seek enforcement of any effluent standard or limitation or to seek any other relief (including relief against the Administrator or a state agency)."[104]

While wording similar to this is found in numerous state and federal environmental statutes, such "savings clauses" are not universal. Therefore, in the absence of such provisions, courts may decide to defer to administrative agencies' "authority" to determine technical issues related to environmental standards and compliance. Perhaps the greatest effect that regulatory standards (either state or federal) may have on common law claims is on the plaintiffs' ability to establish the element of their tort claim related to harm or damage. Arguably, if an agency has blessed the quantity and quality of a discharge either to air or water (i.e., a permit stipulates release of specific pollutants), it may be more difficult for a plaintiff to successfully prove damage. Thus, even if the courts in a particular jurisdiction do not hold that statutory and regulatory standards preempt common law claims, decision-makers hearing such actions (e.g., juries, judges) may, nonetheless, reach decisions which give that effect.

Notes

1. Negligence is a fourth common law theory that can potentially be adapted to pollution and environmental litigation. However, it is only successful in cases in which a plaintiff is able to demonstrate that a "duty was owed" and that "duty was breached by the defendant(s)."
2. *Callahan v. Wasmire*, 13 Ohio Supp. 128.
3. *Village of Euclid v. Ambler Realty*, 272 U.S. 365, 388 (1926).
4. *Capurro v. Galaxy Chem. Co.*, 2 ELR 20386 (Md. Cir. Ct. 1972) (chemical plant emissions enjoinable as public nuisance and also constitute a private nuisance allowing damages for diminution in property values).
5. *Stoughton v. Ft. Worth*, 277 S.W.2d 150 (Tex. Ct. App.—Ft. Worth 1955,

no writ); *Enchave v. City of Grand Junction*, 193 P.2d 277, 118 Colo. 165 (1948).

6. *Ballenger v. City of Grand Saline*, 276 S.W.2d 874 (Tex. Civ. App.— Waco 1955, no writ); see also, *State of N.Y. v. Shore Realty Corp.*, 759 F.2d 1032 (2d Cir. 1985) (owners of waste disposal site must clean-up); *National Sea Clammers Assn. v. City of N.Y.*, 616 F.2d 1222; judgment vacated 453 U.S. 1, 101 S.Ct. 2615 (1981) (plaintiffs brought action against government officials alleging they had permitted discharge of sewage which resulted in damage to their industry); *Philadelphia Elec. Co. v. Hercules, Inc.*, 762 F.2d 303, 315, cert. denied 474 U.S. 980, 106 S.Ct. 384, 388 (1985); *U.S. v. Solvents Recovery Serv. Etc.*, 496 F. Supp. 1127, 1142 (D. Conn. 1980) (property owner sued corporation alleging contamination of groundwater and river as a result of operation of chemical plant. U.S. brought action seeking injunctive relief against two corporations for allegedly polluting groundwater).

7. *Restatement, Second, Torts* '' 821(C).

8. See, *Ozark Poultry Products, Inc. v. Gurman*, 251 Ark. 389, 472 S.W.2d 714 (Ark. S.Ct. 1971) (*rejecting* the argument that a rendering plant was a public nuisance *abatement only at the behest of public officials*, because it inflicted same damage on all homeowners; *Columbia River Fishermen's Protective Union v. St. Helens*, 160 Or. 654, 87 P.2d 195 (1939) (finding a "different injury in kind" to support a public nuisance suit for water pollution which damaged commercial fishermen); *State v. Arizona Pub. Serv. Co.* 2 ELR 20011 (N.M. Dist. Ct., 1972, rev'd, 85 N.M. 165, 510 P.2d 98) (1973) (action against coal-burning elect. generating plant); *Commonwealth v. Barnes & Tucker, Co.*, 455 P.A. 392, 319 A.2d 871 (1984) (discharge of acid run-off from inactive coal mine held to be public nuisance).

9. *Awad v. McColgan*, 357 Mich. at 390, 98 N.W.2d at 573, quoting Thayer, *Public Wrong and Private Action*, 27 Harv.L.R. 317, 326 (1914).

10. *Ryan v. City of Emmetsburg*, 4 N.W.2d 435, 232 Iowa 600; *Hederman v. Cunningham*, 283 S.W.2d 108 (Tex. Civ. App.—Beaumont 1955, no writ).

11. *Vann v. Bowie Sewerage Co.*, 127 Tex. 97, 90 S.W.2d 561 (1936).

12. *U.S. Smelting Co. v. Sisam*, 191 F. 293 (8th Cir. 1911); *Towaliga Falls Power Co. v. Sims*, 6 Ga. App. 749, 65 S.E. 844 (1909).

13. Restatement, Second, Torts ' 826(a), at 3 (Tent. Draft No. 18, 1972).

14. E.g., *Georgia v. Tennessee Copper Co.*, 237 U.S. 678, 35 S.Ct. 752, 59 L.Ed. 1173 (1916) (unavoidable injury); *Reserve Mining Co. v. U.S.*, 514 F.2d 492, 5 ELR 20596 (8th Cir. 1975) (risk of substantial harm); *Spur Industries, Inc. v. Del. E. Webb Development Co.*, 108 Ariz. 178, 494 P.2d 700, 2 ELR 20390 (1972) (en banc, incompatible land use); *Monroe Carp Pond Co. v. River Raisin Paper Co.*, 240 Mich. 279, 215 N.W. 325 (1927) (world must have factories); *Clifton Iron Co. v. Dye*, 87 Ala. 468, 6 So. 192 (1888).

15. See, *People v. New Penn Mines, Inc.*, 212 Cal. App. 2d 667 (1963) (holding that Attorney General's power to sue for public nuisance was pre-

empted by establishment of administrative apparatus to deal with water pollution).

16. Restatement, Second, Torts ' 821(F).

17. *Jost v. Diaryland Power Coop.*, 45 Wis.2d 164, 172 N.W.2d 647 (1970) (held $395 worth of damage to alfalfa crop was "substantial"); *Whalen v. Union Bag & Paper Co.*, 208 N.Y.1, 101 N.E. 805 (1913) ($100 a year damage to farm by paper mill is "substantial").

18. For an interesting discussion of this well known point, see "What Really Pollutes? Study of a Refinery Proves an Eye-Opener," *Wall Street Journal*, March 29, 1993, p. 1.

19. *Board of Commissioners v. Elm Grove Mining Co.*, 122 W.Va. 442, 9 S.E.2d 813 (1940) (public health comes first. Even in as useful and important an industry as the mining of coal, an incidental consequence ... cannot be justified ... if the health of the public is impaired...") (817). *American Smelting & Refining Co. v. Godfrey*, 158 F. 225, 234 (8th Cir. 1907) (installation of bag houses to prevent emissions). See also, *U.S. v. Reserve Mining Co.*, 380 F. Supp. 11, 4 ELR 20573 (D.C. Minn.), injunction stayed, 498 F.2d 1073, 4 ELR 20598 (8th Cir.) stay denied, 419 U.S. 802 (1974), modified and remanded, 514 F.2d 492, 5 ELR 20596 (8th Cir. 1975), further proceedings 408 F.2d 1212, 6 ELR 20481 (D. Minn. 1976) (Court considers rules for absolute abatement, even if it results in plant shut down and widespread economic dislocation, where threatened pollution of public drinking water supplies by asbestos).

20. *Holman v. Athens Empire Laundry Co.*, 149 Ga. 645, 100 S.E. 207 (1919); *Hannum v. Gruber*, 346 Pa. 417, 31 A.2d 99 (1943); *Pawlowicz v. American Locomotive Co.*, 154 N.Y.S. 768 (1915); *Woschak v. Moffat*, 379 Pa. 441, 109 A.2d 310 (1954); *Merkel v. Smith*, 458 S.W.2d 940 (Tex. Civ. App.—Eastland 1970, no writ); *Wilmoth v. Limestone Products, Co.*, 255 S.W.2d 532 (Tex. Civ. App.—Waco 1953, writ ref'd n.r.e.) (concerning a claim that a nearby cement plant constituted a permanent nuisance).

21 *S.A. Gerrard Co. v. Fricker*, 42 Ariz. 503, 27 P.2d 678 (1943).

22. *E.Rauh & Sons Fertilizer Co. v. Shreffler*, 139 P.2d 38 (6th Cir. 1943) (finding nuisance); *Ambrosini v. Alisal Sanitary Dist.*, 154 Cal. App. 2d 720, 317 P.2d 33 (1957) (holding city liable).

23. See, P. Huber, *Galileo's Revenge* at 15.

24. 509 U.S. 579, 113 S.Ct. 2786 (1993)..

25. See, J. Sanders, *Scientific Validity, Admissibility, and Mass Torts after Daubert*, 78 Minn. L. Rev. 1387.

26. *Restatement, Second, Torts* ' 870.

27. Keeton, *Balancing of Equities*, 17 Texas L. Rev. 412 (1940).

28. *Boomer v. Atlantic Cement Co.*, 26 N.Y.2d 219, 309 N.Y. S.2d 312, 257 N.E.2d 870 (1970).

29. See also, *Kentucky-Ohio Gas Co. v. Bowling*, 264 Ky. 470, 95 S.W.2d 1; *Northern Indiana Public Service Co. v. W.J. & M.S. Vesey*, 210 Ind. 338, 200 N.E. 620 (basing a denial of injunctive relief to complainants, in favor of permanent damages, on the public interest in the operation of the gas plant).

30. *Boomer*, at 875.
31. *Madison v. Ducktown Sulfur, Copper & Iron Co.*, 113 Tenn. 331, 83 S.W. 658 (1904).
32. *Hulbert v. California Portland Cement Co.*, 161 Cal. 239, 251, 118 P.928 (1911).
33. *Portland Cement Ass'n. v. Ruckelshaus*, 158 U.S. App. D.C. 308, 486 F.2d 375, 3 ELR 20642 (1973) cert. denied, 417 U.S. 921 (1974).
34. *North Anna Anv. Coalition v. Nuclear Reg. Comm.*, 533 F.2d 655 (1976).
35. *Harrison v. Indiana Auto Shredders Co.*, 528 F.2d 1107, 6 ELR 20179 (7th Cir. 1975).
36. The claimants in *Harrison* also alleged violation of a local air pollution ordinance. (The City of Indianapolis General Ordinance No. 109, 1973). Note also that the Court commented that the district court's findings and conclusions required 114 pages of complicated text.
37. *Harrison*, at 1109.
38. *Harrison* at 1120 fn 20, citing Prosser *Nuisance Without Fault*, 20 Texas L. Rev. 399,410 (1942).
39. *Harrison* at 1125.
40. *American Smelting & Refining Co. v. Godfrey*, 158 F. 255 (8th Cir. 1907); see also, *Stevens v. Rockport Granite Co.*, 216 Mas. 486, 104 N.E. 371 (1914) (court stating that noise control devices could eliminate nuisance to plaintiffs); *Renken v. Harvey Aluminum, Inc.*, 226 F. Supp. 169 (D.C. Or. 1963) (requiring emission control unless defendant could show it was prohibitively expensive).
41. *Georgia v. Tennessee Copper Co.*, 206 U.S. 230, 27 S.Ct. 618, 51 L.Ed. 1038 (1907), 237 U.S. 474, 35 S.Ct. 631, 59 L.Ed. 1054 (1915) (motion to enter a final decree), 237 U.S. 678, 35 S.Ct. 752, 59 L.Ed. 1173 (1915) (decree), modified 240 U.S. 650, 36 S.Ct. 465, 60 L.Ed. 846 (1916); *Fletcher v. Bealey*, 28 Ch.D. 688, 700 (1885) where court refused injunctive relief because "in ten years' time it is highly probable that science (which is now at work on the subject) may have discovered the means for rendering this green liquid innocuous."
42. *Bove v. Donner-Hanna Coke Corp.*, 142 Misc. 329, 254 N.Y.S. 403 (Sup. Ct. 1931), *aff'd* 236 App. Div. 37, 258 N.Y.S. 229 (1932).
43. *Buffalo v. Savage*, 1 Misc.2d 336, 148 N.Y.S.2d 191 (Sup. Ct.), *aff'd mem.*, 309 N.Y. 941, 132 N.E.2d 313 (1955).
44. *Euclid v. Ambler Realty Co.*, 272 U.S. 365, 388, 47 S.Ct. 114, (1926) (Sutherland, J.).
45. *Bartel v. Ridgefield Lumber Co.*, 229 P. 306, 308 (Wash. S. Ct. 1924) (lumber operation damaged adjacent landowners, even though in lumber town).
46. See, *Edwards v. Allouez Mining Co.*, 38 Mish. 46 (1878).
47. See, *Steinmetz v. Kelley*, 72 Ind. 442 (1880).
48. Contributory negligence may be effective in countering toxic tort claims based on negligence of the defendant.
49. *Smith v. Staso Milling Co.*, 18 F.2d 736 (2nd Cir. 1927).

50. *Boomer v. Atlantic Cement Co.*, 26 N.Y.2d 219 (1970).
51. *Spur Industries, Inc. v. Del E. Webb Development Co.*, 108 Ariz. 178, 494 P.2d 700 (1972).
52. See, *Spaulding v. Cameron*, 38 Cal.2d 265, 239 P.2d 625 (1952).
53. See, *United States v. Fixico*, 115 F.2d 389 (10th Cir. 1940); *Love Petroleum Co. v. Jones*, 205 So.2d 274 (Miss. 1967).
54. See, *Union Oil Co. v. Oppen*, 105 F.2d 558, 4 ELR 20618 (9th Cir. 1974).
55. See, *Schatz v. Abbott Laboratories, Inc.*, 51 Ill.2d 143, 281 N.E.2nd 323 (1972).
56. See, for example, *Martin Building Co. v. Imperial Laundry Co.*, 220 Ala. 90, 120 So. 82 (1929); *Hidalgo County Water Improvement District v. Cameron County Water Control & Improvement District*, 250 S.W.2d 941 (Tex. Civ. App.—San Antonio 1952, no writ) (discussing developments in environmental law and injunctions); *Wilmont Homes v. Weiler*, 202 A.2d 576 (1964); *Arizona v. California*, 373 U.S. 546, 83 S.Ct. 1468, 10 L.Ed.2d 542 (1963) (regarding water pollution); *Godard v. Babson-Dow Manufacturing Co.*, 313 Mass. 280, 47 N.E.2d 203 (1943) (regarding vibration pollution).
57. See, *Ryan v. Emmetsburg*, 232 Iowa 600, 4 N.W.2d 435 (1942); Kazan *Trespass-Nuisance-Election of Remedy* 39 Texas L. Rev. 244, 245 (1960) ("the overwhelming majority of courts have held that invasions by dust, soot and cinders, ashes and sawdust, noxious odors and fumes, and gases, if actionable at all, constitute a nuisance").
58. *Martin v. Rental Metals Co.*, 221 Or. 86, 342 P.2d 790 (1959).
59. 3 H&C 774, 159 ENG. REP. 737 (1865), Rev'd, L.R. 1 Ex. 265 (1866), Aff'd, L.R. 3 H.L. 330 (1868).
60. *Id.* at 279-80.
61. A minority of states reject the Rylands doctrine.
62. *Turner v. Big Lake Oil Co.*, 75 A.2d 256 (1950).
63. *Fritz v. E.I. DuPont de Nemours & Co.*, 128 Tex. 155, 96 S.W.2d 221 (1936).
64. See, for example, *Spano v. Perini*, 25 N.Y.2d 11, 302 N.Y.S.2d 527, 250 N.E.2d 31 (1969), where the Court allowed recovery of damage suffered from blasting without proof of negligence or even without proof of actual physical invasion, both of which would have been required to sustain a claim for trespass.
65. Additional discussion of this distinction is provided in the Comments to Restatement, Second, Torts ' 849, which notes, in pertinent part:
Comment on Subsection (1):
a. *Nature of interest protected.* The right to use water is an interest in real property.... The holder of the right is entitled to protection of his use of water from any type of tortious conduct that may be directed at an interest in real property.
b. *Nuisance.* The most common tort arising from an interference with the use of water by an act or conduct not itself a use of water is the commission of a nuisance by polluting the water. Pollution may be the result of a

use of water, as in the case of a manufacturer who withdraws water from a stream, uses it to process materials and then returns it in a polluted condition. It may be the result of a use of land....which involves no taking or harnessing of water but does result in waste or debris being deposited in or finding its way into water. It may be the result of a personal act, as in the case of a person who carries noxious waste or debris to a bridge and dumps it into a stream.... *The rules stated in ''821A-840E governing nuisance are applicable to these interferences with the use of water.*

c. *Trespass.* Some activities that result in an interference with the use of water may call for the application of the rules stated in ' 158 relating to trespass upon land.... The pollution of water may be the result of an act constituting a trespass, as in the case of a foreign substance deposited in the water while it is on the land of the plaintiff or deposited upstream and then brought to the land by the stream.

d. *Negligent, reckless or abnormally dangerous conduct.* A wide variety of other activities that are not themselves uses of water may cause injury to another's use. These may subject the actor to liability under the rules stated in '' 281-499 relating to negligent conduct, in '' 500-503 relating to reckless conduct, or in '' 519-524A relating to abnormally dangerous conduct.

Comment on Subsection (2):

e. *Pollution—No right to pollute.'* There is no riparian right or privilege to pollute water, nor do landowners have rights to pollute surface and groundwater found on or within their land.... The definition of 'use of water' in ' 847 specifically excludes pollution from the category of protected uses....

Many cases arising from water pollution have been couched in terms of the law of riparian rights, probably because most of the plaintiffs have been riparian proprietors and the person most likely to be harmed by pollution of water is one who has a legal right to it.... But riparian law does not encompass the entire problem. Some pollution of water can injure nonriparians and much pollution caused by riparians is the result of acts and activities that cannot be described as exercises of riparian rights even though the acts take place on riparian land. In these cases the law of nuisance is appropriate. It is equally appropriate when the controversy arises between riparians or in the course of an exercise of riparian rights. Nuisance' is a type of harm occurring to plaintiffs, rather than a description of acts or conduct of defendants. The question of whether the pollution is unreasonable and hence a nuisance is the same as the question of whether it is unreasonable and hence a violation of plaintiff's riparian rights, and in both questions a balancing of the interests of the parties is important. It is not surprising to find, therefore, that some courts have put water pollution cases on the basis of riparian rights, some have used the language and principles of nuisance law and some have used the two doctrines interchangeably and even simultaneously.

There is a distinct advantage, however, in selecting nuisance theory rather than riparian doctrine in pollution cases.... The use of nuisance law, on the other hand, reduces the chance of confusion and emphasizes that pollution is a tort and not the exercise of a property right....

f. *Pollution—Right to clean water.* Treating pollution under the heading of nuisance does not impair any right or remedy of a riparian proprietor who complains of harm from pollution. His right to clean water for his use remains untouched. The application of nuisance law operates against the defendant and bars him from claiming that his acts are sanctioned by riparian property rights to pollute. The reason for placing the treatment of pollution in other Chapters of the Restatement is to strengthen and clarify the law of pollution control, not to weaken it. (emphasis added)

66. See, *Atkinson v. Herington Cattle Co.*, 200 Kan. 298, 436 P.2d 816 (1968).
67. See, *Sandusky Portland Cement Co. v. Dickson Pure Ice Co.*, 221 F. 200 (7th Cir. 1950); *Westville v. Whitney Home Builders*, 122 A.2d 233 (1956).
68. *Mann v. Willey*, 64 N.Y.S. 589 (App. Div. 1990), aff'd, 168 N.Y. 664, 661 N.E. 1131 (1901).
69. *Ravndal v. Northfolk Placers*, 91 P.2d 368, 371 (1939).
70. See, for example, *Columbia River Fishermen's Protective Union v. St. Helens*, 160 Or. 654, 87 P.2d 195 (1939).
71. *Bliss v. Washoe Copper Co.*, 186 F. 789 (9th Cir. 1911).
72. *Bliss* at 796.
73. See, *Bove v. Donner-Hanna Coke Corp.*, 258 N.Y.S. 229 (1932) where the court refused injunction to halt the emission of particulates on nearby residents because the plant was deemed to be "modern and up-to-date" and no "reasonable change" to the operation could be made to reduce emissions. Even though the plaintiff resided in the area *before* the industrial facility located nearby, the court believed she should have recognized that the area would eventually be heavily industrialized. See also, *U.S. Smelting v. Sisam*, 191 F. 293 (8th Cir. 1911); *Garland Grain Co. v. P.C. Homeowners Improvement Assn.*, 393 S.W.2d 635 (Tex. Civ. App.—Tyler 1965, writ ref'd n.r.e.).
74. *Mountain Copper Co. v. U.S.*, 142 F.625, 626 (9th Cir. 1906).
75. See, cite at fn 31, *supra.*
76. *McElwain v. Georgia Pacific Corp.*, 245 Or. 247, 421 P.2d 957 (1966).
77. *McElwain* at 959.
78. *Diamond v. General Motors Corp.*, 20 Cal. App. 3d 274, 97 Cal. Rept. 639 (1971) (upholding trial court's denial of injunction, on the basis that such relief would "halt the supply of goods and services essential to life and comfort of persons...represented by the plaintiff;" *Smejkal v. Empire Lite-Roll, Inc.*, 274 Or. 571, 547 P.2d 1363 (1971) (denying injunctive relief and remanding to lower court for calculation of damages); *Borland v. Sanders Lead Co.*, 369 So.2d 523 (Ala. 1979) (balancing value of land when used for agricultural purposes against value of land if put to industrial use).
79. *Ozark Poultry Products, Inc. v. Garman*, 251 Ark. 389, 472 S.W.2d 714 (1971); *Spur Industries v. Del E. Webb Development Co.*, 108 Az. 178, 494 2d 700 (1972); *Bradley v. American Smelting & Refining Co.*, 104 Wash. 2d 677, 709 P.2d 782 (1985).

80. *Sullivan v. Jones & Lughlin Steel Co.*, 208 Pa. 540, 57 A. 1065 (1904). See also, *American Smelting & Refining Co. v. Godfrey*, 158 F.225 (8th Cir. 1907) (enjoining operation of a smelter near farm land and stating "The rights of habitation are superior to the rights of trade, and whenever they conflict, the rights of trade must yield..."); *Dale v. Bryant*, 141 N.E.2d 504 (Ohio 1957) (holding monetary damages would be inadequate to compensate residential landowners for injury caused by industrial neighbor).

81. *Georgia v. Tennessee Copper Co.*, 237 U.S. 474, 35 S.Ct. 631 (1915).

82. *Bradley v. American Smelting and Refining Co.*, 104 Wash.2d 677 (1985).

83. *Vann v. Bowie Sewerage Co.*, 90 S.W.2d 561 (Tex. Com. App. 1936).

84. *Turner v. Big Lake Oil Co.*, 96 S.W.2d 221 (Tex. 1936).

85. *Branch v. Western Petroleum*, 657 P.2d 267 (Utah 1982).

86. *Branch* at 275. See also, *Frank v. Environmental Sanitation Management, Inc.*, 687 S.W.2d 876 (Missouri 1985) (awarding damages to farmers from migrating landfill leachate); *Shutes v. Platte Chemical Co.*, 564 So.2d 1382 (Miss. 1990) (making punitive damages available to plaintiffs able to prove negligent emission of chemicals onto plaintiff's premises).

87. For example, *Hargrove v. Cook*, 41 P. 18 (Cal. 895); *Columbia Ave. Savings-Fund, Safe-Deposit, Title & Trust Co. of Philadelphia v. Prison Commission of Georgia*, 92 F. 801 (1899); *Smith v. Staso Milling Co.*, 18 F.2d 736 (2nd Cir. 1927); *Strobel v. Kerr Salt Co.*, 58 N.E. 142 (N.Y. 1900) (enjoining salt manufacturer from polluting stream); *Jewell v. Hancock*, 175 S.E.2d 847 (Ga. 1970).

88. *Indianapolis Water Co. v. American Strawboard Co.*, 53 F. 970 (D. In. 1893).

89. *Id.*

90. *Sandusky Portland Cement Co. v. Dixon Pure Ice Co.*, 221 F. 200 (7th Cir. 1915).

91. *Ellison v. Rayonier, Inc.*, 156 F. Supp. 214 (W.D. Wash. 1957).

92. *Id.* See also, *Commerce Oil Refining Corp. v. Miner*, 170 F. Supp. 396 (D. R.I. 1959) (court deferring to state regulatory agency's determination of whether discharge of effluent was within permit parameters).

93. *People v. New Penn Mines, Inc.*, 212 Cal. App. 2d 667, 28 Cal. Rept. 337 (1963).

94. *Id.* at 674.

95. *Atlas Chemical Industries, Inc. v. Anderson*, 514 S.W.2d 309 (Tex. Civ. App.—Texarkana) *affirmed* 524 S.W.2d 681 (Tex. 1975).

96. See also, *Stoddard v. Western Carolina Regional Sewer Authority*, 784 F.2d 1200 (4th Cir. 1988) (holding that a state regulatory scheme did not preempt state common law remedies).

97. *Id.* at 319.

98. *Milwaukee v. Illinois & Michigan*, 101 S.Ct. 1784, 451 U.S. 302 (1981).

99. *Middlesex County Sewerage Authority v. Nat'l Sea Clammers Assn.*, 101 S.Ct. 2615 (1981).

100. See also, *New England Legal Foundation v. Costle*, 666 F.2d 30 (2d Cir. 1981) (holding that EPA's approval of use of high sulphur fuel precluded

plaintiff's nuisance action); *Bryski v. City of Chicago*, 499 N.E.2d 162 (1986) (holding that federal regulations preempted plaintiff's claims of noise pollution against airlines at O'Hare Airport).

101.*Tennessee v. Champion International*, 709 S.W.2d 569 (1986); *International Paper Co. v. Ouellette*, 479 U.S. 481, 107 S.Ct. 805 (1987).

102.*Arkansas v. Oklahoma*, 112 S.Ct. 1046 (1992).

103.See, *CAE-LINK Corp. v. Washington Suburban Sanitary Commission*, 602 A.2d 239 (1992).

104.33 U.S.C.A. ' 1365(e); see also, *International Paper Co. v. Ouellette, supra* at fn 103.

5

Rent Seeking on the Legal Frontier

Bruce L. Benson

Government efforts to resolve toxic tort related problems may be far more costly than the problems themselves. One of the less obvious effects of government failure gets to the heart of the explosion in toxic tort litigation and is examined in section I. Government claims the authority to be the source of "the rules of the game" - property rights, liability rules - under which market players pursue their goals. Stability in liability rules characterized tort law up until about thirty years ago. Rapid and sharp changes in liability rules, and variations in such rules among jurisdictions, characterize tort law in recent years. This has created costs due to uncertainty about liability rules, which in turn has made the property rights of producers less secure.

Legal instability means that other parties have incentives to "rush in" and lay claim to the insecure property rights by filing lawsuits. The resulting mass tort litigation means that standard common law tort solutions are less likely to work. The courts are overwhelmed with litigation, causing delays, high administrative costs (e.g., legal fees that can exceed the payments to victims by significant margins), and injustices (e.g., through non-consensual consolidations, coerced settlements, and violations of due process by multiple assessments of punitive damages). Indeed, even though discussion of toxic torts often focuses on alleged health costs or clean-up costs and who should bear them, a substantial portion of the social costs of toxic torts arise from dispute resolution itself, a cost which is relatively high because of government failure.

Consider the Superfund legislation intended to govern the cleanup of the nation's most dangerous hazardous waste sites. As Benanav (1994) explained:

> Passed by Congress in 1980, Superfund was suppose to force polluters to clean up hundreds of old hazardous waste sites across the country.... The reality, however, is that flaws in the Superfund law have created a protracted and adversarial process that has led to legal gridlock - with the result that only a small percentage of waste sites have actually been cleaned. Meanwhile, 30 cents to 40 cents of every Superfund dollar has found its way into the pockets of lawyers and consultants, who, needless to say, haven't lifted a shovel to clean up anything.
>
> The reasons for Superfund's stunning failure are legion, but the greatest problems can be found in its liability system and in its process of determining cleanup standards.

With respect to liability, for instance, the law is not based on a strong negligence standard. Thus, evidence that a firm that unknowingly dumped a small amount of potentially harmful waste at a site used by many other unidentified dumpers can make the firm liable for an entire multimillion dollar cleanup. Further, the law can be applied retroactively, so even if the company did its dumping before 1980 when it was perfectly legal to do so, and even if the firm had a state issued license to do the dumping, it may be held liable for the cleanup. Naturally, a firm sued under these conditions is likely to resist liability through every conceivable means, including litigation: "The predictable outcome of all of this is the mother of all legal nightmares. Typically the Environmental Protection Agency sues the polluters, who in turn pursue everyone in sight in an attempt to unload cleanup costs onto as many others as possible" (Benanav 1994). Companies also sue insurance companies who deny that policies cover such cleanup, and the insurance companies mount vigorous defenses. Both sides have been spending about $1 billion per year in legal costs and cleanups have been delayed for years. But this is only a small part of total litigation costs arising in the toxic tort area.

Tort litigation costs in general are large, even relative to compensation payments. In 1985, for instance, total expenditure nationwide for all tort litigation in state and federal courts was estimated to be between $29 and $36 billion. About $16 to $19 billion of that was spent for various costs of the tort litigation system, including legal fees and court costs, but not including compensation payments (Hensler et al. 1985). Plaintiff's received about $21 to $25 billion, but out of that they had to pay their lawyers roughly a third ($7 to $9 billion), reducing compensation to victims to $14 to $16 billion. Defendants also paid about $4.7 to $5.7 billion in legal fees. So roughly 44 to 48 percent of the money paid out in torts actually went to the victim. About 40 percent went to the lawyers and 4 to 5 percent went to court expenditures and claim processing. The rest of the expenditures, between 6 to 12 percent, were the lost wages for the defendants and plaintiffs that arise because of the time the litigation process takes. Thus, if plaintiffs were exactly compensated for tort harms, it actually cost more to compensate them than the harms themselves cost. Of course, there is no reason to assume exact compensation. Many harms may not be compensated at all, but on the other side, given the growth in punitive damages and awards for intangible harms discussed below, overcompensation seems increasingly possible as well.

The total costs of tort activity are not spread evenly. In product liability cases such as the asbestos cases detailed below, litigation costs are substantially higher than in some other areas of tort law (e.g., auto accident tort). This difference reflects the fact that auto accident tort law is relatively stable so potential litigants know the rules of the game. Product liability law, including toxic torts, is far less stable, however, and as explained below, this instability significantly raises litigation costs.

Direct administrative costs are not the only costs of the tort system. For instance, jurors are rarely fully compensated for their time, so actually court budgets understate the opportunity costs of resources allocated to tort resolution. Furthermore, the growing demands on the courts for tort litigation have created significant spillover costs for disputants in all other areas of civil law, since they suffer from the crowding and longer delays. Common law courts are common access resources allocated on a first-come-first-serve basis (Neely 1982; Benson 1990, 99-101), so the tort explosion adds significantly to the crowding and delay problem: as the judge noted in *In re Ohio Asbestos Litigation*, Nol 83-OAL (N.D. Ohio Gen. Order No. 67, filed June 15, 1983), the resolution rate of eighty asbestos tort cases pending in his court "was so delaying in the progress of unrelated cases" that some new method of settling them was needed.

When liability rules are clear, litigation costs tend to be relatively small. If potential plaintiffs knew with certainty that firms would not be liable (e.g., when the government contractor defense discussed in chapter 3 is clearly established), then the number of suits filed would fall. On the other hand, if it is clear that firms will always be liable (e.g., strict liability), then many claims might be filed but most are likely to be settled without litigation. But if the liability rules are not clear, perhaps because they are changing very rapidly, or because they vary across common law jurisdictions, litigation costs are likely to be high (e.g., even under a strict liability doctrine there can be considerable uncertainty regarding what kinds of defenses might hold and/or how extensively the doctrine might be applied). And under these circumstances, unmeasured costs also can be very high. Uncertainty regarding liability rules means that property rights are relatively insecure, and insecure property rights shortens time horizons, reduces the incentives to invest, innovate, produce, and enter into exchanges. The resulting social costs easily could exceed all of the other costs of toxic torts. Given such potential costs, let us consider the relationship between changing tort liability rules and property rights in more detail.

I. Changing and Varying Liability Rules and the "Rush for Property Rights"[1]

As Epstein explains (1980, 254), in tort law matters, as well as in property and contract, law performs its essential function best only if it remains predictable: "Social dynamism ... is wholly desirable, but not best implemented by Judicial decision.... [P]rivate sources ... should be spared the burden of planning their affairs in an environment filled with unwanted legal uncertainties."

Such predictability, however, no longer seems to characterize tort law. Three required elements once characterized tort actions under common law: (1) a breach of duty owed to the plaintiff by the defendant; (2) a measurable harm suffered by the plaintiff; and (3) evidence that the breach of duty caused the harm. Lack of any of these elements meant that there was no grounds for tort action. All three elements have changed dramatically over the last three and a half decades, particularly in product liability, creating considerable uncertainty about what the law is.

From Breach of Duty to Strict Liability

Manufacturers used to owe a duty of care only to those who purchased a product from the manufacturer. Product liability was a matter of contract law, not tort law (Rubin 1986). Beginning in 1916 with *MacPherson v. Buick Motor Co.*, 217 N.Y. 382 (1916), the duty of care gradually expanded to include those who purchase from others in the contractually-linked distribution chain (e.g., retailers). For instance, an important 1960 Supreme Court of New Jersey decision, *Henningsen v. Bloomfield Motors, Inc.*, 32 N.J. 358 (1960), ruled that the manufacturer's duty was owed to anyone who was likely to use or be exposed to the product. Perhaps even more significantly, the court ruled in *Henningsen*, for the first time in a product case, that the manufacturer was liable under an "implied warranty principle," despite the lack of any convincing evidence of negligence by the manufacturer, or of any defect, and no expressed contractual warranty covering the claimed failure of the product. This implied warranty principle was part of a movement away from the negligence or fault standard of liability (e.g., a breach of duty for care) toward a standard of strict liability.[2]

Now in all states, some producers can be held liable for harms regardless of whether any legal duty to the plaintiff is breached. No fault or negligence is required (if negligence can be demonstrated, cases for the plaintiff are even stronger, however). Considerable uncertainty remains as to how far strict liability will be expanded. Many questions remain regarding which kinds of products are subject to strict liability and which defenses (e.g., assumption-of-risk, unforseen misuse, comparative negligence) might be acceptable in which jurisdictions. Thus, defendants in alleged strict-liability cases are often unwilling to settle.

From Compensatory Damages for Measurable Harms to Damages for Intangibles and Punitive Damages

A successful tort action used to lead to compensatory damages paid to the plaintiff for predictable categories of measurable harms. After *Dillon v. Legg*, 68 Cal. 2d 728, 441 P.2d 912 (1968) in California, courts began to allow plaintiffs to recover for emotional distress. A "dwindling majority" of courts still demand evidence of some physical manifestation of harm,[3] but most courts have abandoned such requirements.[4] For instance, "it is an increasing trend for the courts to award damages for fear of future illness" due to exposure to a potentially toxic product (Pope and Del Giorno 1991, 495). As Pope and Del Giorno (1991, 503) concluded, "While these holdings may appear to be progressive and compassionate, in reality they threaten to contaminate ... tort litigation with frivolous and feigned claims. Allowing recovery for these claims may provide windfalls for plaintiffs who may never be actually injured, thereby depleting resources for those with actual damages." After all, the determination of damages has become much more subjective and uncertain.

Not surprisingly, there is a widespread perception that damage awards have also grown dramatically. The perceptions held by the public are fueled by anecdotal "evidence" produced by the media and by defendant interest groups (e.g., the AMA). Such evidence is inevitably biased by the media's search for the sensationally large or unusual verdict (MacCoun 1993, 138-39) and defendant self-interest, but while empirical research on trial outcomes in general is mixed, trends in trial awards in product liability tends to support the perception.

Hensler et al. (1987) reported that median awards in product liability cases in San Francisco and Cook County, Illinois rose between 1960 and 1984, and that mean awards increased sharply (median awards tend to be much smaller than mean awards because occasional very large awards inflate means [Peterson and Priest 1993]). Moller (1996) reported similar trends in fifteen counties from six states for the 1985-1994 period. There are many other factors that may be more important determinants of growing damage awards, of course, including the increase in medical costs and the increasing ability of the health sector to maintain life for much longer periods, but the perception created by such awards is that producers are increasingly vulnerable to big losses.

Punitive damages exceeding the losses of the plaintiff are also awarded with increasing frequency, *particularly* when the defendant is a business (Moller 1996). When first introduced, punitive damages usually applied only for intentional torts, but such awards are rapidly being extended to other torts, including many involving strict liability where no fault is evident, let alone intent. In some jurisdictions, a firm can be punished when no significant harm has been demonstrated. For instance, an Illinois jury found no credible evidence of physical harm in *Kemner v. Monsanto*, No. 80-L-970 (Cir. Ct., St. Clair City, Ill. 1987), yet it returned a verdict of $1 in compensation and $16 million in punitive damages (Huber 1992, 728).

Punitive damages can be especially devastating for firms involved in large numbers of cases alleging the same harm-causing action in several different common law jurisdictions. Indeed, tort defendants have raised constitutional objections to repetitive punitive damages, but to no avail (Lafferty 1991). For instance, the federal district court of New Jersey held in *Juzwin v. Amtorg Trading Corp.*, 705 F.Supp. 1053 (D.N.J. 1989); 718 F.Supp. 1233 (D.N.J. 1989), *vacating* 705 F.Supp. 1053 (D.N.J. 1989) that imposition of multiple punitive damages may violate the fourteenth amendment but that it was powerless to shape a remedy to this problem, and that it would be unfair to plaintiffs to implement such a ruling retroactively.[5] The perception of these trends in damage awards creates stronger incentives to sue, of course.

From Evidence of "Cause-in-Fact" to Evidence of a Probable Relationship

Historically, a preponderance of evidence establishing that the defendant's action was a "cause-in-fact" of the plaintiff's harm was a necessary but not a sufficient condition for damage awards. An action was considered to have been the cause-in-fact of a harm if the harm would not have occurred "but-for" the action. Many courts now consider evidence that the defendant's product *might be* a substantial factor in producing the harm as sufficient, as evidenced by statistical, epidemiological, or experimental evidence rather than the much more particularistic evidence that used to be required.

In *Jackson v. Johns-Manville Corp.*, 727 F.2d 506, 516 (5th Cir. 1984), *cert. denied*, 106 S. Ct. 3339 (1985), for instance, it was stated that the preponderance rule "provides an 'all or nothing' approach, whereby the plaintiff becomes entitled to full compensation for those ... damages that are proven to be 'probable' (a greater than fifty percent chance)."[6] The statistical, epidemiological, or experimental evidence clearly does not have to be very strong either, at least by any scientific standards. For instance, in affirming the lower court verdict awarding $5.1 million for congenital injuries allegedly from a spermicide in *Wells v. Ortho Pharmaceutical Corporation*, 615 F. Supp. 262, 266-67 (N.D. Ga. 1985), *aff'd*, 788 F.2d 741 (11th Cir. 1986), at 745, the appeals court stated that "it does not matter in terms of deciding the case that the medical community might require more research and evidence before conclusively resolving the question." Yet, the study upon which this decision was based explicitly stated that the results were tentative, and in fact, subsequent research has failed to support the findings (Huber 1992, 724-25). Substantial evidence now suggests that spermicides are not teratogenic. Numerous other examples can be cited, leading some observers to characterize the evidence in such torts as "junk science" (Huber 1992, 724-37; Spyridon 1991, 298-313).

In the past, an inability to link a specific breach of duty to a specific harm would have prevented a damage award, but today if a group of defendants can be identified whose fungible product was the alleged cause of harm, then in some jurisdictions, damages may be apportioned among the group. In *Sindell v. Abbott Laboratories*, 26 Cal. 3d 588, 607 P.2d 924 (1980), the California supreme court ruled that when there are innocent plaintiffs and "negligent" defendants and when the defendants are better able to bear the costs from the accident than the plaintiffs, then the plaintiffs should be able to recover even if they cannot identify who actually caused the harm.[7] The court imposed liability on the defendants in proportion to the companies' market shares of the product at the time that the harmful exposure occurred. *Sindell* was not adopted by many other courts, but the doctrine of joint and several liability (see discussion below), which allows a plaintiff to sue anyone and everyone who might have been the producer, makes market share liability relatively irrelevant anyway.[8] The key point is that high profile cases like this create the impression that courts are willing to structure new rules to fit the circumstances.

An additional condition of recovering damages is that a defendant's breach of duty was the "proximate" cause, but the idea of proximity is a relative one. How close must the connection be for it to be considered proximate? *Henningsen* and subsequent decisions have allowed a producer to be held liable for harms that are increasingly remote from any transaction involving the producer. Thus, when the true proximate cause is immune from liability (e.g., as the government often is, as explained in chapter 3, or as an employer may be under Worker's Compensation), the income of more remote causes (a manufacturer) may be vulnerable. Such changes add to the evolving strict liability doctrine and growing damage awards to make the property rights to producers' income increasingly insecure: these rights are perceived to be "up for grabs" (as in Anderson and Hill 1990, 177]) for any plaintiff who can establish a claim.

While many of the changes in tort law outlined above may sound quite "fair," they clearly have opened up more avenues for pursuing damages and created stronger incentives to file suit.[9] These changes created incentives which have led to an explosion in some types of tort litigations (e.g., toxic torts, medical malpractice), increasing the cost of administering the tort system, including court crowding and delay, and much higher legal fees.

II. Incentive Consequences of Changes in the Tort Law

As Barzel (1989, 64) emphasized, a typical asset is not a homogeneous entity with a single attribute owned by an individual or in common. An asset can have many attributes, and property rights to all of an asset's different attributes are never perfectly defined. The bundle of rights associated with a particular asset can include some rights which assign benefits to particular individuals, while other potential benefits are held in common. Incentives always exist to capture the rents that are generated by attributes held in common. Rents are dissipated through competition as the commons becomes crowded and over-used (e.g., see Johnson and Libecap 1982).[10]

Because of changes in tort law, property rights to income generated by many producers are no longer exclusive. The resulting uncertainty about property rights to producers' income has the effect of transferring the property rights to the use of income generated by productive efforts from the producer to the commons. As Cheung (1974, 58) explains, "When the right to receive income is partly or fully taken away..., the diverted income will tend to be dissipated unless the right to it is exclusively assigned to another individual." In the context of toxic torts, this creates two important incentives.[11]

First, the rights to producers' income are valuable, and because of the uncertainty created by rapidly changing tort law, they are also "up for grabs" (as in Anderson and Hill 1990, 177), for anyone who can establish a claim to them. Thus, individuals have incentives to rush to establish claims to rights to the income of alleged mass tortfeasors. Making such a claim in tort law is not unlike making a claim to oil from federal lands (Libecap 1984), or to shrimp in the Texas Gulf fishery (Johnson and Libecap 1982), or to the public lands as they were privatized (Anderson and Hill 1990).

Libecap (1984, 383) explained that property rights to oil on federal land were assigned only upon extraction during the early years of this century, and as a result, "Excessive wells were dug along property lines to drain oil from neighboring areas; extracted oil was placed in surface storage (open reservoirs as well as tanks), where it was subject to evaporation, fire, and spoilage; and rapid extraction rates reduced total oil recovery as subsurface pressures, necessary for naturally expelling subsurface oil, were prematurely depleted." Similarly, Anderson and Hill (1990, 195), found that "Efforts to give away the public domain created a commons into which squatters and homesteaders rushed to compete for the rents. In the process, pioneers paid for the land in terms of forgone wealth, privations, and hardships, demonstrating that 'there ain't no such thing as free land.'"

Tort claims also require a substantial investment in time and resources (attorneys, expert witnesses, etc.) to meet the requirements of the courts in supporting a claim. The implication of this race for property rights and the resulting rent (or non-exclusive income) dissipation, is that tort plaintiffs are likely to pay dearly for any income they actually capture. Paraphrasing Anderson and Hill (1990), the effort to give income to those who may be suffering from toxic torts creates a commons into which legitimate and illegitimate plaintiffs (homesteaders and squatters) will rush, and therefore, plaintiffs in general will pay very high costs in the form of legal fees, time invested in litigation, anxiety, etc. —there 'ain't no such thing as free relief from toxic harms,' either.

Second, the producers who face the dissipation of what used to be their exclusive income have incentives to compete to retain that income and to reestablish claims to it. They also bear high costs as they invest along every margin that appears to be beneficial in delaying losses or regaining rights (e.g., employing active and innovative defenses in tort litigation, influencing legislators in an effort to redirect the common law courts, using means of limiting liability that are available within the law such as reorganization under Chapter 11 Bankruptcy). Thus, as in Cheung (1974), Libecap (1984), Johnson and Libecap (1982), and Eggertsson (1990, chapter 4), rents will be dissipated along several margins.

Others are also likely to compete for the insecure income. Attorneys who see particularly large pools of susceptible income may actively recruit plaintiffs rather than wait for clients to call upon them. Trial lawyers are a powerful and effective lobby group (perhaps the most powerful in the country), and they will exert a great deal of political pressure to counter demands made by producers for tort reform. In fact, as suggested by McChesney (1987), legislators can also capture part of the insecure income by threatening to pass tort reform legislation. The mere threat of such legislation can lead trial lawyers to make large contributions to appropriate legislators in order to divert such legislation and protect their lucrative source of income.

An Example: Asbestos

Asbestos litigation has been ongoing for long enough to illustrate the nature and potential extent of rent dissipation and other consequences. Asbestos was widely used as insulation prior to the early 1970s, in part because it is fire resistant. It probably saved thousands of lives and huge property damages by preventing large numbers of fires. Asbestos was known to cause health problems, particularly when a person breathed heavy doses of asbestos fibers.[12] However, before the changes in the 1960s, there was no discernable tort liability risk for asbestos producers (Epstein 1982, 18). Tort law would only compensate for injuries to the actual purchasers in the distribution chain, so, for example, employees of purchasers could not make claims against suppliers (in part because they were under Workers' Compensation).

The most proximate cause of the most severe asbestos-related health problems was the United States Government. Millions of tons of asbestos were used to insulate ships built on contract for the Navy under Navy specifications. The way it was applied exposed many thousands of workers to dangerous levels of asbestos, compared to the relatively safe way that asbestos insulation in the construction of buildings was applied—something that Navy officials clearly knew (Castleman 1984, 503-4). Since the government would not admit any liability for harm to the workers who handled or were exposed to asbestos products (Benson 1996), workers were essentially liable for the consequences of not informing themselves of its dangers and guarding against them.[13]

Even more significantly, given the issues addressed here, those who were exposed to asbestos had clear reasons to believe that the property rights to asbestos producers' income were secure. However, as tort law began to change in the early 1960s, producers' property rights to their income began to become less secure. Incentives to litigate developed, culminating in 1973 in *Borel v. Fiberboard Paper Products Corp.*, 473 F.2d 1076 (5th Cir. 1973), *cert. denied*, 419 U.S. 839 (1974). In this case, a worker maintained a product liability claim against an asbestos manufacturer rather than his employer, and the claim was upheld under the theory of strict liability. "This case completely transformed the law" (Epstein 1982, 43). What had been perceived as very secure property rights were suddenly much less secure. Indeed, as Epstein (1982, 44) explained:

> The *retroactive* nature of the duty not only renders the judicial exercise pointless as a matter of deterrence, but also imposes on the firm the impossible task of complying with a liability rule of which it could not have had any knowledge. The standard practices of yesterday have become the source of liability today. Rules, like horses, should not be changed in midstream.[14]

After the *Borel* decision, many additional claims were filed, and as these claims have expanded, plaintiffs have taken advantage of virtually all of the changes in tort law discussed above. Indeed, some of those changes were made in the context of asbestos litigation. The tort system's treatment of asbestos, like its treatment of product liability in general, has been "haphazard" (Epstein 1982, 44). The same evidence can lead to exoneration in one court and punitive damages in another. This uncertainty is at the heart of the litigation explosion. If property rights were clearly defined, there would be much less reason to litigate. When they are not clearly defined the incentives are to try to lay claim to any that are insecure. As Huber (1992, 731) noted, "To a tort lawyer, a single million-dollar verdict that survives all appeals can more than offset a long string of losses."

The Race for Rents

By 1984, the number of asbestos tort cases had passed 30,000 (Hensler et al. 1985), but filings increased every year. Suits are likely to continue to be filed as long as non-exclusive income is perceived to be available from asbestos producers. For instance, prior to 1983, nearly all of the personal-injury suits filed against asbestos firms were by shipyard workers and miners, but in 1983, more that sixty plasterers who had worked on buildings insulated by asbestos filed lawsuits (Chen 1984, 30), suggesting to a whole new set of potential plaintiffs that they might be able to lay claim to the income of former asbestos producers. Similarly, a new area of asbestos litigation began in 1982, known as "rip-out-and-replacement" cases (Chen 1984, 30), in which asbestos suppliers and manufacturers are sued for the cost of replacing material that contain asbestos (initially only schools were involved, but others soon followed). A flood of asbestos suits followed.

In 1992 the president of Keene Corporation reported that there were more than 85,000 cases filed against his firm alone, with an additional 2,000 filings every month, and that they had already settled 75,000 cases (Bailey 1992, 1). The cases all have multiple defendants, so he estimated the total number of cases actually pending for the industry to be 150,000. After all, workers can rarely demonstrate that one particular supplier was the exclusive cause of their harm (Chen 1984, 26). Thus, plaintiffs draw upon the court's willingness to apportion damages among a group of potential tortfeasors. Given that willingness, "the prudent plaintiff will sue every manufacturer that ever supplied any employer for whom he worked over the years" (Epstein 1982, 45).

Rent Dissipation

Some forms of rent dissipation are measurable. Legal fees in asbestos were running at about three times the level of compensation payments for the 30,000 plus claims made before 1985 (Hensler et al. 1985). Asbestos plaintiffs had received about $400 million but they netted $236 million because plaintiffs' lawyers had been paid around $164 million, or 41 percent of the $400 million. In addition, defense lawyers had received over $600 million at the time, including about $205 million in cases that resulted in payments to plaintiffs and $395 million to fight claims by plaintiffs who got nothing. So of the $1 billion that had changed hands, over 75 percent had gone to lawyers. This distribution has persisted. Asbestos firms' payments reached an estimated $12 billion in 1992, with $9 billion going to legal fees (Bailey 1992, 2).

Plaintiff's lawyers generally work on a contingency fee basis, taking from a third to a half of the award received by a client. Since 25 percent of all of the punitive damage awards in tort cases since 1965 have been in asbestos suits (Bailey 1992, 3), the awards can be quite high. Asbestos awards have also included compensatory damages for intangible harms such as fear of cancer. Not surprisingly, lawyers respond to incentives, and many have specialized in the asbestos tort business, flying around the country helping local attorneys pursue thousands of claims. A recent settlement netted two lawyers $125 million, for instance, or about $1 million a day for the time they spent on the case (Bailey 1992, 3). This is somewhat of an outlier, but Bailey estimated that plaintiffs' lawyers' contingency fees in asbestos cases average over $5,000 per hour of work actually done (Bailey 1992, 2). In fact, this is another reason for the high defense fees. Since plaintiff attorneys obtain relatively large fees in asbestos cases, and they can obtain multiple fees by representing multiple clients, they have strong incentives to lead the rush for property rights. The asbestos firms are forced to defend themselves on an ever increasing number of fronts.

There were some 200 asbestos producers, and the average asbestos suit names at least twenty defendants who incur a share of the damages awarded (Chen 1984, 26). These multi-defendant cases are one reason for the huge amount of money being paid to defense attorneys. Typically, a defendant has to hire defense attorneys in every jurisdiction in which it is being sued. Manville, one of the largest asbestos manufacturers, employed sixty different law firms around the country before declaring bankruptcy in 1982.

As courts have become willing to allow awards for such things as fear of disease when no physical manifestation exists, the potential for successful pursuit of premature or illegitimate claims has increased. The race for property rights and the likelihood of depletion of defendants' financial capabilities to pay (or of bankruptcy, as discussed below) encourages plaintiffs to get into the race as early as they possibly can, so the increased willingness of courts to grant damages prior to physical manifestations of exposure simply encourages more claims.

Personal injury lawyers actively recruit plaintiffs, many of whom have little or no discernable asbestos related illness (Chen 1984, 26). The result is analogous to Libecap's (1984, 383) findings of the premature extraction of oil from federal fields, which prematurely depleted subsurface pressures and reduced total oil recovery, and Anderson and Hill's (1990) analysis of premature production on homesteaded lands. For instance, Keene Corporation asked a federal court in 1993 to declare that it had too little money to pay claimants and to order its remaining $100 million in assets to be divided in a class-action settlement for asbestos claimants (*Tallahassee Democrat,* May 14, 1993, p. 8A). Had the suit filings been slower and the awards smaller, Keene might have survived to pay off many of the claimants over time.

Interestingly, in terms of the proximate cause issue, Keene was not in the asbestos business until 1968 when it bought another company, Baldwin Ehret Hill (BEH), which produced asbestos. Only 15 percent of BEH's products contained asbestos. Those products were only 10 percent asbestos and BEH put warning labels on the products. These sales constituted only 2 percent of Keene's sales until 1972 when it stopped all asbestos production. Keene had paid $80 million for BEH, only to find that it was a loser, so they closed the entire operation in 1975. But between 1975 and 1992 Keene paid $400 million on BEH asbestos-related litigation expenses, over five times the value of the BEH assets. With such rapid dissipation of wealth, plaintiffs must rush in if they want to capture any of it.

Another reason for plaintiffs to enter the litigation race for property rights with premature or illegitimate claims is that the courts, overwhelmed by asbestos cases, have begun to consolidate claims to encourage mass settlements. These are non-consensual consolidations (Bailey 1992, 3), and defendants object strongly because the combined evidence from multiple claims can overwhelm a jury with facts, statistics, and anecdotes that would not have been admissible in any single plaintiff's case. Weak or illegitimate claims are able to free ride on stronger claims, sharing in the punitive damages and intangible harm awards. As Huber (1992, 728) explains, "present scientific knowledge" indicates that many of the concerns over asbestos are exaggerated: the risks of high levels of exposure are "grave" but the risks from low levels of exposure (e.g., working in a building insulated with asbestos) are "apparently insignificant." Nonetheless, hundreds of thousands of additional claims are possible because they often need not be legitimate.

Efforts to Defend Insecure Property Rights

The ongoing defensive effort in litigation is but one margin along which asbestos producers have tried to preserve their vulnerable income. They have also lobbied Congress in an effort obtain an Asbestos-Health-Hazards-Compensation bill. Such a bill was introduced in Congress in 1977 by Rep. Millicent Fenwick (whose district included the town of Manville where a large Manville asbestos plant was located), and by Sen. Gary Hart (whose 1980 campaign received contributions from Manville's PAC and several of its officers) in 1980 and again in 1981 (Brodeur 1985, 192, 194-95, 258-62). The bill would have required that compensation for asbestos injuries be obtained from the federal or state no-fault workers' compensation insurance carried by the last employer who exposed a particular worker to asbestos. In other words, the statute would have reestablished the property rights that had been in place before the *Borel* decision by once again placing such issues under workers' compensation. These efforts failed. The Association of Trial Lawyers of America is one of the most powerful interest group in the country and it actively campaigns against tort reform legislation that might protect producers' property rights (Benson 1996, 346).

Some companies have used another defensive tactic in an effort to regain control of the property rights to income: Chapter 11 bankruptcy. Under the Bankruptcy Reform Act of 1978, the court can estimate the value of claims against a company seeking relief, including all legal obligations "no matter how remote or contingent," and develop a plan to expedite that relief. Chapter 11 also produces an automatic stay of all pending litigation and blocks payment of all unsecured debts incurred before filing, including attorneys' fees.

One of the first asbestos firms to declare bankruptcy was Manville. In 1982 it was spending $1 million a month on asbestos-related legal fees (Chen 1984, 30). That year, its insurance company, which had paid out $16 million in asbestos claims during the first quarter of the year alone, informed Manville that it would not pay any more. Manville had seen its asbestos settlements rise form $300,000 in 1975 to $35 million in 1981, and the 1982 figure reached $27 million in August. As a consequence, the company chose to declare Chapter 11 bankruptcy against the 16,500 lawsuits that were pending and the nearly 500 new ones that were being filed every month. The Manville bankruptcy case and reorganization took over six years to accomplish, during which time no claims were paid.

The settlement required Manville to establish a trust that would ultimately be able to pay out $2.5 billion to asbestos claimants and another trust to pay schools, hospitals, and businesses for asbestos removal (Cohn and Bollier 1991, 131). Manville won on the issues of paying only compensation, not punitive damages, and of a mandatory settlement procedure to be followed before litigation against the trust can be waged. Furthermore, their liability was limited so they have no obligation to fully pay all claims that may eventually materialize. The court also approved payments of $102 million to lawyers, accountants, and investment bankers involved in the bankruptcy proceedings.

Bankruptcy was not cheap for Manville, but it secured rights to much more income than it would have been able to otherwise claim. By 1990, 150,000 claims had been filed against the trust rather than the 100,000 predicted, for instance, and the average award was $43,500 rather than the $25,000 predicted (Cohn and Bollier 1991, 132). The trust's commitments to make payments for claims already settled during or before 1990 meant that any new settlements would have to wait at least fifteen years for a payment. There were $6 billion in outstanding claims by 1990.

By August 1992, sixteen asbestos companies had declared Chapter 11 bankruptcy, turning away from the common law of torts to the bankruptcy courts in an effort to clarify their asbestos liabilities (Bailey 1992, 3). There is little doubt that others are going to take this route in an effort to survive with some assets intact unless an alternative arises. Of course, the threat of Chapter 11 adds to the incentives of attorneys to find potential plaintiffs willing to rush in and make claims early.

The same story could be told about many other industries. Recently, for instance, Dow Corning, a company that lost $6.8 million on $2.2 billion in sales in 1994, declared bankruptcy after a $4.2 billion settlement of about 400,000 breast implant cases, not because of the settlement but because about 10,000 plaintiffs chose to pursue separate litigation rather than accept the terms of the settlement (Goodman 1995, 8A). The settlement has subsequently fallen apart.

III. Conclusions

Liability rules that clearly specify who is liable have some very desirable characteristics. Secure property rights eliminate the incentive to litigate which arises when property rights are not secure. Even if theoretically the liability rules in place are not "optimal," the efforts to claim non-exclusive income that arise when liability rules are in a process of change may fully dissipate any efficiency gains that could arise by eliminating the alleged "non-optimal" assignment of liability. Whether the end of traditional negligence standards and their replacement by strict liability (and all of the other changes that have ocurred) have been movements in the direction of the theoretical "optimum" or not the movement itself has generated enormous investments in socially wasteful rent-seeking.

But are the changes producing better rules? It is not clear that the demise of traditional tort standards is appropriate even when the producer is the low-cost provider of information. The individual user of a product is in many cases likely to be the least-cost accident avoider. Thus, a negligence standard is likely to be the most efficient standard, and the dissipation of rents that we are observing may actually be arising in the context of a move away from the actual optimum.

Schuck (1986, 13) noted that "In reality, the Agent Orange litigation [and we can add, the Asbestos, DES, Dalkon Shield, etc.] prefigures a grim dimension of our future; it is a harbinger of mass toxic torts yet to come." He suggested that future disputes may involve pharmaceutical, food additives, industrial compounds, pollutants, toxic waste landfills, radiation, or any of an unpredictable number of unknown technological advances. While he may be right, it will not be because of the technological advances and the "toxic age"; it will be because the government has failed to clarify and secure property rights. Indeed, the same trends are also likely to accelerate in the non-toxic areas of product liability (Benson 1996).

When courts assert that they must "fashion new standards" rather than hold to traditional tort principles, they signal that they are ready to entertain new reasons for redistributing producers' income. If potential plaintiffs do not see the signal, tort lawyers will, so continued recruitment of clients can be expected in the evolving race to capture increasingly less secure property rights to producers' incomes. Of course, this process ultimately may undermine incentives to seek and produce technological advances (or to produce at all), unless the pressure that is being put on state and federal legislatures forces them to assign the insecure property rights to someone (producers, consumers, or workers) in order to end the continual dissipation of non-exclusive income through the litigation process.[15] As Epstein (1982, 46) concludes: "there is not an endless supply of water in the trough. We must somehow undo the unsystematic and unthinking judicial activism. Otherwise—as more and more cases work their way through the legal system, and more and more firms take the bankruptcy route—the only doors left will be closed, and marked 'No Exit.'"

Notes

1. This section draws from Benson (1996).
2. See *Greenman v Yuba Power Products, Inc.* 59 Cal. 2d. 57, 377 P.2d 897 (1963).
3. For instance, see *Payton v. Abbott Labs* 386 Mass 540, 437 N.E.2d 171, 181 (1982), and *Ball v. Joy Manufacturing Co.* 55 F.Supp 1344 (S.D. W.Va. 1990), *aff'd sub nom. Ball v. Joy Technologies Inc.*, 1991 WL 146815 (4th Cir. 1991).
4. *Potter v. Firestone Tire and Rubber Co.* 274 Cal. Rptr. 885 (Ca. App. 1990), *review granted*, 278 Cal. Rptr. 836, 806 P.2d 308 (Cal. 1991); *Herber v.*

Johns-Manville Corp. 785 F.2d 79 (3d Cir. 1986); *Wetherill v. University of Chicago* 565 F.Supp. 1553 (N.D. Ill. 1983); *Wells v Ortho Pharmaceutical* 615 F. Supp. 262 (N.D. Ga. 1985); *Laxton v. Orkin Exterminating Co.* 639 S.W.2d 431 (Tenn. 1982); *In re Moorenovich* 634 F.Supp. 634 (D. Me. 1986); *Sterling v. Velsicol Chemical Corp.* 855 F.2d 1188 (6th Cir. 1988).

5. The same arguments were rejected in *Man v. Raymark Industries* 728 F. Supp. 1461 (D. Haw. 1989), however, and the defendant's actions in *Tetuan v. A. H. Robins Co.* 241 Kan. 441, 738 P.2d 1210 (1987) were viewed to be a series of different wrongs rather than a single wrong, thereby avoiding the constitutional claims.

6. Courts are divided as to the treatment of this approach (Brenner 1989, 789). Under the "strong" version, statistical correlation indicating that the probability exceeds fifty percent is not sufficient - some particularistic proof of direct and actual knowledge of the causal relation is required (*Namm v. Charles E. Frosst & Co.*, 178 N.J. Sper. 19, 427 A.2d 1121, 1125 [App. Div. 1981]). The weak version allows a verdict for the plaintiff based solely on statistical probability (*In re Agent Orange Product Liability Litigation.*, 597 F. Supp. 740, 835 [E.D.N.Y. 1984]).

7. The idea that a firm is >better able to bear the cost' has been justified in the California courts by arguments about insurance costs or risk spreading through slightly higher prices (see for example, *Greenman v Yuba Power Products, Inc.*, 59 Cal. 2d. 57, 377 P.2d 897 [1963]), implying that placing liability on firms is the efficient allocation of these costs, but actually a "deep pocket" criteria is probably being applied.

8. In *Namm v. Charles E. Frost & Co.* the court held that the *Sindell* theory of enterprise liability would "result in total abandonment of the well settled principle that manufacturers are only responsible for damages caused by a defective product upon proof that the product was defective and that the defect arose while the product was in the control of the defendant." In *Payton v. Abbott Labs*, 386 Mass. 540, 437 N.E.2d 171 (1982), the court was willing to use market shares as long as each defendant was held liable for the portion of damages caused to the total injured population. Other states have also adopted modified versions of enterprise liability or market share theories - see *George v. Parke-Davis*, 107 Wash. 2d 584, 733 P.2d 507 (1987); *Collins v. Eli Lilly & Co.*, 116 Wis. 2d 166, 342 N.W. 2d 37 (1984); and *Abel v. Eli Lilly & Co.*, 48 Mich. 311, 343 N.W. 2d 164 (1984).

9. The issue of why these rapid changes in tort law occurred is interesting. The changes coincide with the evolving political power of environmental groups, for instance, and with rapid changes in statute and administrative law as responses to such interest groups, so it may well reflect the same political forces. After all, the courts clearly are subject to and respond to interest group demands (Benson 1990, 112-17; Neely 1982). The changes also correspond in terms of timing with changes in the ethical rules of the legal profession. State bar associations virtually eliminated all of the long-

standing restrictions on lawyers' freedom to advertise, to solicit clients, and to finance clients, especially in personal injury, civil rights, and "public interest" cases (Schuck 1986, 26). Thus, lawyers are now free to conceive, initiate, and sustain litigation themselves without being constrained by identifiable clients (e.g., as in a class action suit). With these reduced constraints, lawyers may be pursuing cases which would not have been tried before, and therefore forcing judges to consider significant issue for which no clear precedent existed (see Benson [1990, 62-71] for an example of this kind of transformation in law arising from a sudden increase in lawyer involvement in criminal trials). Nonetheless, no effort is made here to either determine the source of these changes in tort law or discuss the literature that has developed on that subject (e.g., see Rabin 1988; Schuck 1986, 32; Cooter and Ulen 1988, 438-39).

10. Indeed, crowding in the case of toxic torts spills over into other litigation, as noted above, and see *In re Ohio Asbestos Litigation*, Nol 83-OAL (N.D. Ohio Gen. Order No. 67, filed June 15, 1983).

11. In a broader context, other impacts include the reduced incentives to invest in producing any product that might be misused and result in injury, higher insurance costs, higher product prices, smaller entities than might be dictated by scale economies (in order to limit potential liabilities), and so on.

12. Castleman (1984, 31) provides a detailed chronology of asbestos related medical research. Among other things, he concluded that "asbestosis was by 1935 widely recognized as a mortal threat affecting a large fraction of those who had regularly worked with the material." Furthermore, the most severe exposures were in Navy shipyards, and the Navy published "Minimal Requirements for Safety and Industrial Health in Contract Shipyards" in 1943 (the first of several reports) recommending segregation of dusty work, use of exhaust ventilation and respirators, and periodic medical exams; yet such precautions were not mandated and only some shipyards advised the use of respiratory protection (Castleman 1984, 166). Some effects of asbestos were not confirmed until the mid-1960s, of course (Epstein 1982, 18), and some alleged effects are still not confirmed, but the point is that the knowledge of asbestos-related problems that existed was widely disseminated when the exposures that have generated most of the asbestos cases arose.

13. Note that a good deal of information about potential health consequences of asbestos exposure was availableCsee note 12. Also note that the government still will not admit liability for alleged asbestos harms. See also chapter 3 for a discussion of potential consequences of the general approach that governments have taken to toxic liability issues.

14. Epstein (1982, 44) emphasized the forum change from Worker's Compensation to the common-law courts. Here the focus is on the reduced security of property rights to income.

15. Indeed, the "uniform practice in every other industrialized country" is to bar all tort action against suppliers (Epstein 1982, 46). Thus, shifting such liability and its accompanying litigation costs onto U.S. suppliers may be part of the reason for the widely perceived weakening of America's com-

petitive position in world markets—another negative externality of government action.

References

Anderson, Terry L., and Peter J. Hill. 1990. "The Race for Property Rights." *Journal of Law and Economics* 33 (April): 177-97.

Bailey, Glenn W. 1992. "Asbestos Litigation Monster Rewards Plaintiffs' Lawyers While Devouring Jobs and Economic Growth." *Legal Backgrounder* 24 (August 28): 1-4.

Barzel, Yoram. 1989. *Economic Analysis of Property Rights*. Cambridge: Cambridge University Press.

Benanav, Gary G. 1994. "At Last, Congress May Finally Clean Up Superfund." *Wall Street Journal* (May 3): A14.

Benson, Bruce L. 1990. *The Enterprise of Law: Justice Without the State*. San Francisco: Pacific Research Institute for Public Policy.

————. 1996. "Uncertainty, the Race for Property Rights, and Rent Dissipation due to Judicial Changes in Product Liability Tort Law." *Cultural Dynamics* 8 (November): 333-351.

————. "Toxic Torts by Government." Chapter 3 in this volume.

Brenner, Jeffrey S. 1989. "Alternatives to Litigation: Toxic Torts and Alternative Dispute Resolution – A Proposed Solution to the Mass Tort Case." *Rutgers Law Journal* 20 (Spring): 779-821.

Brodeur, Paul. 1985. *Outrageous Misconduct*. New York: Pantheon Books.

Castleman, Barry I. 1984. *Asbestos: Medical and Legal Aspects*. New York: Harcourt Brace Jovanovich Publishers.

Chen, Edwin. 1984. "Asbestos Litigation is a Growth Industry," *Atlantic* 254 (July): 24-32.

Cheung, Steven N. S. 1974. "A Theory of Price Control." *Journal of Law and Economics*. 17 (April): 53-71.

Cohen, Henry S., and David Bollier. 1991. *The Great Hartford Circus Fire: Creative Settlement of Mass Disasters*. New Haven, CT: Yale University Press.

Cooter, Robert and Thomas Ulen. 1988. *Law and Economics*. New York: Harper Collins.

Eggertsson, Thrainn. 1990. *Economic Behavior and Institutions*. Cambridge: Cambridge University Press.

Epstein, Richard A. 1982. "Manville: The Bankruptcy of Product Liability." *Regulation* (September/October): 14-19, 43-46.

————. 1980. "The Static Concept of the Common Law," *The Journal of Legal Studies*. 9 (March): 253-75.

Goodman, David. 1995. "Are Companies Abusing Chapter 11?" *Tallahassee Democrat* (May 16): 8A.

Hensler, Deborah R., William L. F. Felstiner, Molly Selvin, and Patricia A. Ebner. 1985. *Asbestos in the Courts: The Challenge of Mass Toxic Torts*. Santa Monica, CA: Rand Institute for Civil Justice.

Hensler, Deborah R., Mary E. Vaiana, James S. Kakalik, and Mark A. Peterson. 1987. *Trends in Tort Litigation: The Story Behind the Statistics.* Santa Monica, CA: Rand Institute for Civil Justice.

Huber, Peter. 1992. "Junk Science in the Courtroom." *Valparaiso University Law Review* 26 (Summer): 723-55.

Johnson, Ronald N., and Gary D. Libecap. 1982. "Contracting Problems and Regulation: The Case of the Fishery." *American Economic Review* 72: 1005-22.

Lafferty, David. 1991. "*Juzwin v. Amtorg Trading Corp.*: Multiple Assessments of Punitive Damages in Toxic Tort Litigation." *Pace Environmental Law Review* 8 (Spring): 647-65.

Libecap, Gary D. 1984. "The Political Allocation of Mineral Rights: a Re-evaluation of Teapot Dome." *Journal of Economic History* 44 (June): 381-91.

MacCoun, Robert. 1993. "Inside the Black Box: What Empirical Research Tells Us about Decision-making by Civil Juries." In *Verdict: Assessing the Civil Jury System*, ed. by Robert E. Litan. Washington, DC: The Brookings Institute.

Madison, James. "The Federalist No. 44." In Alexander Hamilton, John Jay, and James Madison, *The Federalist: A Commentary on the Constitution of the United States.* New York: The Modern Library, 1937.

McChesney, Fred S. 1987. "Rent Extraction and Rent Creation in the Economic Theory of Regulation." *Journal of Legal Studies* 16 (January): 101-18.

Moller, Erik. 1996. *Trends in Civil Jury Verdicts Since 1985.* Santa Monica, CA: Rand Institute for Civil Justice.

Neely, Richard. 1982. *Why Courts Don't Work.* New York: McGraw-Hill.

Peterson, Mark A., and George L. Priest. 1983. *The Civil Jury: Trends in Trials and Verdicts, Cook County, Illinois, 1960-1979.* Santa Monica, CA: Rand Institute for Civil Justice.

Pope, Michael A., and John F. Del Giorno. 1991. "Novel Damage Theories May Contaminate Toxic Tort Litigation." *Defense Council Journal* 58 (October): 495-503.

Rabin, Robert L. 1988. "Tort Law in Transition: Tracing the Patterns of Sociolegal Change." *Valparaiso University Law Review* 23 (Fall): 1-32.

Rubin, Alvin B. 1986. "Mass Torts and Litigation Disasters." *Georgia Law Review* 20 (Winter): 429-53.

Schuck, Peter H. 1986. *Agent Orange on Trial: Mass Toxic Disasters in the Courts.* Cambridge, MA: Belknap Press.

Spyridon, Gregg L. 1991. "Scientific Evidence vs. 'Junk Science'—Proof of Medical Causation in Toxic Tort Litigation; The Fifth Circuit 'Fryes' a New Test (*Christophersen v. Allied-Signal Corp.*)." *Mississippi Law Review* 61 (Fall): 287-313.

6

Toxic Torts:
Problems of Damages and Issues of Liability

David D. Haddock and Daniel D. Polsby

Science News (February 6, 1993) brought an alarming report from Iowa City: sleep is hazardous to one's health. More precisely, REM sleep, a normal phase in the sleep cycle, appears to be highly correlated with, and may possibly contribute to, the risk of ischemic heart disease, often a killer. The same issue of *Science News* also brought the dismaying intelligence (from the University of Wisconsin) that plastic cutting boards, long thought to be more sanitary in food preparation than the wooden kind, and thus required by many public health departments, are in fact more hospitable environments to potentially deadly coliform bacteria.

Plastic cutting boards and dream-filled sleep thus join a lengthening list of health risks: Listerine (throat cancer); dietary fat (cardiovascular disease; cancer); aerosol sprays (skin cancer); untreated pajamas (burns); flame retardants in pajamas (cancer); aspirin (stroke; gastrointestinal disease); failure to take aspirin (heart disease; colon and rectal cancer); hot dogs (attention deficit disorder; cancer); asbestos (cancer; pulmonary disease); the fuel tanks of certain pickup trucks and small cars (burns); solar radiation (skin cancer); treated food (cancer); untreated food (salmonella and coliform contamination); refined sugars (cardiovascular disease; periodontal disease; attention deficit disorder); saccharine (cancer); aspartame (headache; vertigo); caffeine (cardiovascular disease); beer (liver disease); birth control pills (cancer, heart disease); intrauterine devices (uterine hemorrhaging); silicon prostheses (immune system impairment); cathode rays (eye and brain disease; epilepsy); halogens in the drinking water (cancer); inadequately treated drinking water (cholera).

To the above one might add any number of general environmental degradations, sometimes called toxic torts when they have an attributably human origin: hydrocarbon combustion residues of all kinds, and hydrocarbon precipitations such as gasoline vapors (cancer); formaldehyde vapors (cancer); mercury, lead, and other heavy metals (brain, liver, and kidney damage); plutonium from leftover atomic weapons (cancer); industrial nuclear waste products (cancer); risks posed by habitat loss to the ecosystems and biodiversity of the planet, from the spotted owl and snail darter to the ongoing clear cutting of rain forests by greedy landowners (who thus threaten the survival of aerobic life forms in the known universe).

We could go on indefinitely cataloguing disasters impending and in progress, even more of which may eventually become the subject of legal liability claims in an American courtroom. But longer lists are not required to make our point: namely, that the environment seems to have become so remarkably full of jeopardy that one should be surprised to live long enough to be able to read all the way through this introduction before being struck down by some insidious newfangled doom. Yet the national mortality statistics regularly report a counter result: people are living longer and longer all the time, and seemingly with greater vigor, too, if the golf courses of Florida and Arizona offer a trustworthy indication. How dare the populace louse up a perfectly good model? Is there a paradox at work, or just a word game?

The confusion comes from misframing the matter to be analyzed. The occasional newsworthy calamity is a misleading representation of modern life, even though as our analytical and statistical tools get better, inevitably we can ferret out more and more products and procedures that are somehow occasionally connected to this or that problem. But by proceeding in such a fashion we may be learning more about how clever academically and politically ambitious investigators can be than we are about anything important about the world. In order to understand the relationships between legal institutions and real risks, we should be less interested in particular members of the set, and more interested in the common properties of the entire class of things to which these risky-but-beneficial entities belong. We observe that all the members of the set have benefits, and all have risks; and we observe that, in gross, the world in which we live today, though it is filled with un-countable risks of which our ancestors were innocent, is a far safer, more comfortable, and more predictable place than was true even one hundred years ago.

Nevertheless, that realization does not imply that the legal system should be indifferent to hazards. Thus, the question remains what the legal system ought to do about risk. Strict liability with a defense of contributory negligence, in which defendants bear plaintiffs' losses unless it can be shown that the loss would not have resulted except for the plaintiff's misbehavior, continues to gain in acceptance. The prominence of strict liability rules reflects an intuition that such rules furnish superior incentives to actors to take account of the harms that their activities impose on the people around them. Professor William Jones (1992), relying on arguments first set forth by Shavell (1980; 1987), has recently put the argument thus: If actors' conduct were completely transparent, so that it was apparent (or would become so) whether the actor should or should not have done a certain thing, then there would be little to choose between the rules of negligence on the one hand, and strict liability with certain defenses on the other. Imposing liability on whoever is found to have been behaving in a careless fashion would induce all the actors to "internalize" the costs of injuries that are apt to occur in the course of life, and therefore to undertake the appropriate level of caretaking. But actors' conduct is not completely transparent in that way. Indeed, some kinds of behavior—Shavell's example is activity levels—will be quite opaque, at least to a court.

For instance, every time an airline flies a plane bound for Europe there will be some chance of various kinds of harm resulting even if it acts with due care. Two trips per day will create substantially more risk than one. Transparent to analysis and straightforward for a court to adjudicate are such questions as: were flight officers and engineers selected and supervised properly? was the airframe maintained correctly? and so on. But opaque to analysis, and therefore difficult for a court to factor into a behavioral standard against which to compare the company's conduct, will be such questions as whether the airline ought simply to have run one (careful) flight per day instead of two.

Because so large a proportion of social harms result from such unmonitorable behaviors, it behooves the legal system to seek a convention for assigning liability that gives proper incentives irrespective of whether one can accurately characterize an actor's behavior as careful or careless, correct or incorrect. It is on that ground that Jones recommends that the legal system adopt a strict liability rule for hazardous activities. If a defendant must bear the costs of injuries whether or not he was behaving negligently, he will have incentives to undertake all precautions that cost less than their expected benefits. That will be true even if a court cannot accurately decide whether or not particular precautions were cost-justified.

Jones's argument is fine so far as it goes; but it goes only half-way through Shavell's analysis, and thus only half-way to making a perfectly valid point. True enough, there are many aspects of the behavior of potential defendants that are (for one reason or another) unmonitorable. But there is a correlative point about the behavior of potential plaintiffs that calls into question the automatic superiority of strict liability rules. Plaintiffs, after all, have some responsibility for their own safety. Some plaintiffs' behavior (like some defendants' behavior) is monitorable and thus usable for fabricating defenses against a strict liability judgment. But some plaintiffs' behavior (in parallel with some defendants' behavior) is unmonitorable. And so the unasked and unanswered questions in the Jones position are: What is the effect of defendants' strict liability on plaintiffs' non-monitorable behavior? Why doesn't the logic of the argument lead one to the *inverse* of defendant's strict liability, namely, plaintiff's strict liability, what is usually called negligence, in which plaintiffs bear their own losses unless they can show that the loss would not have resulted except for the defendant's misbehavior? The essay that follows is meant to explain the limitations of theory in deciding between negligence and strict liability. A proper comparison of the two doctrines, it will be seen, ultimately must be an empirical task.

1. The Nirvana Fallacy

Economic theory predicts, and much practical experience confirms, that the more nearly an owner's entitlement to an asset approaches "complete" property rights, the more productively that asset will be used. The owner of a complete property right has an incentive to use the asset in its most valuable way, or to sell it to someone who will.

If property rights are incomplete, meaning that an asset owner will not be able to appropriate all its value to himself, the asset may on occasion be diverted away from its highest value use to a lower value one. Suppose the ostensible owner can't successfully forbid someone from appropriating (some of) the value of his (incompletely owned) asset. Party A's actions interfere with party B's plans even though the benefit to party A is less than the cost to party B, and B cannot make A pay for the loss. In such an instance one has what is called a *Pareto-relevant externality* (Buchanan and Stubblebine 1962).

That much is generally accepted among economists. Similarly, economists understand why property rights may be left incomplete though that predictably leads to costly externalities. The reason is that *defining and enforcing entitlements is costly.* It would be a Pyrrhic transaction to reduce externalities by $1 at a cost in other resources of $1.01 or more (Demsetz 1967).

In practice, therefore, a world devoid of all externalities would be an inefficient world, or, at least, could be efficient only if defining, enforcing, and transacting entitlements were costless. Thus, one does not establish the existence of a real-world inefficiency simply by discovering an externality. Such a discovery, rather, may suggest a potentially fruitful comparative institutional issue, namely, whether the externality is being, not eradicated, but *optimized*—cut down to as small a size as is desirable given the transaction and learning (or information) costs (Demsetz 1969).

It does not follow from the recognition that some externalities are efficient that it is pointless to consider a system of complete property rights enforceable at zero cost. Thinking about things in that way can provide a baseline against which alternative real-world arrangements can be evaluated. In other words, the concept of "costlessly complete property rights" provides a sort of zero on a gauge of efficiency just as "zero-Kelvin" establishes a baseline on a thermometer. It is no aspersion on the concept of "zero-Kelvin" that such a temperature is never actually observed, and it is no criticism of the concept of "complete property rights with zero transaction costs" that in the real world there can be no such state.

No one should be surprised if economists, doing field research in the real world, discover numerous divergences from perfect arrangements; externalities are everywhere. Sometimes, as is true of oil spills, they are conspicuous and obnoxious. Much modern torts scholarship seems to take the position that once one discovers an externality, it axiomatically follows that there should be a change in liability rules that would give each implicated agent an incentive to internalize the externality. That conclusion is a *non sequitur.* Whether compulsory internalization "should" happen or not depends entirely on the values assigned to two factors: (1) what a given proposed rule will cost, and (2) the cost of the residual externality. That sum will always be a positive number and the best anyone can do is to discover the arrangements that will *minimize* the sum of costs. When rules are costly to implement and/or residual externalities are likely to be large (as may be optimal) compulsory internalization is not warranted.

That often-repeated point hardly introduces a novelty to economic analysis. Yet in practice, most economists ignore it more often than not. The typical analysis of oil spills and other "toxic" torts is but one example: researchers strive to discover a legal arrangement that will exactly replicate costless, complete, property rights. Demsetz (1969) tried to call attention to the otherworldliness of analyzing existing institutions from the timeless, unconstrained standpoint of "nirvana":

> The view that now pervades much public policy economics implicitly presents the relevant choice as between an ideal norm and an existing "imperfect" institutional arrangement. This *nirvana* approach differs considerably from a *comparative institution* approach in which the relevant choice is between alternative real institutional arrangements. In practice, those who adopt the nirvana viewpoint seek to discover discrepancies between the ideal and the real and if discrepancies are found, they deduce that the real is inefficient. Users of the comparative institution approach attempt to assess which alternative real institutional arraignment seems best able to cope with the economic problem; practitioners of this approach may use an ideal norm to provide standards from which divergences are assessed for all practical alternatives of interest and select as efficient that alternative which seems most likely to minimize the divergence. (Demsetz 1969, 1)

In other words, an external cost is like any other cost in one important way—it is to be optimized, and that will rarely mean obliterated. We are not dismayed that automobiles can only be produced if a certain amount of real resources (such as workers' leisure time) must be sacrificed via market transactions. Why need we be dismayed if automobile production similarly requires some sacrifices via external costs? The sacrifices should be dismaying only if they exceed those that would be required under plausible alternative arrangements.

The nirvana fallacy involves the usually implicit and often unconscious assumption that some real-world costs of a given endeavor are zero, and that certain liability rules will cause the model's actors to optimize all the remaining costly activities. Or, researchers may instead discover that the combined costs of externalities and what one must pay to get rid of them cannot altogether be extinguished regardless of institutional arrangements, and then declare hopelessly that every such arrangement is "inefficient."

At various times, two well-regarded scholars of law and economics, Professors Polinsky and Shavell have committed each of these *faux pas* (Polinsky 1982; Shavell 1982; Polinsky and Shavell 1992). Though Demsetz' proposition is widely known and cited by economists, Polinsky and Shavell (1992) cling to the nirvana fallacy in their assertion that strict liability beats negligence as a means of assigning liability for toxic environmental torts (oil spills and the like), because it causes producers to internalize the costs of injury. For that more limited set of injuries, then, Polinsky and Shavell's position resembles that of Jones. Under negligence rules, defendants (whom we will for clarity sometimes refer to as producers), bear none of the external costs of their actions so long as the marginal benefit of omitted care-taking behavior does not exceed the marginal cost of those actions. But the external costs imposed by a producer's actions are real costs to society. If the producer ignores a part of the costs arising from its endeavor, it will choose to produce too much.

2. Simulations

Our argument is that the choice between liability rules must ultimately rest on empirical findings. Not even the greatest of theoreticians could possibly answer this question unless they were in possession of the relevant data. We progress by stages. First, we discuss a simulation that enables us to gain some *a priori* perspective on the magnitude of the injuries that individuals suffer as a result of prevalent hazardous production techniques present in our economy. This initial simulation incorporates a number of strong simplifying assumptions. For instance, in this model there are only two commodities whose production generates a hazard to bystanders. Second, each has a demand elasticity of zero. In other words, the number of units sold does not vary with changes in price, such as those that might be induced by changes in the liability exposure of the producers. Third, we assume away litigation costs. Finally, we assume away any behavioral response by either the potential plaintiffs or the potential defendants to a change in the liability rule. In other words, the production and consumption of the commodities is the same under either a negligence or a strict liability rule.

Given these assumptions, it turns out that strict liability does produce some anomalies: under strict liability people who have functionally identical experiences would be treated differently by the legal system. A number of individuals who have not been injured by the *set* of beneficial-but-hazardous activities are endowed with causes of action against given *members* of the set. In effect, they are allowed to sue the system for injury when actually they have been net beneficiaries of its workings. Thus, some people who have been extremely lucky in life will receive more solicitude in court than many others whose more modest fortunes might be thought to entitle them to more judicial concern.

In the course of subsequent discussion we relax all of these assumptions. First, we consider the impact of expanding the number of beneficial-but-hazardous products. In the course of that discussion a striking point begins to emerge: the more products one adds to the simulation, the stronger the principal conclusions of the initial simulation become. Next, we ask whether the pessimism about hazards that we have built into our model accounts for the results we have obtained. Evidently, the high level of hazard we assume does account for an extraordinarily large number of litigation opportunities. But reducing the hazard also makes trivial a problem with which advocates of strict liability should certainly be concerned, namely, whether real (as opposed to merely formal or legal) injury has been imposed on potential plaintiffs.

Finally, we assume that when the liability rule changes, so will people's behavior change. As Jones would have it, that includes a behavioral response by producer-defendants. Even then, however, we show that surpluses aggregated over the economy can be reduced (dramatically in our simulation) if plaintiffs also respond to changed liability rules (and if litigation costs are positive).

A. A Pessimistic Two-Product Simulation

Beginning with the strong set of assumptions outlined above, assume that the surplus or injury realized by any individual as a result of the ongoing production of one product is independent of any surplus or injury that individual may realize as a result of the production of the other product. Stated differently, the observation that Hylton is much better off as a result of consuming product A tells one nothing about the advantage or injury Hylton is realizing as a result of B's production.

Assume that for each product, 10 percent of the population realizes an advantage ("consumer's surplus") of +20, 20 percent realizes a surplus of +12, 40 percent a surplus of +4, 20 percent an injury of -4, and 10 percent an injury of -12 as illustrated by Figure 6.1. Thus, we have a discontinuous ersatz "Normal" distribution with few enough points for the reader easily to follow the discussion. The average, and most common, realization from either product is +4. So the product is beneficial on net, but because its production leads to some people being injured, hazardous as well. (Indeed, either product injures 30 percent of the population, so the model's products are very hazardous in comparison to most real world ones.) Suppose we examine a community of 10,000 families. Three thousand individuals are being injured by product A (1,000 of them quite severely), and 3,000 by product B. Life seems hard in our imaginary community, but how hard is it really?

FIGURE 6.1

Distribution of Outcomes Across Population: One Product

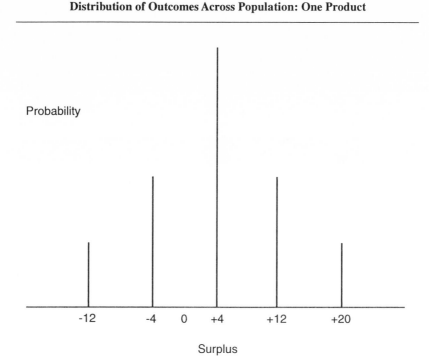

Probability

-12   -4   0   +4   +12   +20

Surplus

Suppose we consider "hazardous production" as the unit of analysis, rather than considering each product seriatim. Table 6.1 summarizes such a prospect. Consider the thousand individuals most severely injured by product A, those shown along the bottom row of the table. Because 10 percent experience a benefit of 20 from product B, 100 of the 1,000 individuals along the bottom row actually realize a net benefit from hazardous production, because 20 minus 12 sums to 8. That is shown in the lower left-hand cell of Table 6.1.

TABLE 6.1

Aggregates Surpluses: Two Products

		Product B				
		+20 N=1,000	+12 N=2,000	+4 N=4,000	-4 N=2,000	-12 N=1,000
Product A	+20 N=1,000	20 + 20 = 40 N = 100	20 + 12 = 32 N = 200	20 + 4 = 24 N = 400	20 – 4 = 16 N = 200	20 – 12 = 8 N = 100
	+12 N=2,000	20 + 12 = 32 N = 200	12 + 12 = 24 N = 400	12 + 4 = 16 N = 800	12 – 4 = 8 N = 400	12 – 12 = 0 N = 200
	+4 N=4,000	20 + 4 = 24 N = 400	12 + 4 = 16 N = 800	4 + 4 = 8 N = 1,600	4 – 4 = 0 N = 800	4 – 12 = –8 N = 400
	-4 N=2,000	20 – 4 = 16 N = 200	12 – 4 = 8 N = 400	4 – 4 = 0 N = 800	-4 – 4 = –8 N = 400	-4 – 12 = –16 N = 200
	-12 N=1,000	20 – 12 = 8 N = 100	12 – 12 = 0 N = 200	4 – 12 = –8 N = 400	-4–12=–16 N = 200	-12–12=–24 N = 100

Notice that 8 is the average net return from the two products together for the community's 10,000 individuals as a whole, yet under strict liability the individuals in the lower left-hand cell will have a cause of action against the hazardous-but-beneficial marketplace. Now consider all the 2,600 individuals along the lower-left to upper-right diagonal, in other words, all those whose aggregated benefit is 8 (which, of course, includes the 100 individuals we have already begun examining). Under strict liability, 1,000 of those individuals are entitled to sue, 200 claiming recompense of 12, 800 claiming 4. In contrast, the 1,600 individuals in the central cell of the table, those with an average experience (+4) from each product, have the same net experience of 8, but no cause of action. If the hazardous-but-beneficial marketplace is the unit of analysis rather than one component product in isolation, a strict liability rule seems to treat functionally identical individuals quite differently.

Those two phenomena—a cause of action in the hands of parties who are uninjured on net, and diverse treatment of previously identical parties—are not isolated occurrences. For example, the 200 individuals in the cell to the immediate right of the lower-left cell have suffered *no* injury, yet similarly have a cause of action against the producers of A. A comprehensive examination of the table reveals that only those individuals in the nine cells in the lower-right corner suffered an injury if one aggregates over both products. That sums to just 1,700 of the 10,000 families, even though either product considered alone injures 3,000. The individuals with a cause of action, in contrast, are all those in either the two bottom rows or the two right-hand columns, which amounts to 5,100 of the 10,000 individuals under observation. Indeed, 1,400 individuals with a net realization equal to or better than the average, nearly as many as those actually injured in net, have a cause of action enabling them further to improve their lot relative to—and ultimately at the expense of—other individuals in the community.

When one considers the hazardous market *as a whole*, the number of aggregate injuries falls. This is the normal result of pooling; anyone moderately familiar with statistics or finance would expect such a result, and would find the contrary surprising. Less obvious is how the number of potential lawsuits balloons as new products are added to the mix. *It is particularly ironic that an increasing proportion of those lawsuits belong to people who are actually better off as more and more hazardous-but-beneficial products find their way into circulation.* Moreover, even those who recognized and disdained the economic wastefulness of suits in such a situation would be pushed in the direction of suing nevertheless, for the situation is pervaded with a prisoner's dilemma, in which the dominant rational strategy will always be to behave "unsocially" no matter what other people do, because one will pay indirectly for lawsuits brought by others (in the form of higher prices for hazardous-but-beneficial products), whether or not one initiates lawsuits of one's own.

B. N-Product Simulation

It is a straightforward exercise, if tedious, to extend the analysis to marketplaces with a larger number of beneficial-but-hazardous products. Suppose we consider four such products. When the calculations are made, the number of individuals injured falls from 1,700 to 1,269, but the number with a cause of action increases from 5,100 to 7,599, roughly six times as many as were actually injured. Those with a cause of action who also have an average or better realization also increases, from 1,400 to 3,498, close to three times as many as were injured; 534 of those individuals would actually have two separate causes of action.

Even with this simple model, then, two contrasting tendencies are foreshadowed. As the number of hazardous-but-beneficial products in the market increases, the number of individuals injured by the marketplace (as opposed to a particular product) falls, but the number with at least one cause of action increases. Although we relax an additional number of our presently strong assumptions below, those tendencies, it will be seen, will not disappear.

Consider a world with twelve beneficial-but-hazardous products, for instance. Less than 10 percent (9.32 percent) of the population will have suffered an aggregated injury. But virtually everyone (99.78 percent of the population) will have at least one cause of action. To get an idea of the scale of that effect some readers may find it useful to compare these numbers with the populations of real places. It would be as if, out of the entire United States, only those who live in New York state had been injured, but only those who live in El Paso lack even one cause of action. Everybody else in the whole country, it appears, would have benefited from participation in the marketplace for hazardous-but-beneficial stuff but, at the same time, would have at least one good cause of action against something in that marketplace, and many of those people would have multiple causes of action.

C. An Optimistic Two-Product Simulation

As noted, to this point the probability distribution of the simulations has been a pessimistic one wherein 30 percent of the population was injured by any individual endeavor. That, of course, leads to an unusually large number of individuals with a cause of action. But by the same token, it also leads to an unusually large number of unfortunate souls who have realized an aggregated loss from beneficial-but-hazardous activities. In other words, if the assumption is altered in a way that reduces the magnitude of resource dissipation through litigation, the same alteration simultaneously reduces the number of individuals who have any reason to claim they have been injured by the system.

The following alteration in the probability distribution can be used for a brief illustration.

Probability	Realization
.1	+28
.2	+20
.4	+12
.2	+4
.1	-4

The distribution is still an ersatz Normal distribution, and the variance has not been changed. But now only 10 percent of the population is injured by any one beneficial-but-hazardous endeavor.

Consider Table 6.2, which shows the outcome after the innovation of two beneficial-but-hazardous activities. Now only 19 percent have a cause of action, as opposed to 51 percent in Table 6.1 above. Hence, the proportion of the populace with a cause of action has been reduced by nearly two-thirds. But only 1 percent of the population has suffered an aggregated injury in Table 6.2, as contrasted with 17 percent in Table 6.1. The number of people with any *real reason* to complain, in other words, is plummeting toward zero, while the number of people with a *legal right* to complain is descending, but very slowly in comparison with the number who have been injured by the beneficial-but-hazardous marketplace. The pessimism of our original probability distribution may have ballooned the litigation problem, but that is the only way to prevent the fundamental problem of interest—true, as opposed to legal, injury—from becoming rapidly trivial.

TABLE 6.2

Aggregated Surpluses with More Optimistic Distribution

Product B

		+28 N=1,000	+20 N=2,000	+12 N=4,000	+4 N=2,000	-4 N=1,000
Product A	+28 N=1,000	28 + 28 = 56 N = 100	28 + 20 = 48 N = 200	28 + 12 = 40 N = 400	28 + 4 = 32 N = 200	28 – 4 = 24 N = 100
	+20 N=2,000	28 + 20 = 48 N = 200	20 + 20 = 40 N = 400	20 + 12 = 32 N = 800	20 + 4 = 24 N = 400	20 – 4 = 16 N = 200
	+12 N=4,000	28 + 12 = 40 N = 400	20 + 12 = 32 N = 800	12 + 12 = 24 N = 1,600	12 + 4 = 16 N = 800	12 – 4 = 8 N = 400
	+4 N=2,000	28 + 4 = 32 N = 200	20 + 4 = 24 N = 400	12 + 4 = 16 N = 800	4 + 4 = 8 N = 400	4 – 4 = 0 N = 200
	-4 N=1,000	28 – 4 = 24 N = 100	20 – 4 = 16 N = 200	12 – 4 = 8 N = 400	4 – 4 = 0 N = 200	-4 – 4 = -8 N = 100

D. Behavioral Responses to Changed Remedies

So far our discussion seems to point decisively away from strict liability because it endows most people, who are made better off by the world of beneficial-but-hazardous products, with a right to complain about and obtain redress from producers, which is to say ultimately from themselves. But the critique is incomplete; only straw men are thought to argue for completely getting rid of dangerous products. The serious problem that has been noticed is simply that without virtuous liability incentives, too much of a good-and-also-bad thing will be produced.

Jones, for example, argues that under a negligence standard, the costs of the injuries that occur in the normal course of business are not internalized, which causes the prices of hazardous products to be too low, which leads to the consumption of too many of them. We can include such a point in our model by altering some of its assumptions, and if we do, we shall find that overconsumption does not necessarily follow from the premise that injuries have not been internalized by producers. Two changes in the model, in particular, will make the point: first, like Jones, alter the model so that demand elasticities are negative; but second, unlike Jones (and unlike Polinsky and Shavell in their most recent guise), relax the assumption that the behavior of plaintiffs does not change to take account of a change in the expectation that they will be compensated for their injuries. If we do that we observe that a strict liability rule will motivate defendant producers to lower their output in reaction to a price-induced feedback from the customers. Consumer surplus will diminish, although whether injuries go up or go down will depend on the strength of plaintiffs' response to changes in liability rule relative to the strength of defendants' responses to such changes. If, in addition, we relax the assumption that litigation is costless, it becomes evident that surplus is reduced below whatever net it would otherwise have reached.

In order to implement these changed assumptions, suppose that for each product there are 4,000 individuals with demand $P=14-2q$ as shown by D_x on Figure 6.2; 2,000 with demand $P=14-(4/6q)$ as shown by D_y; and 1,000 individuals with demand $P=14-(4/10q)$ as shown by D_z. Those assumptions are depicted in Figure 6.2. The average cost of the product (neglecting the cost of injuries) in the hazardous-but-beneficial industries is assumed to be equal to marginal cost, which is set equal to 10. There are 3,000 individuals injured by the production of each product; 2,000 suffering an injury of -4; and 1,000 suffering an injury of -12. For computational simplicity, it is assumed that those injured consume none of the product that is responsible for their injuries. The industries are assumed to be perfectly competitive. (This set of assumptions leads, incidentally, to Table 6.1 above as a special case.) Assume that each of the individuals in the model expends one unit of surplus endeavoring to reduce the probability that he will be injured by one of the production processes. Table 6.3 shows the surplus that various consumers receive from a market that sells two beneficial-but-hazardous items under a system of negligence. As before, there are 2,000 people being injured to the extent of -4, by the existence of Product A; and another 1,000 being injured by -12.

FIGURE 6.2

Range of Demands for Hazardous but Beneficial Product Across Population

---

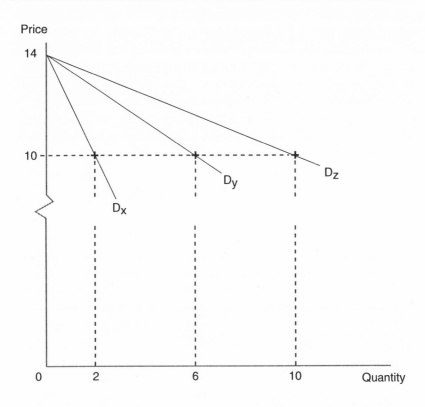

Identical statements are true of Product B. If a system of negligence is replaced by one of strict liability, then that cost will be added to the price (because by assumption the industry is perfectly competitive).

Assume that litigation costs equal one-third of the amount at issue, a low-ball and therefore conservative estimate so far as we can ascertain (Tullock 1985; Danzon and Lillard 1986). Also note that if fewer hazards are present in the environment, then potential plaintiffs will take fewer precautions against being injured by those hazards. Indeed, if they are to be fully compensated, literally, for all the injuries they suf-

TABLE 6.3

Aggregates Surpluses Net of Costs of Precaution Under Rule of Negligence

Product B

		+20	+12	+4	-4	-12
	+20	20 + 20 = 40 Precaution -1 Net 39 N = 100	20 + 12 = 32 Precaution -1 Net 31 N = 200	20 + 4 = 24 Precaution -1 Net 23 N = 400	20 – 4 = 16 Precaution -1 Net 15 N = 200	20 – 12 = 8 Precaution -1 Net 7 N = 100
	+12	20 + 12 = 32 Precaution -1 Net 31 N = 200	12 + 12 = 24 Precaution -1 Net 23 N = 400	12 + 4 = 16 Precaution -1 Net 15 N = 800	12 – 4 = 8 Precaution -1 Net 7 N = 400	12 – 12 = 0 Precaution -1 Net -1 N = 200
Product A	+4	20 + 4 = 24 Precaution -1 Net 23 N = 400	12 + 4 = 16 Precaution -1 Net 15 N = 800	4 + 4 = 8 Precaution -1 Net 7 N = 1,600	4 – 4 = 0 Precaution -1 Net -1 N = 800	4 – 12 = -8 Precaution -1 Net -9 N = 400
	-4	20 – 4 = 16 Precaution -1 Net 15 N = 200	12 – 4 = 8 Precaution -1 Net 7 N = 400	4 – 4 = 0 Precaution -1 Net -1 N = 800	-4 – 4 = -8 Precaution -1 Net -9 N = 400	-4 – 12 = -16 Precaution -1 Net -17 N = 200
	-12	20 – 12 = 8 Precaution -1 Net 7 N = 100	12 – 12 = 0 Precaution -1 Net -1 N = 200	4 – 12 = -8 Precaution -1 Net -9 N = 400	-4 – 12 = -16 Precaution -1 Net -17 N = 200	-12 – 12 = -24 Precaution -1 Net -25 N = 100

fer, then tautologically there is no reason for them to take any precautions at all. For the moment, make the relatively strong assumption that decreasing carefulness by potential plaintiffs exactly offsets the decreased hazard induced by falling consumption. That assumption means that the same amount of injury occurs, the only change is in the mechanism through which it is borne. Monotonous arithmetic computation leads one to conclude that putting the cost of injuries into the price of each product results in an increase in the price of each product of 2, from 10 to 12. Given the demand assumptions set forth in the previous

FIGURE 6.3

Changed Surplus when Liability Rule Change Induces Price Change

---

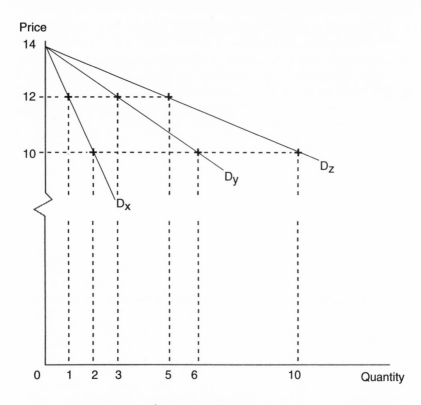

paragraph, the quantity demanded by every buyer is cut in half; which is shown in Figure 6.3 along the line indicating price of 12.

Table 6.4 shows Table 6.3 with the addition of the surpluses that are realized by various categories of individuals once strict liability is introduced. If one aggregates net surpluses under the two systems of liability, one discovers that total surplus has fallen from 70,000 to 30,000. Of course, the money that consumers no longer spend on hazardous-but-beneficial products will be spent on other things that are less hazardous. The increased surplus in those markets will offset the shrinkage occurring in the hazardous industries. Whether that offset is important or trivial depends on the demand and supply elasticities in the markets for less hazardous products. One cannot simply assume that it is important; it may or may not be. Jones' *prima facie* claim is therefore refuted: the world is not necessarily better off under strict liability rather than negligence. Of course that doesn't mean that the world is necessarily better off under negligence than strict liability—anyway, we do not claim to have proved such a thing. Our point is simply that the problem boils down to an empirical issue, which cannot be resolved by musing about it.

It is striking to observe how, as one moves from Table 6.3 to Table 6.4, variance diminishes in the surpluses of individuals; the winners don't gain as much as they did before, but the *comparative* losers are no longer *absolute* losers. Most of the variance reduction, however, comes by imposing large decrements on the big winners, while the big losers

TABLE 6.4

Aggregated Surpluses Inclusive of Awards Under Rule of Strict Liability

Product B

Product A		+5	+3	+1	-4	-12
	+5	Prior Net 39 5 + 5 = 10 N = 100	Prior Net 31 5 + 3 = 8 N = 200	Prior Net 23 5 + 1 = 6 N = 400	Prior Net 15 5 – 4 = 1 Award 4 Net 5 N = 200	Prior Net 7 5 – 12 = -7 Award 12 Net 5 N = 100
	+3	Prior Net 31 5 + 3 = 8 N = 200	Prior Net 23 3 + 3 =6 N = 400	Prior Net 15 3 + 1 = 4 N = 800	Prior Net 7 3 – 4 = -1 Award 4 Net 3 N = 400	Prior Net -1 3 – 12 = -9 Award 12 Net 3 N = 200
	+1	Prior Net 23 5 + 1 = 6 N = 400	Prior Net 15 3 + 1 = 4 N = 800	Prior Net 7 1 + 1 = 2 N = 1,600	Prior Net -1 1 – 4 = -3 Award 4 Net 1 N = 800	Prior Net -9 1 – 12 = -11 Award 12 Net 1 N = 400
	-4	Prior Net 15 5 – 4 = 1 Award 4 Net 5 N = 200	Prior Net 7 3 – 4 = -1 Award 4 Net 3 N = 400	Prior Net -1 1 – 4 = -3 Award 4 Net 1 N = 800	Prior Net -9 -4 – 4 = -8 Award 8 Net 0 N = 400	Prior Net -17 -4 – 12 = -16 Award 16 Net 0 N = 200
	-12	Prior Net 7 5 – 12 = -7 Award 12 Net 5 N = 100	Prior Net -1 3 – 12 = -9 Award 12 Net 3 N = 200	Prior Net -9 1 – 12 = -11 Award 12 Net 1 N = 400	Prior Net -17 -4 – 12 = -16 Award 16 Net 0 N = 200	Prior Net -25 -12–12 = -24 Award 24 Net 0 N = 100

gain only modestly. Indeed, more than a quarter of the people with causes of action under a system of strict liability have had their surpluses reduced by the switch away from a negligence rule. Look at the lower left-hand cell, for example: under strict liability those represented there are on balance worse off though each gains a cause of action against one product that would not have been available under a negligence system. That results because the individual is induced to diminish his consumption of the other product to half its prior level. Bear in mind that as the number of beneficial-but-hazardous products goes up, the number of individuals who are apt to be made better off by a switch from negligence to strict liability goes down, and the number of people who are apt to be made worse off goes up.

Our argument so far has assumed that the decrease in consumption of a hazardous product will lead to an *exactly offsetting increase* in the carelessness of those who are exposed to the hazard. It is hardly believable that such an assumption would hold for all products and in all circumstances, but it is not crucial to our argument that the offset be total. Any substantial offset, even if it is less than complete, will point in the same direction: the change in surplus will still be negative, but perhaps not so strongly negative as would be observed if the offset were total.

Admittedly if there is no offset at all, or if it is weak, surplus will increase by a switch to the rule of strict liability. But which of these assumptions comes closer to modeling any particular category of injury cannot be established through pure theory. Indeed, there is no theoretical reason why the offset in behavior could not actually be *more* than total. That would be the case, for example, if plaintiffs, being relieved of *all* the costs of accidents in which they are involved, were to undertake *no* precautions even though their precautions were crucial to the avoidance of injury.

3. Unobservable Negligence and the Optimal Placement of Liability

Legal scholars continue to be preoccupied with the economics of tort law thanks to John Prather Brown's application of the Coase Theorem more than two decades ago (Brown 1972). Crucial aspects of Brown's model can be illustrated as in Figure 6.4. The hypothesis of marginal analysis is that, beginning with any given level of a particular behavior, the crucial question is whether an increase in that behavior will bring more good than harm or vice versa. If the good exceeds the harm, the law would, ideally, seek to encourage that behavior. But if the harm exceeds the good, the law would seek to discourage it.

In the Brown model, that behavior seemed to be merely common "caretaking." If relining one's brakes for $100 would lead to a $110 reduction in expected injuries to others, tort law ideally would seek to encourage that investment. By the same token, an ideal tort law would not seek to inspire one to reline one's brakes at a cost of $100 if expected injuries to others were less than $100, even if those injuries were greater than zero.

FIGURE 6.4

Optimal Caretaking

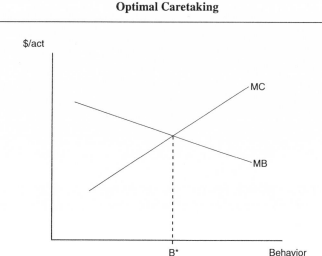

Shavell (1980; 1987) introduced a complicating refinement to the Brown model by extending the behavior at issue to something that he called an "activity." Suppose one's brakes have been so maintained that one can take a trip as carefully as the law expects, which, again, does not mean "with zero probability of causing injury." Suppose that for any ten-mile trip the *pro rata* expected injury to other parties is 90 cents. Ideally, the law would seek to encourage potential trip-takers to stay home unless taking the trip was worth at least 90 cents to the driver (net of all other costs).

The distinction between "caretaking" and "activity level," though it has become conventional in the law and economics literature, seems to us to be misplaced. The distinction that ought to be important is between *behavior that will be observable to a court and characterizable by it as "negligent" or "not negligent,"* and *behavior that will not be so observable and subject to characterization.* Undoubtedly some kinds of "caretaking" behavior are more easily perceived and characterized in relation to a standard of care than is true of some aspects of "activity levels." To recur to the previous example, it may be easier for a judge and jury to figure out how often one should reline brakes than to know or generalize about how often and how far "reasonably careful" people travel.

But one can readily observe and evaluate some sorts of items that look to be activity levels. For example, truckers in interstate commerce are required to keep logs that tell how many hours they have worked, which they must display upon a police officer's demand. And by the same token, some caretaking is difficult to observe after the fact. For example, in a serious accident much evidence is typically destroyed, see, *e.g.*, *Henningson* v. *Bloomfield Motors, Inc.*, 161 A.2d 69, 75 (N.J. 1960). Moreover, the observability of either caretaking or activity level changes over time with the advent of new methods of monitoring.

Accordingly, the crucial question is whether and how far a person's behavior can be observed, monitored, and characterized as consistent or inconsistent with some legal standard (such as due care). The law may well regard it as *attractive* to utilize self-executing rules (such as the rule that an actor must pay for the injuries done by his behavior), when attempting to regulate hard-to-observe behavior. But hardly ever can the law utilize self-executing rules to regulate the behavior of *all* the parties to a hazardous interaction. If the purpose of the law is to maximize the value of society's resources, it is a mistake to look only at the behavior of a potential defendant, taking no account of incentives that might lead potential plaintiffs to avoid or minimize injuries to themselves. In principle, plaintiffs and defendants are alike in their ability to vary both their caretaking and activity levels. In principle, coordination between potential defendants and plaintiffs has social value. In principle, it would be easier to deploy society's material resources at their greatest value by laying incentives on both actors rather than only on one of them.

The four-cell matrix illustrated in Figure 6.5 recapitulates the above analysis of tort liability conditions as first set forth by Brown. The model regards all aspects of the behavior of both parties as transparent to the law; there is no such thing as uncertainty whether a given behavior, by either plaintiff or defendant, was legally required; and no uncertainty whether such behavior as was required had or had not occurred. Scholars have found such a model to be a useful introduction to the analysis of tort law.

The vertical axis graphs the potential behavior of a defendant according to whether it was negligent or non-negligent. The horizontal axis does the same for a plaintiff. The northeast and southwest cells are the analytically easy boxes, for each of them covers a situation where one party was behaving negligently but the other party was not. As Brown showed, the negligent party in those cells will have the incentive to move its behavior to the standard of due care if it is required to pay for the losses that occur.

FIGURE 6.5

Outcome of Efficient Liability Rules when All Caretaking is Transparent

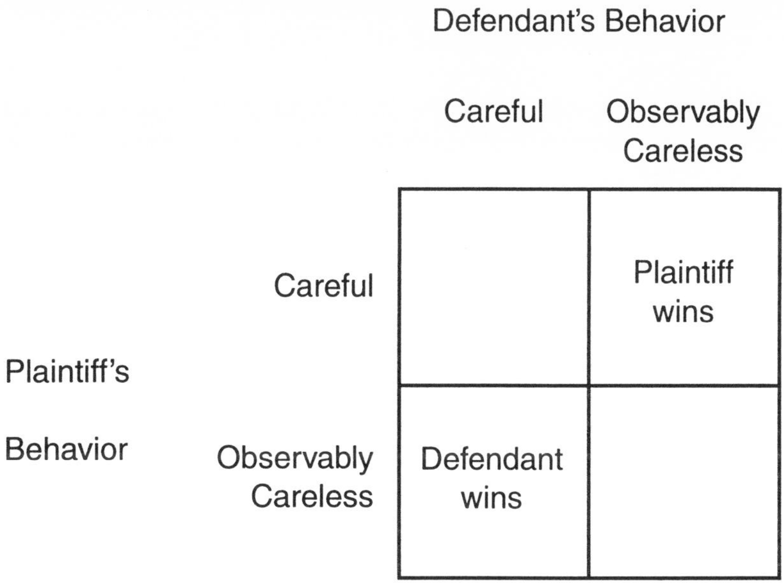

		Defendant's Behavior	
		Careful	Observably Careless
Plaintiff's	Careful		Plaintiff wins
Behavior	Observably Careless	Defendant wins	

The southeast and northwest corners are the hard cases in which either both parties have been negligent or both have been non-negligent. Brown's argument does not elucidate a principle for deciding whether negligence on the one hand, versus strict liability with appropriate defenses on the other, should govern such "tie" cases. Indeed, either rule seems to work in such circumstances.

Figure 6.6 is essentially a modification of Figure 6.5 to reflect the fact that some behavior that is negligent (in the sense that its value to the actor is less than the expected costs that it imposes on other parties) is not, for one reason or another, observable, at least not by a court. For example, it may in practice be difficult to ascertain the value to an actor of the last trip per week that he makes to the store, and thus to find out whether one trip fewer might, on the margin, represent less of a loss to him than a gain to other people who are at risk of injury every time such a trip is made. Other examples can be given in which a court may find it impractical to decide whether the value of a defendant's precautionary behavior was less than or greater than the expected costs his behavior would impose on others. Suppose a traffic accident occurs because a tire blew out, and a subsequent examination of the evidence leaves one uncertain whether or not the tire blew because it was overinflated. If it was overinflated, the driver was "negligent"; if it blew because of a structural weakness the driver generally would not be negligent, because people who are being "reasonably" careful (and hence not-negligent) will not ordinarily discover latent defects in tires.

In Figure 6.6, the four cells of Figure 6.5 retain their respective places in the northeast, northwest, southeast, and southwest corners of the diagram. Instances in which the defendant is observably negligent and the plaintiff not (northeast corner) remain easy cases; similarly easy are cases in which the plaintiff is observably negligent and the defendant not. Instances in which both are either non-negligent or both are observably negligent remain "tie" cases where it is not obvious that either rule systematically yields inappropriate results. But Figure 6.6 has five extra cases. Anywhere in the middle column the plaintiff is negligent but unobservably so; anywhere in the middle row the defendant is negligent but unobservably so. In the middle cell both are unobservably negligent.

In respect to these five added boxes, no one has yet proposed a liability rule that systematically dominates all competitors. That is because when one imposes a legal liability on one party, the other party,

FIGURE 6.6

Outcome of Efficient Liability Rules when Some Caretaking is Unmonitorable

---

Defendant's Behavior

		Careful	Unobservably Careless	Observably Careless
Plaintiff's	Careful		Plaintiff would win in Nirvana	Plaintiff wins
Behavior	Unobservably Careless	Defendant would win in Nirvana		
	Observably Careless	Defendant wins		

being off the hook, can be expected to relax his precautions or self-protective measures. The losses caused to a plaintiff by a defendant are always an externality to a defendant who is not required to pay for them; but by the same token, plaintiffs who can get their injuries paid for by other people have no incentive to take precautions, including cost-justified precautions, in their own defense.

Given that one of the parties will inevitably be off the hook depending upon the choice of liability rule, an ideal, or nirvana, solution to the problem is out of the question. Whoever is relieved of the residual liability—that which will be affixed when no observable negligence by either party can be found—will be less inclined to control his unobservable behavior in a way that equally weighs the benefits to him with the costs imposed on others. If a perfect rule is beyond reach, the task becomes selecting that one of the imperfect alternative rules that least deviates from the ideal. In other words, the question to be answered is whether the overall harm done by the unobservable negligence that has been induced by relieving a defendant of residual liability will be greater than or less than the overall harm done if it is the plaintiff that has been relieved of residual liability. If greater, then a rule of strict liability will avoid the greater harm; if less, then a rule of negligence is to be preferred. Meditation can take one no further.

At this point in the argument the researcher's task becomes wholly empirical. One must estimate what changes one may expect, and with what probability, in the behavior of defendants and plaintiffs as a result of changing their incentives. Upon inquiry, furthermore, one would be nonplussed to find one size fits all; instead one would expect to see different answers to the question depending on time, circumstance, and subject matter. Presuming these variables do matter, it is understandable that the law has always entrusted common law courts, with their specialty of particularistic empirical inquiry, with deciding which rule ought to govern which category of case.

4. The Impact of Liability Rules on the Rate of Innovation

It is incongruous to treat hazardous-but-beneficial activities as though they were "states of nature," that would occur at some given level regardless of the institutional mechanisms under which they are governed. If the innovation of the process and its regulation are interrelated, then one must take account of the interrelationship; the alternative is confusion. The way beneficial-but-hazardous endeavor A was regulated in 1980 would be expected to have had an impact on whether or not beneficial-but-hazardous endeavor B was undertaken in 1990. Courts, and whoever would advise courts concerning the selection of liability rules, must possess at least an implicit theory concerning the optimal rate of innovation in society.

Yet precisely here a serious problem appears. The optimal rate of innovation is uncertain (Dasgupta and Stiglitz 1982; Mortensen 1980). The difficulty may well go beyond simply that one lacks necessary data; the problem may be formally insoluble. Many people, however, do seem to have quite settled views on this question (*e.g.*, Buchanan and Faith 1981; Reich 1991). Those views should lead to predictable opinions on what liability rule the law should employ, whether injuries arise from toxic torts, or from some other endeavor accompanying innovation: If the observers believe that innovation is proceeding too rapidly, then all else being equal they ought to prefer more strict liability protection and less negligence protection; and conversely if they believe that the pace of innovation is too slow. Defendants under both rules have equal entitlement and ability to get compensation for the social *benefits* produced by their actions. But with respect to social *harms*, they are different. The negligence defendant must pay only for injuries inflicted *wrongfully*; the strict liability defendant for injuries he caused whether wrongfully or not. The negligence defendant is "subsidized" (so to speak) relative to the strict liability defendant whenever those two liability domains are distinct. But, because plaintiffs also make decisions that affect the likelihood of an injury, the alternative here is not *whether* to subsidize, but *whom*. The alternative to "subsidizing" defendants is "subsidizing" plaintiffs. The issue is, which subsidy is to be preferred. If "not enough innovation" is the problem, then all else being equal subsidizing defendants—that is, producers—is preferable.

5. Environmental Torts

A. *The Argument for Strict Liability as a Control on Unobservable Behavior*

Speaking specifically of the Exxon Valdez oil spill, Polinsky and Shavell (1992) have recently proposed that strict liability should govern in cases of serious environmental torts. Figure 6.7 illustrates the argument. MC_p shows producers' direct production costs (in other words, labor costs, material costs, administrative costs, and, particularly, the costs of taking "due care" to avoid injuries). In a competitive industry, producers will choose to produce Q units. MC_t shows MC_p with the addition of all costs of non-negligently inflicted injuries "caused" by the product. Polinsky and Shavell see Q* as the socially optimal amount of production for firms in this industry. Under negligence rules it will seem to the firms in the industry that producing units between Q* and Q is worthwhile, because market price P exceeds the marginal *private* costs of producing those units. Under strict liability rules they will see it differently, because they will be obliged to take into account the costs of the injuries associated with producing those additional units. Under negligence rules, therefore, a deadweight loss accrues, in an amount corresponding to the area of the cross-hatched triangle.

No mention is made of plaintiffs' conduct in Polinsky and Shavell's latest take on the problem, although in his 1980 piece and his 1987 book, plaintiffs' conduct played a major role in Shavell's thinking; indeed, in circumstances where plaintiffs could get out of the way of harm more cheaply than defendants could avoid creating the risk of harm, negligence would dominate strict liability.

FIGURE 6.7

Deadweight Loss if Product does not Internalize External Costs

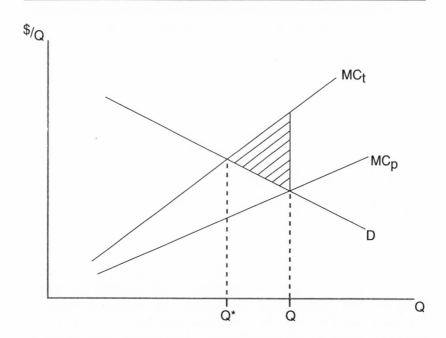

B. Refutation of the Proposition that Strict Liability is Always an Appropriate Control on Unobservable Behavior

Polinsky and Shavell are recycling a disguised replica of Pigou's (1932) argument from early in this century, Pigou in a poke, so to speak, in which a tax is to be levied on a socially costly externality in order to discourage its creation. Actually, because in Polinsky and Shavell's version the "tax collector" is a private plaintiff rather than a public revenue department, their solution yields a *worse* run of results than a Pigouvian tax would do rather than improving upon it. In instances where courts are unable to detect contributory negligence, plaintiffs are, under the Polinsky-Shavell rule, able to remain in the way of a continuing harm and yet recover, when having them get out of the way might be the least costly solution. A Pigouvian tax, at least, would not underwrite such a perverse result.

It was the defect of Pigouvian taxation that motivated Coase to criticize Pigou; Polinsky and Shavell seem not only to have missed the Pigou-specific point of *The Problem of Social Cost*, but the larger point as well, that the concept of "damages" does not have a life of its own. It takes two parties, at a minimum, for "damages" to exist. "Damages" arise because of mutually inconsistent, competing desires for the use of a given resource. If party A uses the resource, then party B cannot, and that is an injury to party B. On the other hand, if party B is privileged to forbid party A's use, then that is an injury to party A. Neither one of these injuries amounts to "damages," in a legal sense, until it has been decided who is entitled to the resource and on what terms. The Polinsky-Shavell theoretical assertions may well be correct on the empirical issue, that ultimately Exxon ought to be strictly liable for oil spill damages in Prince William Sound (that the non-Exxon "owners" of the resources damaged by the oil spill were "entitled" not to have their "property" spoiled by spilled oil).

That Exxon is the proper party to bear these costs cannot simply be assumed, however: in order to reach that result one would have to know (on the one hand) the costs of additional precautions by Exxon (or of doing with less of the resource), and (on the other hand) the *value* of the resources placed in jeopardy by the endeavor of shipping oil by sea, and the cost of additional precautions by the owners of *those* resources. To neither of these empirical questions do Polinsky and Shavell speak, nor is the correct answer obvious. For all one can tell, though, Polinsky and Shavell regard getting the right answer as unimportant. *Plus Pigouvian que Pigou*, their views seem antiquated in the Age of Coase.

Coase (1960) pointed out that in high transaction cost cases, of which coastal oil (or other obnoxious cargo) spill cases are a plausible example, entitlements were less likely to pass from current to new owners than in cases where the costs of transactions were lower. Hence, unlike low transactions costs cases (where only small harm is done by mis-placing initial entitlements, because claimants who value them more highly than their current owners can bid them away), in high transactions costs cases it is important for courts to lodge entitlements properly in the first place.

Elsewhere, Shavell (1980; 1987) and Polinsky (1980) each appears to recognize that the conduct of potential plaintiffs plays a potentially crucial role in accidents. Yet their Exxon discussion relies on an odd elasticity assumption, namely, that the liability rule will affect the number of potential defendants and how they behave, but not the number of potential plaintiffs and how *they* behave. One's intuition would seem to be otherwise, namely that (for example) the amount of capital invested in fishing Prince William Sound will be a function of whether the owners of fishing boats and equipment can recover against non-negligent polluters, or whether, rather, they would be allowed indemnity only if they could prove that their losses had been inflicted negligently. The liability rule ought to be expected to affect investments *both* in fishing gear *and* in tankers, not one *or* the other. A negligence rule *will* induce "too much" investment and activity by petroleum shippers in a nirvana sense, but strict liability will by the same token induce "too much" investment by fishermen. Though in each instance, each investment is "too much" by a nirvana standard, only one, the more expensive one, is "too much" in a Coasian sense. Clinging to a nirvana standard leads inevitably to hopelessness and frustration; a comparative institutions approach leads to empirics and hence to an answer.

6. Insurance Considerations

We have suggested that a court in search of a tort liability rule should not necessarily prefer strict liability and ignore its accompanying Pigouvian question about whether and how much the customers of a producer are affected when the producer has to pay for (and hence charge for) injuries attributed to the use of that product. Even if efficiency were the only object of a court's endeavor, there are other-than-Pigouvian features that also have to be taken into account. Perhaps the most prominent of those is what could be called the insurance problem.

One way to characterize the distinction between negligence and strict liability is that only with the latter is there embedded in the price of a product a "premium" for "insurance" against injuries "caused," but non-negligently, by the use of the product. In the event of such an injury, the producer will call upon the quasi-reserve thus created; each potential defendant will, in addition to its primary endeavor, operate as a mini-insurance company (re-insuring risks, of course, as other insurance companies do, where and as appropriate). Among the other iniquities discussed above, it is often inefficient to ask producers to function in this way. Several points should be made here.

1. There are, to begin with, economies of scope in the purchase of insurance. Consumers will generally find it cheaper to purchase one-stop first party insurance for all the risks of non-negligent harm that matter to them—Blue Cross or life insurance or disability insurance or fire insurance—rather than obtaining coverage in minuscule driblets from every widget manufacturer, seller, and shipper with whom they come into contact.

If an individual loses a foot, for example, he is unlikely to think it a serious loss if suffered while using a lawn mower but a trivial loss if suffered as a result of a fall in the woods. Thus, consumers will (in many cases) recognize a need to purchase first party insurance anyway, to cover risks from legally unattributable causes. Accordingly, there will be a certain amount of "double coverage" with the resulting potential for litigious hassling among insurance companies concerning which of them ought to be primarily liable.

And to the extent that consumers do not purchase double coverage, there is a further and related cost that ought to seem ironic to those who crave the insurance coverage aspects of the strict liability system. Because compulsory third party insurance in the price of widgets reduces the number of residual risks that are covered by first party insurance, third party insurance makes first party insurance generally less valuable to consumers, and therefore less likely to be bought, assuming that there is overhead in the provision of insurance. Although every consumer injured by a widget may be covered by the quasi-insurance in the price of the product, on some margin some other consumer will calculate that obtaining first party insurance is no longer worth while. What if that margin-dwelling person is then overtaken by some injury for which no defendant can be found to answer? To our knowledge, nobody has even tried to ascertain whether one actually gains more universal insurance coverage through strict liability than one loses. It seems unlikely, though, that the marginal effect of strict liability on a consumer's disposition to own first-party coverage is zero.

2. A market distortion results from the opposite of the foregoing

effect as well. Say the competitive price of a widget, by itself, is $10; of course in a strict liability world the widget manufacturer cannot offer widgets coupled with product-specific insurance at that price but must demand a higher price, say a premium of $2. The consumer, of course, is faced with one single number, $12, leading to a problem, from the buyer's point of view, of bounded rationality. Assume the consumer is rationally ignorant about the price of the insurance component, but believes that, whatever it is, it is "fair" (meaning the amount of the premiums will equal the amount of the benefits paid out). The consumer is knowledgeable about the value *to himself* of the widget component of the widget-insurance compound good—knows, let's say, that if the widget component is priced at $11 or less, the consumer should buy it (the consumer thinks it's worth that), but if more than that, the consumer should not. The (rationally ignorant) consumer is left in a quandary, because he cannot parse the $12 price into its insurance component and its widget component to ascertain whether the latter is greater or less than $11. Indeed, the consumer would have to become irrationally knowledgeable in order to be sure not to miss buying widgets that are worth more (to that consumer) than its (obscured) price, while purchasing gidgets that are worth *less* than they actually cost. In either case, some deviation from efficient market distribution will occur; the consumer will buy too many overpriced and too few underpriced products.

3. A system of strict liability assumes there will be, and encourages there to be, *more* insurance policies in force, than a negligence system. If there are scale economies in insurance, that leads to resource dissipation. For example, society may pay extra costs of administration. Here again, the magnitude of the cost has never been ascertained, but it certainly cannot be written off as zero.

4. If one ties insurance compulsorily to a product, one automatically treats every consumer identically with respect to the risks of product malfunction. But consumers are not fungible, and indeed may not even be very similar, on a number of dimensions that matter to the efficient management of risk. For example:

A. People differ in their tastes for risk. Although (most) all individuals are said to be "risk averse," they are not identically so, and indeed, it is well known that a person's background culture, financial position, and psychological makeup may significantly affect his or her taste for risk (and risk assumption).

B. Even if all people were identical in their fundamental taste for risk, they are differently situated in regard to what they have to worry about if a bad thing happens. A widower with five children would feel the need for a larger life insurance policy than a man with identical inborn tastes for risk but a young working wife, or no wife but grown-up children.

C. Even if A and B were assumed away, people are not identically situated in their capacities to adapt to risks, to learn the idiosyncrasies of widgets, and to avoid the dangers inherent in them. Some people are simply more competent with tools than others, and some people are better trained than others. Those who are more skillful should be encouraged to greater levels of care, but product-linked insurance is too blunt an instrument to accomplish this objective.

5. Finally, coupling a product with insurance against injuries that may occur during the course of the product's use is regressive (Priest 1981); and it seems everywhere to be understood that regressivity is either a bad thing or at least not a good thing. The injured parties are legally entitled to be compensated for the value of time lost as a result of the injury. All else being equal, people with higher valued time are wealthier than those with lower valued time. Thus, though injuries themselves may be physically identical, high-wage buyers are entitled to larger recoveries than low-wage buyers. But it is the buyers *as a group* who pay for those recoveries through the enhanced prices that must be paid for the product. People who are appreciably poorer or richer than the average member of the group cannot be assumed to have the same risk consumption characteristics as people who are near the average. If sellers are not able, as typically they will not be, to unbundle the product and insurance components and price them separately, then every buyer of a given product—wealthy, poor, or average—will pay the same insurance premium mark-up.

The result is perverse. Poor and wealthy buyers will both pay identical insurance premiums, but poorer consumers will receive a smaller benefit, and wealthier ones a larger, should the quasi-underwritten loss— a product-related injury—occur. If they ignore the other drawbacks of product-tied insurance that we noted above, wealthy people should prefer to have insurance coverage wrapped in the price of products they buy, since they could then pass on a part of the cost of their insurance coverage to others. But poor people should find it cheaper to buy their insurance in the form of a conventional insurance policy, where there can be a reasonably close fit between the premiums they pay and the benefits they can expect to receive.

Law students confronted with such an argument characteristically reply that poor people cannot afford to buy insurance as such, and therefore need to have insurance coverage wrapped in the price of the milk. This proposition will not do. If a poor person cannot afford to buy insurance, then he cannot afford to buy insurance. One hardly changes that circumstance by selling the insurance in one way rather than another, that is, tied to a product instead of freestanding. And of course, if the insurance a poor person cannot afford is made a mandatory part of the milk he used to be able to afford, then inability to afford insurance will spread out into an inability to afford milk-plus-insurance, and his children must then learn to drink the compassion of the misguided in lieu of milk. This seems a harsh benevolence to visit on relatively disadvantaged people.

7. Conclusion

The business of adjudication is exacting but hardly exact. Academic theorizing is capable of adding elegance and symmetry to the work of judges, but not, unfortunately, to solving the most difficult problems of judging. Theory shows that strict liability rules do save certain resources relative to negligence rules—sometimes. Yet a more complete theory, resting on nearly identical assumptions, makes clear that strict liability rules waste other resources relative to negligence rules at other times. What no theory tells or can tell is whether (or, more properly, in what settings) the savings exceed the waste and which rule, therefore, should be preferred. From time out of mind, common law courts have struggled to weigh the savings and waste against each other, deciding for strict liability in some instances to and negligence in others, and changing the rule sometimes as the technology of caretaking, and the forms in which consumer wealth are generated and held have evolved. Academics, it seems, are fated to spectate rather than officiate as the story unfolds.

References

Buchanan, James M., and Roger L. Faith. 1981. "Entrepreneurship and the Internalization of Externalities." *Journal of Law and Economics* 24 (April): 96-111.

Buchanan, James M., and William Craig Stubblebine. 1962. "Externality". *Economica* 29 (November): 371-84.

Coase, Ronald H.. 1960. "The Problem of Social Cost." *Journal of Law and Economics* 3 (October): 1-44.

Danzon, Patricia Munch, and Lee A. Lillard. 1983. "Settlement Out of Court: The Resolution of Medical Malpractice Claims." *Journal of Legal Studies* 12 (June): 345-77.

Dasgupta, Partha, and Joseph E. Stiglitz. 1980. "Uncertainty, Industrial Structure, and the Speed of R & D." *Bell Journal of Economics* 11 (Spring): 1-28.

Demsetz, Harold. 1969. "Information and Efficiency: Another Viewpoint." *Journal of Law and Economics* 12 (April): 1-22.

———. 1967. "Toward a Theory of Property Rights." *American Economic Review: Papers and Proceedings* 57 (May): 347-59.

Jones, William K. 1992. "Strict Liability For Hazardous Enterprise." *Columbia Law Review* 92: 1705-79.

Mortensen, Dale T. 1982. "Property Rights and Efficiency in Mating, Racing, and Related Games." *American Economic Review* 72 (December): 969-79.

Pigou, A. C. 1932. *The Economics of Welfare*. London: Macmillan.

Polinsky, A. Mitchell. 1980. "Strict Liability vs. Negligence in a Market Setting." *American Economic Review: Papers and Proceedings* 70 (May): 363-67.

Polinsky, A. Mitchell, and Steven Shavell. 1992. *Optimal Cleanup and Liability After Environmentally Harmful Discharges*. Stanford Law School, John M. Olin Working Paper No. 99 (September).

Priest, George L. 1981. "A Theory of the Consumer Product Warranty." *Yale Law Journal* 90 (May): 1297-1352.

Reich, Robert B. 1991. *The Work of Nations: Preparing Ourselves for 21st Century Capitalism*. New York: Simon & Schuster, Pounds.

Shavell, Steven. 1987. *Economic Analysis of Accident Law*. Harvard Univ. Press, Cambridge, Mass.

———. 1980. "Strict Liability versus Negligence." *Journal of Legal Studies* 9 (January): 1-25.

Tullock, Gordon. 1986. "Negotiated Settlement." In *Law and Economics and the Economics of Legal Regulation* , edited by J. Matthias Graf v. d. Schulenberg and Göran Skogh. Dordrecht: Kluwer Academic Publishers.

7

Insurance, Information, and Toxic Risk

Donald N. Dewees[1]

I. The Problem

Liability for the harm caused by toxic substances may arise in at least three situations. First, workers may be exposed to toxic substances in the workplace giving rise to workers' compensation claims and in some cases to product liability suits against the supplier of the toxic substance itself. Second, purchasers and users of consumer products may be exposed to toxic substances in those products giving rise to product liability suits against the manufacturer. In recent years, product liability suits by consumers have been outnumbered by product liability suits arising from workplace hazards; the asbestos personal injury litigation is predominantly workplace-related. Finally, the discharge of toxic substances into the environment may give rise to lawsuits by individuals who have suffered personal injuries arising from exposure to the substance, by individuals whose property or business has been injured by the substance, or by individuals or organizations claiming damages from injury to specific ecosystem or from more general injury to the environment. An important category of environmental litigation is law suits to recover the cost of decontaminating land that has been polluted by one or more toxic substances.

This paper examines the effectiveness of tort law in deterring behavior that gives rise to toxic injuries and in compensating victims of that behavior. It also explores the extent to which insurance assists or impedes the deterrence of harm caused by toxic substances. It then examines the influence of increasing tort liability on the operation of the market for insurance. The first examination is theoretical, reviewing the implications of economic theory for these questions. I then turn to the empirical evidence regarding these questions. The empirical analysis is performed separately for the three types of injury identified above.

Why should this analysis focus on toxic substances, when the questions raised here could be raised with respect to any number of causes of tort litigation? Three factors distinguish the problems caused by toxic substances from other causes of harm, at least to some degree. First, the risk caused by toxic substances is often not apparent to potential victims. Unlike the risk of injury in accidents such as motor vehicle accidents which are intuitively apparent to motorists and pedestrians alike, or the obvious risk of accidental injury in the mining and construction industries, the risk of injury from exposure to a toxic substance is often not apparent. Frequently, victims are not aware that they are exposed to the substance, much less that it may cause harm. This lack of information on the part of potential victims reduces the victim's inclination to take precautions or to demand that the owner or person in control of the substance take precautions. Sometimes the injurer is not aware that his behavior is dangerous and fails to take precautions out of ignorance.

Second, it is often difficult for a victim to determine or to prove to a court that his or her injury was caused by exposure to a toxic substance that was the responsibility of the defendant. Where the outcome of the exposure is an increase in the risk of contracting a common disease such as lung cancer, it is impossible to prove that a particular person's disease was caused by the defendant's toxic substance, or indeed by *any* toxic substance. It may be difficult or impossible to establish that the victim was exposed to the substance for which the defendant is responsible, much less to prove the extent of the exposure. Even in the infrequent cases in which the outcome of the exposure is unique to a particular toxic substance, the source of the substance may not be easy to establish.

Third, many toxic substances give rise to harm that manifests itself long after the exposure occurs. Some cancers have a latency period of ten, twenty, or thirty years, or more. Birth defects will not be manifest until the exposed person gives birth, and in some cases not until many years after the birth. When the harm arises long after the causative event, it is far more difficult to establish the facts that prove causation simply because of fading memories and lost or destroyed physical evidence. In addition, the discounting of future costs and risks causes individuals and firms to take less seriously the risk of injury and lawsuits that may occur decades in the future than the risk of immediate injuries that may give rise to a legal claim quite quickly. For all of these reasons the incentive to take care is attenuated.

Two major objectives may be attributed to the tort system: deterrence of harmful activity and compensation of victims. The evaluation performed in this paper focuses on the achievement of economically efficient or optimal deterrence, defined as precautions that minimize the sum of all costs including precautions taken by all parties, harm to the victim, and transactions costs (Shavell 1987; Polinsky 1989; Posner 1986). This definition of an efficient or optimal outcome is specifically an economic definition referring to cost minimization; it is quite different from a requirement of complete or total deterrence, since in many cases the cost of complete deterrence would be so high that it could not be justified in economic terms. In addition, some consideration is given to the compensation objective, viewed both as the achievement of corrective justice, in which the wrongdoer is required to restore the victim to his pre-injury status (Weinrib 1989), and as the achievement of distributive justice, in which victims generally are compensated for the losses that they suffer irrespective of the moral culpability of the defendant.

II. Theoretical Evaluation

A. The Tort System Without Insurance

Shavell (1987, chapter 2) distinguishes between unilateral and bilateral accidents, a distinction that is important as among similarly situated parties, in cases such as auto accidents. In the case of toxic substances, however, the victim is rarely in a position to exercise much care; the textbook examples of the laundry locating next to the smoky factory are not useful for evaluating toxic pollution of air, land, and water that create small risks for large numbers of persons or for the environment. Our primary interest is therefore in models that focus on unilateral care, with only a secondary interest in care taken by the victim.

There is also the question whether a liability regime influences activity levels to an economically efficient extent. In some cases, reducing toxic exposures may require substantial reductions in the activity giving rise to the exposure; the best way to reduce pesticide runoff is to reduce pesticide use. In other cases, the technology of pollution control may be sufficiently effective that high levels of control may be achieved with minimal changes in the underlying activity level. Chlor-alkali plants in Canada reduced their discharge of mercury by 99 percent between 1970 and 1973 while increasing production costs by only a few percent; the output reduction arising from this control is negligible (Dewees 1990). I believe that in the majority of cases reducing the activity level is a minor factor in controlling toxic pollution; so it is permissible to analyze policies with models that ignore the effect on activity levels.

The general results of theoretical analyses of some simple situations are easily summarized. Assume first that the liability regime affects the care taken by the victim and the injurer, and that it does not affect the activity levels of the parties. Then for unilateral accidents, in which only the injurer's care affects the outcome, strict liability *or* negligence achieves economically efficient deterrence, minimizing the sum of precautions, harm, and transactions costs. (Shavell 1987, 8). For bilateral accidents, in which the outcome depends on the care taken by each individual, strict liability with a defense of contributory negligence *or* any form of the negligence rule (negligence, negligence with contributory negligence, comparative negligence) achieves economically optimal precautions by both parties (Shavell 1987, 16).

In some cases, the economically optimal control of the problem requires that one or both parties reduce their activity level as well as taking precautions. For example, if the chemical industry discharges toxic wastes, the optimal solution may involve some reduction in chemical production as well as some control of the discharge of toxic wastes from chemical plants. If the optimal solution requires both parties to reduce their activity levels and to take precautions, then no single liability rule achieves optimal deterrence. Strict liability with a defense of contributory negligence creates the appropriate incentives for the "injurer," but it does not adequately deter activity by the victim because the victim does not bear the cost of accidents for which he or she is not contributorily negligent (Shavell 1987, 29). If the excessive activity of victims as a result of this defect is not large, then this rule is the best choice. If the activity of victims is more important than the activity of injurers in causing the harm, then one might prefer a negligence rule. Because I believe that in the case of harm caused by toxic substances the activity levels of victims are not often a major factor in optimal control of the problem, strict liability with a defense of contributory negligence is the preferred rule.

The results just presented are based on some strong assumptions. The parties must be perfectly informed about the risks arising from their activities and about the reduction in risk associated with all relevant precautions. This means that each must know what precautions the other party is taking. Furthermore, the courts must correctly apply rules of causation and due care. The parties are assumed to be risk-neutral. Of course, with risk-neutral parties, there is no incentive to purchase insurance.

Consider relaxing some of these assumptions. First, suppose that one of the parties is misinformed about risk levels. If the injurer or the victim underestimates (overestimates) the risks that he causes or faces, he will take insufficient (excessive) precautions and engage in excessive (insufficient) activity levels. If a party correctly understands current risk levels but underestimates (overestimates) the effect of his own precautions on risk reduction then he will undertake less (more) precautions than if he were perfectly informed.

Second, suppose that courts do not reach the correct result in every case. If court errors are random, this should not affect the behavior of risk-neutral parties. If courts generally find excessive (insufficient) liability, then injurers will be induced to take excessive (insufficient) precautions and engage in insufficient (excessive) activity, while victims will undertake insufficient (excessive) precautions and excessive (inadequate) activity.

Third, suppose that some injurers have limited assets so that they cannot pay large liability awards. To the extent that large awards are a part of the expected cost of a given level of activity and precaution, the injurer's incentive to reduce risks and activity will be attenuated because his expected *actual* liability is reduced.

With respect to risks that are well understood and observable such as an acute rare disease or an acute poisoning of fish or wildlife, *and* when the full value of the loss is recognized by the courts, then tort liability with no insurance should achieve economically optimal deterrence and compensation. But we should expect under deterrence in the common cases where the injury does not affect private property or a person, as with general ecosystem harm; where the discharge or the necessary causal linkage is not apparent; where the harm is in the form of a disease of long latency or another latent injury such as chemicals percolating unobserved through the soil; or where the defendant is judgment-proof. Conversely, over deterrence might be expected if courts tend to find causation erroneously, or if costly clean-up actions are mandated where they are not warranted.

B. The Tort System with Insurance

In contrast to the assumptions in the previous section, suppose that parties are not risk neutral. Instead, assume that potential victims are risk-averse so that they prefer to be insured against losses. In addition, while some injurers, such as a large corporation, may be risk-neutral, others, such as a small corporation or an individual, may be risk-averse. The risk-averse injurer may choose to purchase liability insurance to reduce its risk of financial loss.

Shavell (1987, chapter 9) outlines the basic implications of these assumptions for economically optimal behavior under various additional assumptions. First, if both injurers and victims are risk-averse but insurance is not available, the relative appeal of strict liability and of negligence rules depends on the relative risk-aversion of the parties. If victims are more risk-averse than injurers, then strict liability is the more attractive rule, since injurers are forced to bear most losses that are inflicted on victims, shifting losses to the less risk-averse party (Shavell 1987, 210). Deterrence is achieved to the extent that the injurer bears the costs of the injuries.

When liability insurance is available to injurers and accident insurance is available to victims, the economically optimal regime depends once again on information available to the insurer. If insurers can observe the care taken by injurers, then liability insurance is provided and optimal precautions are taken under either a strict liability or a negligence regime (Shavell 1987, 213). If insurers cannot observe the care taken by injurers, then insurance companies provide less than full coverage, to induce injurers to take some care. Care taken by injurers is suboptimal but better than if full coverage was provided, and welfare is higher than if there were no insurance.

When insurance is available to both insurers and victims, the choice of liability rule is less important than in the situation in which no insurance is available. If the insurers are able to observe care taken by the injurer and the injured, then economically optimal deterrence should be achieved through risk-rating of premiums. If insurers can observe care levels imperfectly, they tend to increase deductibles and policy limits in order to share risks with the insured, but still achieve suboptimal deterrence.

This discussion has assumed that litigation to enforce legal rights is costless, yet we know that litigation is very expensive. Some injuries are not worth litigating when litigation costs are high, in that the deterrence achieved by the plaintiff's winning the suit may have a value less than the administrative cost of the litigation itself. Shavell (1987, 162-70) shows that we cannot assume in general that litigation will be either excessive or insufficient when litigation is costly. Where injuries are small or the deterrent effect of litigation is likely to be small, it may be socially optimal to leave the losses on the victim(s) rather than to use costly litigation to shift them to injurers.

A few additional conclusions regarding the economics of insurance are relevant to the evaluation of toxic liability. First, under plausible assumptions, economically optimal insurance would cover pecuniary losses but not nonpecuniary losses, so if the goal of the liability system is to provide optimal insurance for uninsured victims, no awards should be made for pain and suffering (Rea 1982). On the other hand, pain and suffering represents a real social cost of an accident, so optimal deterrence may demand that the injurer be faced with this cost. Thus there is a tension between the optimal liability rule for compensation purposes and for deterrence purposes. Second, if losses are large, some injurers have insufficient assets to satisfy a court judgment. In such cases, the deterrent effect of the tort system in the absence of insurance falls short of optimal deterrence. Liability insurance provided by a fully-informed insurer could give rise to a risk-rating system that induced the injurer to take full precautions. Unfortunately, the injurer who has limited assets has a diminished incentive to purchase liability insurance and therefore a diminished incentive to take precautions (Shavell 1987, 240).

C. The Role of Information

The preceding section shows that information plays a crucial role in determining the behavior of the parties to toxic injuries and in determining the role played by insurance. Knowledge of the type and extent of harm caused by the injurer's action is an important determinant of the extent to which insurance is available and of the incentives for the injurer to take precautions. In the case of toxic substances, this requires knowledge of the harm caused by exposure to the substance, the extent of the victim's exposure to the substance and the extent to which the defendant is responsible for that exposure. The next section of this paper briefly reviews one of these information problems: what is the state of our scientific knowledge regarding the harm caused to humans, to other species, or to the environment generally by exposure to different quantities of the hundreds of toxic substances that are currently a matter of concern?

III. Scientific Knowledge Regarding Toxicity of Toxic Substances

A. General Issues

The lack of data on the toxicity of chemicals is striking. There are four methods of testing substances for toxicity: short-term molecular assays that test for mutagenicity, of interest because 90 percent of known carcinogens are also mutagens; animal bioassays that test the substance on animals in a laboratory setting; epidemiological studies that may generate statistical estimates of a dose-response function; and cluster analysis of diseases shared by members of a group exposed to a toxic substance (Brennan 1988, 502-9). Molecular assays are relatively inexpensive and reasonably good at identifying mutagens, but one cannot be certain that all mutagens are also human carcinogens. Nor can we be certain that a substance that causes cancer in laboratory animals when administered in high doses will cause cancer in humans at low doses, although this is often assumed. Therefore, positive results in the first two types of studies do not necessarily mean that a substance is harmful to humans in the low doses generally received through environmental exposure. Thus, while the first two types of studies are widely used as an input to occupational and environmental regulation, they do not provide a solid basis for quantitative risk assessment. Epidemiological studies can provide quantitative estimates of the relative risk for an exposed population, but this requires quantitative data on the intensity and duration of exposure to the substance and careful medical diagnosis of the condition of the patient, performed for large numbers of exposed individuals, often many years after the exposure occurred if there is a long latency period. High costs and the requirement of a large study population severely limit the number of reliable epidemiological studies.

Uncertainties in dose-response models complicate the assessment of harm, as different dose-response models can alter cancer predictions arising from low exposures by up to four orders of magnitude (Paustenbach 1989, 402). This issue is of vital importance, since epidemiological studies are often performed on workers with high exposures, but the environmental risks arise, if at all, from far lower exposures. A model is required to extrapolate risks in the low exposure region where there are no data. The U.S. Environmental Protection Agency (EPA) has attacked this problem by outlining a specific manner in which these models must be constructed. The EPA's model has been criticized as being too conservative, predicting health risks much higher than those that actually exist (Paustenbach 1989, 382).

As a result, accurate human dose-response functions currently exist for few pollutants. In a study of chemicals regulated by occupational health and safety standards, which also represent potential environmental hazards, only three out of the twelve substances studied (asbestos, benzene, and mercury) yielded sufficient evidence to support a reasonable estimate of the dose-response function (Dewees and Daniels 1988, 57; see also Mendeloff 1988, chapter 4). More recently, the EPA reported that only 10 percent of 2,800 known atmospheric pollutants have been tested for mutagenicity or carcinogenicity, of which ninety-seven have tested positive in whole animal bioassays (EPA 1990, 237). There is debate whether animal data provide a reliable basis for estimating human cancer risks,[2] yet human dose-response data reliable enough to support a reasonable estimate of the dose-response function currently exist for relatively few pollutants. The EPA found that of seventy-two toxic air pollutants, nine are proven carcinogens and fifty-one are probable carcinogens; but of the fifty-one, there is insufficient *human* evidence to estimate the potency for forty-seven. While the EPA provides "unit risk factors" which measure carcinogenic potency for many of these substances, the actual risk may be higher, lower, or even zero (EPA 1990, ES-4). With all these uncertainties, the EPA estimates that the air toxics that it has studied may cause less than one-half of one percent of all U.S. cancer deaths, not a significant portion of the cancer problem. Courts cannot determine with any confidence that there has been harm to a plaintiff with such facts.[3] Portney (1990a) estimates that control of air toxics in the U.S. will generate benefits worth between zero and $4 billion per year, the lower value representing the possibility that there may in fact be *no* adverse health effects arising from the low ambient exposure to these substances.

The limitations of existing data may be seen by briefly reviewing the state of the evidence regarding health and environmental effects of asbestos, a well-known and highly regulated substance, and by examining the general state of evidence regarding water-borne toxic substances.

B. Air Toxics: Asbestos

Medical evidence regarding disease risks should be better for asbestos than for almost any other substance, since millions of workers have been heavily exposed to the fibre since the 1930s and more than a dozen epidemiological studies have traced the effects of this exposure for thousands of these workers (ORCA 1984, chapter 4; Mossman et al., 1990). The principal causes of premature death among those who have worked with asbestos are asbestosis, a chronic restrictive lung disease caused only by exposure to asbestos fibers; lung cancer; and mesothelioma, a rare cancer of the surface lining cells of the lung and abdomen. There is evidence that asbestosis does not arise at low exposures, including any environmental exposure (ORCA 1984, 9). With respect to lung cancer and mesothelioma, the evidence on risks from low exposures is the subject of continuing debate; risks at low exposures have not been proven, yet the generally-accepted linear model would predict them. Most regulatory agencies have accepted that asbestos is a carcinogen at high exposures, and have assumed that the risk of contracting lung cancer and mesothelioma is proportional to the intensity and duration of the exposure to airborne asbestos fibers, recognizing that this may overstate the actual risk at low exposures (ORCA 1984, 283; OSHA 1986).

There is debate whether the three commercially important types of asbestos—chrysotile, amosite, and crocidolite—differ in their potency for causing cancer. The Ontario Royal Commission on Asbestos answered in the affirmative (ORCA 1984, 8), and Ontario regulations have been more strict for crocidolite than for chrysotile, while in the U.S. both the EPA and the Occupational Safety and Health Administration (OSHA) have not accepted that risk differences are sufficiently great to justify differential regulation (OSHA 1986, 22637; EPA 1989). Mossman et al. (1990) argue that chrysotile asbestos does not increase cancer risks at current *occupational* exposure levels, much less at environmental exposures. There is also debate whether different manufacturing processes lead to different risks, and whether risks depend on fibre dimension (ORCA 1984, 260-61).

Surprisingly, then, for this much-studied substance, debate continues on virtually all risk assessment issues including the fundamental question whether there is any risk at all arising from low level exposures, especially to chrysotile.[4] While regulatory actions often rely upon risk assessment data, these are subject to considerable uncertainty.

C. Water Toxics

Concerns regarding toxic substances in water include concerns about the heavy metals and the thousands of organic chemicals that are or may be toxic to humans. Colborn et al. (1990, 134) list some of the symptoms thought to arise from these metals and chemicals, including poor reproduction in bald eagles; developmental abnormalities, birth defects, and population declines in some species of fish-eating birds and animals; failure of some species of fish to reproduce *in situ,* and failure of other species to survive. But while these aggregate effects have been observed, it is often difficult to relate them to a specific pollutant, much less to determine the quantitative relationship between dose and response. After expressing grave concern about the dangers posed by toxic chemicals in the Great Lakes, and noting that laboratory experiments have demonstrated associations between some toxics and some health effects in animals, Colborn et al. (1990, 133) conclude:

> Despite the *association* between health effects and concentrations of contaminants, it is very difficult to prove that individual contaminants have *caused* specific effects in wildlife populations... [L]aboratory experiments have only limited application to the natural ecosystem... Despite unexpected and significant population declines among many Great Lakes species, the only specific chemicals to which a cause-and-effect relationship in wildlife has so far been assigned unequivocally are DDT and the product to which it decays, DDE; these have been shown to cause eggshell thinning (and therefore reproductive failure) in bird species.

The "Virtual Elimination Task Force" of the International Joint Commission (IJC) presents one table showing effects in eleven species of wildlife that are *associated* with exposure to a mixture of many contaminants in the Great Lakes (IJC 1991, 9), and a separate table showing established cause and effect relationships between eight substances and harms caused to eight species, but it is not clear that these relationships have been established at environmental concentrations of the substance (IJC 1991, 10).

Humans are warned to limit their consumption of some fish in some locations in the Great Lakes to avoid ingesting excessive amounts of these chemicals, yet while some of these chemicals have been shown to disrupt the normal functioning of cells in animals, "clinical effects of residues on human health are not yet understood" (Colborn et al. 1990, xxi). The primary pathway of exposure to these chemicals is through food, particularly fish and fowl at the top of the food chain. After expressing serious concern about the possible effects of toxics on human development, Colborn et al. (1990, 181) conclude:

> Although human tissue undoubtedly contains an array of toxic substances, little is known about the effects of such residues on human health. There may be no measurable health effect at all associated with toxic substances at the levels generally found in human tissues.... However, nine of the IJC's critical pollutants have been associated with adverse effects in the human nervous system. These chemicals appear to pose more of a risk as influences on human development than as carcinogens. But the effects of exposure during development are often subtle; they may not be readily evident at birth and in early childhood unless they are the focus of carefully designed studies.

The Agent Orange litigation, which concluded with a settlement of $180 million for the exposed veterans and a statement by the court that the plaintiffs had failed to present credible evidence (meaning epidemiological evidence) of a causal link between their exposure and their diseases, thoroughly reviewed the harm to human health arising from dioxin exposure (Schuck 1986; Brennan 1989, 49). Agent Orange was contaminated by dioxin, which animal studies have linked to a number of injuries including cancer, neurological damage, and hematological effects. The human evidence is far less clear. A major survey of this evidence concluded that the existing data neither proved nor disproved a link between dioxin exposure arising from Agent Orange and cancer in humans (Brennan 1989, 51).

D. Additional Issues

Most analyses of pollution control benefits focus on human health effects or on short-term environmental harm. Yet some of those most seriously concerned about environmental damage believe that ecosystems are not linear and stable, but non-linear and unstable, so that many types of pollution may cause harm that is irreversible, or may initiate a process of ecological change that would be catastrophic. Others believe simply that human activity is now so massive that the resulting environmental harm can be disastrous, whether or not the ecosystem is in principle stable.[5] Because reliable models to estimate these effects are not generally available, damage estimates usually ignore long-term and major ecological effects, although arguably these effects might be a major consideration in assessing the harm caused by the release of toxic substances.

E. Conclusions

We have reliable information about the human dose-response function for only a small number of substances out of the thousands that may cause significant harm. Estimated dose-response functions that rely upon debatable assumptions exist for another small number of substances. For the rest there may be evidence of toxicity arising from short-term bioassays or animal studies employing massive doses of the substance, but there is often no solid basis for estimating the quantitative human risks associated with low exposures to a given toxic substance. The state of evidence regarding effects on plants, animals, and fish is no better; while there are experimental data these results tend to be highly specific to a species and high dose, and observations in nature can rarely determine that specific substances caused a particular harm. Often we have some reason to worry that a substance may cause harm to species or to an ecosystem, but we have no basis for quantifying that harm or even determining whether in the long run it would be significant or insignificant.

The deficiencies in this factual underpinning create terrible problems for the use of the tort system to deter harmful behavior because assessments of the harm arising from a pollutant discharge are generally subject to enormous uncertainty and error. The possibility of tort liability may increase deterrence compared to a regime of no liability, but it cannot be argued that *in general* this represents a movement toward the optimum. The uncertainty of liability gives rise to a vigorous demand for liability insurance by potential polluters and to a reluctance on the part of insurers to supply it.

IV. The Occupational Health and Safety Experience

In general, no-fault workers compensation regimes have replaced tort actions in North America as a means of compensating victims of workplace accidents, while occupational health and safety regulation has largely replaced the deterrent function of tort. The empirical evidence on the efficacy of the tort system and of liability insurance, therefore, is to be found largely in: (1) studies of the functioning of the tort system before workers' compensation was introduced; (2) studies of certain specific regimes where tort survives, such as that for railway workers in the U.S.; and (3) studies that focus on industrial disease/ product liability claims which in the U.S. are not excluded by the workers' compensation law. Conclusions drawn from such evidence must be qualified to the extent that they may reflect; historical conditions no longer relevant the specific nature of the sector that has adopted the tort system rather than workers' compensation or the peculiar problems associated with industrial diseases.

A. Deterrence

When workers' compensation insurance was introduced early in the century, the tort law applicable to workplace accidents was in a process of rapid evolution (Friedman and Ladinsky 1967), and workers were winning an increasing number of cases (Schwartz 1981). However, causation was often difficult to prove, workers had some reservations about bringing a lawsuit or testifying in favor of a plaintiff for fear of impairing their relationship with the employer, and fatalities were undervalued by the standard tort award. Occupational diseases were largely ignored. Furthermore, low income workers would have had difficulty financing a lawsuit that was not reasonably likely to succeed unless contingent fees were available. These input considerations suggest that tort law could have caused substantial but far less than economically optimal deterrence. If tort were to replace workers' compensation today, the limitations regarding proof of causation and concern about the employment relationship would remain. In addition, tort awards for premature fatalities, based on foregone earnings, are a small fraction, perhaps one-tenth, of estimates of the "value of life" based on individual willingness to pay to avoid risks of fatality (Dewees 1986, 305-6). Indeed, one study of the choices available to an asbestos manufacturer whose workers suffered "an occupational health disaster" concluded that the expectation of full tort liability for all the disease that actually occurred would not have induced the manufacturer to reduce exposures below those that led to the disaster (Dewees 1986). A tort regime would therefore provide seriously suboptimal deterrence to employees exposed to toxic substances.

What deterrent effect might we expect from the workers' compensation system? Most employers pay premiums based on the average loss experience in an industry, where industries are rather narrowly defined (ALI 1991, 122-23). This by itself would provide a modest incentive to take precautions because only a fraction of the claims cost would be imposed on a specific employer. Some medium and large firms are risk-rated, with premiums depending on claims rates, although the risk-rating is only partial, since full risk-rating would no longer be insurance. Very large firms may be self-rated or may self-insure. For them, the full cost of all workers' compensation claims are paid by the firm itself. Some insurance carriers conduct loss-management programs for groups of firms with particular risk problems, but I have seen no data on the prevalence of such programs for occupational diseases.

There has been little empirical study of how care and activity levels have been affected by tort liability for occupational injury and still less dealing with occupational health. The evidence that does exist does not clearly establish whether tort liability or workers' compensation gave rise to greater deterrence of accident-causing activity (Chelius 1976; Fishback 1987; Graebner 1976). More relevant to the problem of toxic exposures is a comparison of the exposure of workers to asbestos since 1970 in the U.S., where workers exposed to asbestos have brought an avalanche of product liability lawsuits against asbestos producers and suppliers, with exposures of similar workers in Ontario where, because of differences in the workers' compensation legislation, there has been no such litigation.[6] There was no significant difference in supplier warnings about the hazards of the product, and no significant difference in worker exposure levels, but there was a greater reduction in asbestos use in the U.S. than in Canada (Dewees and Daniels 1988, 63-66). The reduction in asbestos use may be a consequence of the litigation, but it may not have been economically efficient, given the low worker exposures prevailing at the time. While this limited evidence is not conclusive, it suggests that even massive tort litigation directed at product suppliers who were insured may not achieve ends substantially different from those achieved through collective bargaining, the regulatory system, and the operation of the workers' compensation system.

The most recent study of the effect of workers' compensation on injury rates examined fatality rates arising from workplace accidents (Moore and Viscusi 1990). They found that increasing benefits causes a *decrease* in the injury rate such that worker fatalities would have been 20 to 27 percent higher in the U.S. in the absence of workers' compensation. This means that the effect of workers' compensation in protecting workers from fatalities arising from workplace accidents has been far greater than the effect of OSHA. Nevertheless, this finding applies only to fatal accidents, not to fatalities arising from exposure to toxic substances. In the case of toxic exposures that give rise to diseases with long latency periods the deterrent effect of workers' compensation liability will be greatly attenuated compared to the deterrent effect of liability for a similar risk of fatalities from accidents. Because workplace health risks are far less obvious to workers than accident risks, the effect of occupational health regulations is likely far greater than the effect of occupational safety regulations. Thus, I expect that workers' compensation systems deter worker exposures to toxic substances to some extent, but the effect is probably small compared to the deterrence of workplace accidents.

The deterrent effect of workers' compensation would be greater to the extent that carriers, whether private or public, operate risk management or risk rating programs with respect to the exposure of workers to toxic substances. I have not found data indicating whether such programs are more or less common than accident prevention programs among U.S. workers' compensation carriers.

I conclude that a tort regime alone would be a blunt and relatively ineffective instrument for controlling worker exposures to toxic substances. The primary barriers to reducing worker exposures are incomplete scientific information regarding the risks posed by toxic substances, incomplete information regarding actual worker exposures (both of which impede effective regulation as well), the long latency period of many occupational diseases, and the difficulty in determining whether particular health outcomes should be attributed to exposure to toxic substances in the workplace. I conclude from this analysis that reliance on workers' compensation insurance in the absence of tort, that is assuming that workers could not bring work-related product liability suits, would not increase worker exposure to toxic substances above the level which would occur under a tort regime alone.

B. The Effect of Liability on Insurance

Workers' compensation insurance is required by law in most states, although large firms may sometimes self-insure. In some states this insurance is provided either exclusively or alternatively by a state fund. In this situation there is little risk of insurers abandoning the field when claims costs increase; employers must pay any reasonable premium increases required to provide adequate coverage. The dramatic improvement in eligibility rules for occupational disease during the 1970s and the simultaneous increase in occupational disease claims has imposed strains on the workers' compensation system, but it has not caused the same sense of crisis that has been felt in the tort liability insurance regime. Yet during the same period, an explosion has occurred in product liability claims and payouts, resulting in an insurance crisis. The greater predictability of liability under workers' compensation has been described as an advantage that this no-fault compensation regime enjoys in comparison to the fault-based tort regime (ALI 1991, 107). In short, while workers' compensation premiums have risen more rapidly than inflation for several decades, this has not caused a crisis in the price and availability of this form of insurance.

C. The Effect of Insurance on Compensation

Before 1970, workers who suffered from occupational diseases were poorly compensated for their illness and disability (Barth and Hunt 1980, 2-6, 256). Frequently, limitation periods for filing claims would expire before the disease manifested itself, and in many cases it was difficult for the worker to establish that a disease arose out of a workplace exposure (Barth and Hunt 1980). While only 15 percent of occupational injury claims were contested by employers in the mid-1970's, almost 90 percent of occupational illness claims were contested (ALI 1991, 120). Benefits, generally set at about two-thirds of the pre-injury wage, were subject to limits on total or annual payout in some jurisdictions, disadvantaging workers with long-term disabilities such as asbestosis. Worse yet, pensions were usually based on the worker's age when last exposed, which could be one or more decades before the disease manifested itself, and were not indexed for inflation. A worker with chronic lung disease could receive a pension decades after his exposure that was a tiny fraction of current wages in his occupation (ALI 1991, 115).

All of these deficiencies in compensation contributed to the under deterrence achieved by the workers' compensation system at that time.

During the 1970s and 1980s, most workers' compensation systems were modified to deal more fairly with industrial disease, raising compensation levels closer to the theoretical two-thirds of lost wages (Larson 1988; Moore and Viscusi 1990, 5-6). Victims of occupational disease still receive compensation that is a smaller fraction of their pre-injury wage than do victims of occupational accidents, and while this may reflect the difficulty of proving that the disease arose from a workplace exposure, it does not prove that claimants are in fact under compensated for *workplace-related* illness. Permanent disability pensions are still not inflation-indexed in many cases, so under compensation of chronic disease claims continues. It is not possible to provide an accurate estimate of the adequacy of workers' compensation coverage of industrial diseases arising from exposures to toxic substances but most of the problems identified here would contribute to under compensation. The opposite result would occur only if courts or carriers generally paid benefits to workers whose disease did not arise out of a workplace exposure.

How does this compensation compare to that which would arise in the absence of workers' compensation insurance? Because this insurance is pervasive, it is difficult to predict what a tort-based system would be like. It is likely that in a tort-based system many employers would voluntarily purchase liability insurance to handle their workers' claims. If the tort system relied on negligence, then injured workers would not be compensated when there was no employer negligence, while if it relied on strict liability fewer workers would fail to recover. However, the difficulties involved in proving that the disease arose out of the workplace exposure, the much higher cost of litigation compared to handling insurance claims, and the limited assets of many employers who might not insure suggest that workers would be less fully compensated under a tort regime than they are under the current workers' compensation regime. That workers' compensation is superior to tort is confirmed by the fact that despite extensive criticism of the workers' compensation system, there has been no general demand to replace it with a pure tort regime.

A contrary argument could be made based on the experience with asbestos claims. Injured asbestos workers have had the opportunity to file workers' compensation claims or to bring product liability suits against the suppliers who sold asbestos to the employer. Up to 1980, only 29 percent of long-term asbestos workers whose illness had been caused by asbestos filed claims before death (Barth 1981). At that time a flood of personal injury claims had been filed against the manufacturers. It is thought that in general workers chose to pursue product liability claims rather than workers' compensation claims because the expected award was higher for the former.

This limited evidence does not establish conclusively whether a tort-based system or a no-fault workers' compensation system comes closer to providing full compensation to those whose disease actually results from workplace exposure, but it suggests that workers' compensation is superior.

V. Product Liability and Insurance

A. History

During the mid-1980s, most U.S. jurisdictions faced a crisis in the availability and affordability of liability insurance: premiums skyrocketed, sometimes by several hundred percent in one year; in some cases, coverage itself became unavailable (ALI 1991, Vol. 1, 55). During the decade from 1968 to 1978, premiums for general liability insurance, which includes product liability, increased almost five-fold, equivalent to 8.8 percent per year after adjusting for inflation, and over the ensuing decade they almost tripled, increasing over 4 percent per year after inflation (Viscusi 1991, 27-28). Between 1975 and 1989 the number of product liability claims filed annually in federal district courts increased from 2,393 to 13,408, an increase of 460 percent. Average product liability awards increased from $195,000 in 1971 to $1,536,000 in 1988, while median awards increased from $71,500 to $405,000 (Viscusi 1991, 96-97). Since the early 1970s, the applicable legal doctrine has been strict liability, and litigation has centered on whether a product was defective and whether the manufacturer adequately warned consumers of its risks. A study of accidental injuries found that in product-related workplace accidents, 24 percent of victims consider claiming but only four percent go so far as to hire a lawyer, while for non-occupational accidents, only eight percent consider taking action and only one percent hire a lawyer (Hensler et al. 1991). This difference is likely attributable to the greater ease of filing workers' compensation claims compared to launching a civil suit for non-workplace accidents. The U.S.

Consumer Product Safety Commission found that of consumers injured by products less than three percent filed claims (Fleming 1977), although the significance of this figure is unclear since there is no information regarding the number of these accidents in which product defects rather than consumer negligence was responsible for the injury.

B. Deterrence with Insurance

The deterrent effect of the tort system in the presence of liability insurance depends on the premiums being risk-rated or the insurer engaging in loss management programs. Evidence from the 1970s suggests that product liability insurance was rarely risk-rated, as most insurers ignored data on the number and size of product liability claims (Inter-Agency Task Force 1978, I-22). Indeed, insurers typically combined product liability coverage with other forms of coverage and did not even keep separate records of product liability claims experience. Yet 86 percent of all firms surveyed had some form of product liability insurance coverage (Inter-Agency Task Force 1978, III-9). This coverage declined rapidly during the ensuing decade, however (Priest 1989). While only one-quarter of all policies provided deductibles, which would afford some deterrent effect, the level of the deductibles increased fourfold between 1971 and 1976 (Inter-Agency Task Force 1978, III-12). In addition, 11 percent of firms reported exclusions of coverage on some products. Furthermore, the extent of coverage declined during the 1970's and 1980s, increasing the extent of self-insurance.

Turning to empirical evidence, we find some evidence that the emergence of strict product liability doctrine has reduced the level of socially beneficial innovations, reduced the availability of socially beneficial products and services, and caused consumer substitution toward more dangerous products. Huber (1988) provides anecdotal evidence of reduced research on contraceptives during the 1970s and the virtual collapse in the number of manufacturers of vaccine in the two decades to 1986. He also reports the removal from the market of beneficial products, such as swine flu vaccine, small aircraft production, and a drug used to control eye twitching (Botulinum), although it is not proven that these products in fact yield net social benefits.

Viscusi and Moore (1991) perform a statistical analysis linking product liability costs with various measures of innovation. They find that firms holding patents on products face product liability insurance premiums higher than firms that do not, while firms that introduce new patents also have a higher product liability burden. This demonstrates that product liability is a greater burden on product innovators.

A 1986 survey by the Conference Board found that product liability led to the discontinuation of product lines, decisions against introducing new products, and discontinuation of product research, although it also led to improved safety of certain products or product lines and improved safety warnings (McGuire 1988).

Ashford and Stone (1991) review liability and its effects on innovation and safety in the chemical industry. They find under deterrence in both chronic and acute hazards, since firms do not pay the full social cost of injuries caused by their negligence. They find that the firms pay only a small fraction of the social cost of injuries caused to consumers and innocent bystanders (Ashford and Stone 1991, 389-91). They conclude that new and safer products are developed when there is modest deterrence from liability, and suggest that if injury costs were more fully internalized by firms, new technologies would be stimulated (Ashford and Stone 1991, 417). Their study does not state clearly the role of insurance in this deterrence effect.

Johnson analyzes the impact of unwarranted product liability suits on innovation in the chemical industry by reviewing spending on R&D and by reviewing new patent trends. He finds that R&D spending has increased or remained constant in real dollars in all years between 1975 and 1989 (Johnson 1991, 432). In addition, patents in the chemical area continue to account for the majority of new U.S. patent registrations. While several companies were forced to close due to liability risks, he concludes that tort suits have not caused unmanageable problems and that product liability has not reduced its innovation rate. He is not clear on the extent to which the chemical industry is covered by product liability insurance.

One substance that has been the subject of major litigation is asbestos. The principal reductions in risk arose from banning the use of friable asbestos in building insulation in the early 1970s and reductions in the allowable workplace exposure to asbestos, also during the 1970s. Both of these improvements arose from government regulation, not from tort litigation, although it might be argued that the publicity generated by personal injury litigation, and the factual information that it revealed, might have accelerated somewhat the advent of regulations. In any event the prospect of litigation apparently did not deter exposures prior to the 1970s, and the fact of liability in the 1970s may have had little effect since regulations were adopted to control risks during this time. It seems unlikely that the existence of insurance prior to the 1970s reduced deterrence, which was minimal in any case. Perhaps the disappearance of insurance against asbestos risks during the 1970s increased the caution of manufacturers and users, but it seems likely that insurers would have put in place rigorous risk-management programs at that time, so the existence of insurance might have made little difference in behaviour.

I conclude that product liability, with insurance, has significantly affected manufacturers, reducing the flow of new products. It appears that insurance has not been carefully risk-rated, so the deterrent effect has been imprecise, causing beneficial products to be curtailed along with risky products. The decline in the availability of insurance has increased manufacturer caution. The evidence does not establish whether the benefits of this litigation exceed or fall short of the costs.

C. The Effect of Liability on Insurance

What was the effect of the rapid growth of product liability claims in the 1970s and 1980s on the insurance availability? Priest (1989, 9-10) found that one type of self-insurance, the use of captive insurance companies, increased from 200 in 1970 to 2,300 in 1986. In addition, total self-insurance costs increased from about 5 percent in 1970 to about 52 percent in 1979, and it is estimated that this growth continued during the 1980s. The level of deductibles increased four-fold between 1971 and 1976 (Inter-Agency Task Force 1978, III-12). This evidence suggests that the rapid increase in product liability during the 1970s greatly reduced the insurance coverage that was available to manufacturers for this form of loss.

D. Compensation with Insurance

Input analysis suggests that the tort system is unlikely to provide economically optimal compensation. Although a strict liability regime provides broader victim entitlements than a negligence regime, compensation remains seriously incomplete since, for a victim to receive any damage award, a product defect must still be proven. Claims initiation rates are vastly less than unity—probably less than 3 percent of product-related accidents. In addition, damage rules do not reflect optimal insurance: non-pecuniary damages are inconsistent with optimal insurance, although the recent relaxation of the collateral source rule in many jurisdictions can be seen as reflecting economically efficient insurance considerations.

Output analysis suggests that the tort system generates high transaction costs in that victims' net compensation is less than half of total tort expenditures. In addition, the compensation received by victims is far less than the economic losses they face as a result of an accident: the reimbursement rate is between 54.1 percent and 69.5 depending on whether or not the injury occurred at work, and the tort system is only responsible for 5.3 to 7.5 percent of this recovery (Hensler et al. 1991, 107). Furthermore, damages for pain and suffering account for between 30 and 57 percent of total damage awards for products liability claims in which bodily injury payments have been received. In addition, at least for asbestos-related claims, long delays occur before claims are closed, with the average time between filing a claim and its closure being over two and one-half years, and about 11 percent taking six years or more (Kakalik et al. 1984).

Viewed from a corrective rather than distributive justice perspective, the tort system does not perform substantially better, albeit often for different reasons. Strict liability enables victims to obtain compensation even when manufacturers have not acted negligently. When liability is found, damage rules often result in victims receiving inadequate compensation, especially where rules provide for ceilings on non-pecuniary losses, collateral offsets, and limits on awards for fatal injuries. High transaction costs and delays, even in successful claims, mean that very few meritorious victims are actually made whole by the tort system. Low claims initiation rates, even for meritorious claims, are a product of the costs and delays of litigation and result in non-compensation of many meritorious victims.

Thus from both distributive and corrective justice perspectives, the empirical evidence suggests that the tort system fails in appropriately compensating victims of product-related accidents.

VI. Environmental Harm

A. *Liability for Toxic Environmental Torts*

The "environmental" problems considered here include injuries to persons, property, or the environment caused by the discharge of air and water pollution and by the discharge and disposal of solid and liquid wastes. There are four classic environmental tort actions—private nuisance, trespass, public nuisance, and negligence—as well as civil actions based on liability created by statutes such as the Comprehensive Environmental Response Compensation and Liability Act, CERCLA.[7] Such actions represent a rapidly growing area of environmental civil litigation. Other papers have traced the history of litigation regarding environmental toxic torts, so that is not to be repeated here.

Tort doctrine excludes recovery in many environmental cases. Private nuisance requires that the plaintiffs have an interest in real property, and excludes recovery where the harm to any individual is not substantial, where the harm is not unreasonable (U.S.), where the pollution is consistent with the neighborhood, where the victim is unreasonably sensitive, and where the plaintiff has come to the nuisance (Keeton et al. 1984, at §§ 87, 88, 88A, 88D). The doctrine of riparian rights applies only to injury to riparian property, while the doctrine of prior appropriation applied in the western U.S. limits the rights of junior appropriators, excluding many harms from tort protection (Grad 1985, at §2.02). Where the harm is not associated with private property, individuals can rarely sue, and governments rarely do, so injury to public lands and general ecosystem damage are poorly protected, as are the interests of travelers and of visitors in public places. The public trust doctrine has expanded to protect environmental and recreational interests in navigable waters (Austin 1989). Damage rules exclude recovery of pure economic loss or for especially sensitive plaintiffs, and courts have been reluctant to award damages for aesthetic or recreational loss or for the risk of future disease (Keeton et al. 1984). In the limited set of cases where tort doctrine supports an action, proof of causation requires that the plaintiff establish four elements: that the plaintiff suffered actual damages; that the harm arose from a specific pollutant; that the pollutant is of a type discharged by the defendant; and that the defendant in fact discharged the pollutant that harmed the plaintiff (Development 1986, 1617-30). This chain of causation is often impossible to prove, especially with multiple polluters or with substances causing a risk of future disease (Brennan 1988). Finally, the cost of litigation precludes lawsuits for pollutants causing small losses per victim, even where the aggregate damage may be quite large.

These considerations suggest that tort will be useful primarily for local pollution problems involving a single polluter and very substantial damage,[8] and will be of little significance for pollutants dispersed over a large area, or discharged in a developed area with many other pollution sources, including most air and water pollution problems.

If we focus on environmental toxic injuries, tort law appears to be of limited value. In the early 1980s a study group mandated in CERCLA concluded that most plaintiffs face "substantial substantive and procedural barriers in a personal injury action for hazardous waste exposure, particularly where individual claims are relatively small" (Grad 1985, 718). The barriers include causation, joinder of parties plaintiff, apportionment of damages, and statutes of limitation. During the 1980s, litigation over damage from hazardous waste expanded, with much of the expansion related to property damage cases where it is relatively easy to prove that the presence of the toxic waste has reduced property values or required costly remedial measures. Many successful environmental toxic tort cases involve contaminated groundwater, which gives physical evidence of the exposure, and which causes clusters of cases that may be found statistically significant (Brennan 1989, 34). Still, with respect to toxics as with other environmental personal injuries, most pollution discharge does not attract litigation, most polluters pay no compensation, and deterrence will be minimal (Menell 1991).

Responding in part to the great difficulty that environmental plaintiffs face, the Congress, and many states, have passed laws giving explicit rights to compensation to some pollution victims. CERCLA or Superfund authorizes the EPA to arrange for the cleanup of designated hazardous waste sites and then to seek reimbursement from "responsible" parties: present or past owners or operators of the waste facility, generators of the waste, and any party transporting the wastes.[9] Superfund cleanup at a site may lead to a lawsuit to recover the costs from responsible parties who then sue each other to apportion liability. The facts revealed in the cleanup provide the basis for further lawsuits for property damage and personal injury by those owning property or living in the vicinity of the site. Early in the program, it was estimated that these transactions costs amount to between 24 percent and 44 percent of total clean-up expenditures (Butler 1985). More recently, Acton and Dixon (1992) conclude that administrative costs account for 88 percent of total insurance expenditures, while the GAO (1992, 7) reports that administrative costs account for almost 50 percent of federally funded costs.[10] It has been alleged that clean-up costs themselves have often been far greater than necessary to achieve low or negligible risks to human populations (Yandle 2000, ch. 1, this volume).

The courts have imposed: strict liability[11], retroactive liability[12] holding parties responsible for actions that preceded the 1980 enactment, and joint and several liability.[13] These rules diverge considerably from what is required for economically efficient deterrence because some parties may be held liable for far more than the full social cost of their activities, while others pay for little or none of the damage they have caused (Developments 1986, 1513). A number of states have enacted laws dealing with toxic substances incorporating provisions similar to those in CERCLA (Huber 1988, 140). These laws probably impose on those responsible parties with relatively substantial assets, such as large corporations or smaller corporations with extensive insurance coverage, costs considerably greater than the social cost of their activity, thus causing excessive deterrence activity.

Before CERCLA the problem of oil spills at sea gave rise to legislation because of the poor prospects for private common law suits to recover for damage. Congress provided in Section 311 of the 1977 Clean Water Act[14] that persons responsible for oil spills in the marine environment may be liable to the government for any natural resource damages that result, including the cost or expense of restoring or replacing the natural resources. More recently, the Oil Pollution Act of 1990[15] (OPA) regulates the transportation of oil to reduce the likelihood of an accident and the amount of oil spilled should an accident occur. Victims are compensated for natural resource damage, injury to property, lost revenues, profits, or subsistence use from natural resources, and the net cost of providing public services during cleanup. This compensation is financed by liability of responsible parties, up to specified limits, and by a five cent per barrel tax on oil. This is a form of compulsory liability insurance provided by the government.

B. Availability and Risk-Rating of Environmental Insurance

Comprehensive (or Commercial) General Liability (CGL) insurance coverage, purchased by many businesses, historically covered liability for damage arising from many causes including pollution. During the 1960s, rising liability for some types of pollution damage led some insurers to insert in their policies a pollution exclusion clause that excluded liability for pollution *unless* the damage is caused by a discharge that is "sudden and accidental." These policies were intended not to cover normal or intentional pollution discharges, nor did they cover discharges that were unlawful (Abraham 1988, 953). By 1973 the industry produced a standard pollution exclusion clause. Still, insurers were exposed to a long trail of occurrence-based policies written in earlier years without this exclusion (Foster 1990, 31).

In 1975, one insurer offered environmental liability insurance (ELI) specifically to cover gradual pollution discharge. The policy was limited to $4 million per occurrence and $8 million per year. In 1980 other firms began to offer this coverage, and by 1982 this was an established line of business (Foster 1990, 33-38). The high potential risks inherent in this insurance were controlled by limiting coverage to specific sites for any insured, covering only claims made during the insured year, and by requiring a site inspection and risk assessment prior to approving the policy. The risk assessment covered the hazard of the substance, site suitability, potentially exposed population, facility design, and management practices including training, state inspection practices, and past regulatory compliance (Foster 1990, 45). Coverage was limited by excluding certain facilities of the insured that were thought to pose high risks, excluding any accidents where the responsible party had knowingly ignored statutory law or governmental regulations, and by imposing large deductibles and low limits (Foster 1990, 85-87). Premiums for environmental liability insurance were never tailored precisely to the extent of the insured's risk because of the difficulty of estimating the magnitude of that risk when there was little historical loss data to provide the basis for estimation and because of the enormous variability of losses that may arise from toxic discharges.

In the 1980s two dramatic developments imposed massive environmental claims on the insurance industry. First, CERCLA's retroactive strict joint and several liability created unpredictable but potentially enormous liabilities where none had existed before. The unpredictability was heightened because strict liability imposes liability for risks that are unknown and perhaps unknowable when the acts occur, precluding any estimation of risk. Second, many courts interpreted the "not sudden and accidental" exclusion as merely a repetition of the "not expected or intentional" clause, thereby extending coverage to incidents that were in layman's terms not sudden and accidental. The courts have also included clean-up costs in the insurance coverage although the policies on their face apply only to liability to third parties, not to damage to the insured's property (Abraham 1988, 964). The industry responded by excluding all environmental liability during the 1980's in the United States. ELI coverage peaked in early 1984, and by 1985 it had become virtually unavailable (Foster 1990, 115).

Since Canada has no legislation comparable to CERCLA, and since Canadian courts have not yet altered the meaning of "sudden and accidental," the complete exclusion of pollution coverage did not begin until the mid-1980s, culminating in 1990 in part because of fear that the U.S. experience could be repeated in Canada. Since 1985 a very restricted rider for "limited pollution liability" has been available in Ontario (Poch 1989, 85).

C. The Deterrent Effect of Liability with Insurance

Did the existence of liability insurance dampen the deterrent effect of tort liability for environmental discharges? There are several reasons to believe that the dampening effect of insurance was modest. First, only a fraction of all releases of toxic substances during the 1970s and 1980s would have been covered by any form of insurance, given the pollution exclusion and the relative small market for ELI insurance. Second, ELI policies were restricted, covering limited facilities and limited events, with high deductibles and low limits. The exclusion of accidents involving violations of law or regulations created a powerful incentive to avoid violations thus reducing the risk of discharges. Perhaps most important, the risk assessments required for ELI coverage greatly improved the insured's understanding of the risks that he faced and the possibilities for reducing those risks. Indeed, one could argue that insureds would have been more careful when carrying ELI coverage than without it. Foster (1990, 113) concludes that while ELI insurance cannot substitute for regulation, it can supplement regulation by encouraging safe practices.

There is limited empirical evidence of the deterrent effect of tort liability during the 1980s when very limited insurance was available. The Rand Corporation studied corporate responses to, and the economic consequences of, expanded civil liability in all areas (Reuter 1988). The study is based on a series of interviews with senior corporate officials from the chemical industry, the pharmaceutical industry, the semiconductor industry, and small firms.[16] Of all the varieties of liability, the expansion of environmental liability, including that arising from CERCLA, is identified along with product and wrongful dismissal liability as having the most significant influence on corporate behaviour

(Reuter 1988, vi). All of the participants agreed that large potential environmental liability had made their firms more sensitive to environmental consequences (Reuter 1988, 19-21). Both the chemical and pharmaceutical groups indicated that a determination of potential environmental liabilities is critical to decisions regarding corporate acquisitions. Each participant had rejected a candidate acquisition on this basis. In the semi-conductor group environmental liability was a major concern and a source of delay in the site acquisition process. In the chemical industry liability was considered a second-order concern in site acquisitions ranking behind restrictive land-use permits. Waste disposal decisions have clearly taken on greater significance. Two firms reported centralizing waste disposal decisions previously left to individual plants, and some firms, fearing exposure to liability stemming from negligence on the part of other firms, have decided to avoid mixing their wastes with those of other firms. All of the small firms indicated that they had pulled up underground tanks, because of a mandate from their insurance company or the fear that land values would decline if there were contamination.

This analysis indicates that environmental liability has had its greatest effect with respect to land contamination by toxic wastes, where liability is governed by CERCLA, and do not indicate that there has been a significant effect on traditional air and water pollution discharge.

D. The Effect of Insurance on Compensation

An empirical analysis of the efficacy of the tort system in promoting the goal of corrective justice depends on the ability to compare the compensation actually paid to victims by parties at fault with a valuation of the total damage suffered because of their wrongful acts. Such an analysis is problematic for environmental injuries. Even in the study of individual occurrences it is often very difficult to identify all of those who may have been affected, to ascertain the full extent of the damage, and to value that damage. It is also often difficult to identify those who are at fault and to apportion responsibility accurately in cases involving multiple sources. Analysis of distributive justice is somewhat less difficult, in that it is not necessary to determine whether there was wrongful conduct, nor to apportion harm among multiple wrongdoers. Still, I have found no comprehensive, aggregate empirical analysis of the corrective or distributive justice performance of the tort system in the field of environmental harm.

The serious difficulties faced by plaintiffs attempting to pursue environmental litigation arising from problems in establishing causation are disabling with respect to achieving the compensation goals of corrective and distributive justice. Adequate compensatory performance for personal injuries is arguable only with respect to corrective justice, and then only in the small subset of cases in which there is highly localized pollution that significantly increases the incidence of a disease, so that the victims can establish causation from a known polluter. With respect to property damage, the almost unprincipled liability under CERCLA is inconsistent with both corrective and distributive justice.

Turning to output measures, while some victims do recover from polluters, their numbers are small in comparison to the number of people who may be affected by pollution, and their recoveries often fall short of full compensation (ELI 1980). Worse yet, the costs of environmental litigation are probably greater than for tort claims in general, further reducing plaintiffs' net recoveries. Even in the area of property damage by toxic pollutants, actual tort recoveries have covered only a small fraction of clean-up costs, reflecting in part the EPA's admirable reluctance to exercise fully its authority to pursue deep pockets.

VII. Conclusions

The theoretical analysis of this paper provides three principal conclusions regarding the role of insurance in toxic tort liability. First, if insurers are well informed about the risks posed by their insureds, and if they risk-rate their premiums accordingly or engage in optimal loss management programs, if all victims of toxic torts could establish the facts necessary to secure a recovery, and if litigation were costless, then the tort system with insurance would yield deterrence that approached the economic optimum, and would achieve corrective justice goals with respect to compensation. Second, if as seems likely only a small fraction of toxic tort victims sue, and if few insurance policies are rigorously risk-rated, then both deterrence and compensation will be poorly served. Third, when risks are uncertain, as in a period of rapid changes in tort doctrine, insurers will prefer not to issue new policies.

The literature on the actual experience with insurance and toxic torts tends to support these theoretical conclusions. First, as a general matter, tort liability appears to perform poorly with respect to the deterrence of harm arising from toxic substances whether in the workplace, in products, or in the environment, particularly when there is a long delay before the harmful effect is realized. Where the ultimate harm is a risk to human health or the environment, the deterrent effect of tort is particularly attenuated, in part because of very difficult problems of proof. This poor performance with respect to deterrence is matched by poor performance with respect to compensation; the failure to deter generally arises from a failure of the victim to secure full compensation. Where the harm arises from the contamination of land, tort liability is considerably more effective because liability may arise soon after the contamination is discovered, and the likely requirement to fund clean-up costs avoids many problems of proving that humans or the environment have suffered actual damage.

What has been the effect of insurance on the deterrent effect of tort litigation? As noted above, I do not believe that in the case of toxic torts there is much deterrence to be blunted, except in the case of the contamination of land. On the other hand, the insurance companies have imposed some significant loss management programs which require insureds to take precautions and to document these precautions. Thus the surveillance of the insurance companies over their insureds may give rise to greater deterrence than would have occurred in the absence of an insurance regime. This assumes that in the absence of insurance, the individual defendants would not have reacted in a similar way to the risks of litigation. This is plausible in a world of limited information, where the insurance company provides an information-gathering role; its broad exposure alerts it to the risk of litigation and it can provide information regarding effective loss-management strategies to a number of similar clients. The issue is: Who is better able to gather and utilize the relevant information—the insured or the insurer. The assertion that insurance has increased deterrence assumes that there are economies of scale in gathering this information.

On the other hand, insurance has aided in victim compensation. While the volume of toxic litigation has been relatively modest, many defendants did not have assets to cover the full value of claims against them, so there would have been less litigation if insurance coverage had not been present.

The increase in toxic tort liability during the 1980s appears to have had a significant effect on the provision of insurance coverage. Asbestos liability claims have had a devastating impact on those insurance companies whose clients have been the principal defendants. More generally, as toxic tort liability has arisen in a particular field, the insurers have tended to leave that field until the magnitude of the new liability could be estimated with some accuracy. Thus in the mid-1980s, it was very difficult to secure environmental liability coverage, and product liability coverage was much less widespread and subject to greater loss-sharing with the insured than previously. In the case of workers' compensation, the liability for workplace diseases arising from exposure to toxic substances has imposed high costs on some workers' compensation carriers, but has not led to reductions in coverage, in part because coverage is often mandated by state law.

Finally, what are the policy implications of these findings? For tort to cause economically efficient deterrence the risk of liability must be predicted. This is unlikely for many toxic substances in the near future, so the expansion of toxic tort litigation would appear not to serve either deterrence or compensation goals. Where there is liability, the ability of insurance companies to establish loss-management programs may be greater than that of smaller organizations, so insurance with risk-rating and loss management may help control environmental harm. But it is essential that courts create a climate of stability and predictability in interpreting insurance policies, or no insurance will be offered.

Notes

1. The research reported here grows out of a study for the American Law Institute conducted jointly with Michael Trebilcock of the Faculty of Law at the University of Toronto. I wish to thank the ALI and Michael for their contribution to this work. I am grateful for funding provided by the Social Sciences and Humanities Research Council of Canada and for research assistance by Ariana Birnbaum.
2. See, e.g., Ames (1987) arguing that only human epidemiology is useful for predicting human risks and Allen, Crump, and Shipp (1988), arguing that there is a close correlation between the potency estimates of animal studies and human epidemiological studies.
3. Regulatory agencies must confront the same evidentiary problems, which has caused slow progress in toxic regulation. For a discussion of the California program for regulating air toxics, which relies on risk assessment, see Ohshita and Seigneur (1993).
4. These issues were argued in the debate regarding EPA's rule banning most asbestos products and in the subsequent litigation of that rule. Asbestos; Manufacture, Importation, Processing and Distribution in Commerce Prohibitions, 54 *Fed. Reg.* 29,507 (July 12, 1989), 40 *C.F.R.* Part 763, Subpart I, §§ 763.160-179. On October 18, 1991, the 5th Circuit Court of

Appeals vacated this regulation. *Corrosion Proof Fittings v Environmental Protection Agency*, 947 F.2d 1201 (5th Cir. 1991).

5. For examples of these concerns see the Brundtland report (WCED 1987), or chapters in Mathews (1991).

6. The Ontario Workers' Compensation Act prohibits an employee from suing not just his own employer but *any* employer in Schedule 1, which includes most private companies in the Province, including Johns-Manville, the largest supplier of asbestos. R.S.O. 1980, c. 539, s. 8(9).

7. Pub.L.No.96-510, 94 Stat.2767, 42 U.S.C.§§ 9601-9675 (1982 & Supp IV. 1986).

8. Many reported cases are of this type. See, Grad (1985); Brennan (1989, 57) discussing the Woburn case; and Environmental Law Institute (1980).

9. *Id.* § 9607.

10. For a discussion of the complexity of litigation between insureds and insurers over coverage for CERCLA liability, see Abraham (1993).

11. *New York* v *Shore Reality Corp.*, 759 F.2nd 1032, 1042 (2d Cir. 1985).

12. *Jones* v *Inmont Corp.*, 584 F. Supp 1425 (S.D. Ohio 1984).

13. Pub.L. No. 99-499, 100 Stat. 1613, 42 U.S.C. § 9613(f).

14. Pub. L. 95-217, 91 Stat. 1566, 33 U.S.C. § 1321.

15. Pub. L. 101-380, U.S.C. § 2701-2761. See Dunford (1992) for a discussion.

16. The authors conclude that the study interviewed firms and executives who were particularly concerned about liability. For this reason it is said that the groups would be an inappropriate sample from which to draw representative conclusions. For the purpose of this study, which was to learn about a range of opinions and behavioral responses, biases against recent liability trends were not considered to be of great significance (Reuter 1988, 52).

References

Abraham, Kenneth S. 1988. "Environmental Liability and the Limits of Insurance." *Columbia Law Review* 88: 942.

Abraham, Kenneth S. 1993. "Cleaning Up the Environmental Liability Insurance Mess." *Valpariso University Law Review* 27: 601-36.

Acton, Jan Paul, and Lloyd Dixon. 1993. *Superfund and Transaction Costs: The Experience of Insurers and Very Large Industrial Firms*. Santa Monica, CA: RAND Corporation.

Allen, B., K. Crump, and A. Shipp. 1988. "Correlation Between Carcinogenic Potency of Chemicals in Animals and Humans." *Risk Analysis* 8: 531.

American Law Institute. 1991. *Enterprise Responsibility for Personal Injury: Reporters' Study*. Philadelphia, PA: American Law Institute.

Ames, Bruce N. 1987. "Six Common Errors Relating to Environmental Pollution." *Regulatory Toxicology & Pharmacology* 7: 379-83.

Ashford, Nicolas, and Robert Stone. 1991. "Liability Innovation, and Safety in the Chemical Industry." In *The Liability Maze: The Impact of Liability*

Law on Safety and Innovation, edited by Peter Huber and Robert Litan. Washington,DC: The Brookings Institution.

Austin, Brent R. 1989. "The Public Trust Misapplied: *Phillips Petroleum v. Mississippi* and the Need to Rethink an Ancient Doctrine." *Ecology Law Quarterly* 16: 967.

Barth, P. S. 1981. "Compensation for Asbestos-Associated Disease: A Survey of Asbestos Insulation Workers in the United States and Canada." In *Disability Compensation for Asbestos-Associated Disease in the U.S.* New York: Environmental Sciences Laboratory, Mt. Sinai School of Medicine.

Barth, P. S., and H. Hunt. 1980. *Workers' Compensation and Work Related Illnesses and Diseases*. Cambridge, MA: MIT Press.

Brennan, Troyen A. 1988. "Causal Chains and Statistical Links: The Role of Scientific Uncertainty in Hazardous-Substance Litigation." *Cornell Law Review* 73: 469.

——. 1989. "Helping the Courts with Toxic Torts." *University of Pittsburgh Law Review* 51: 1.

Butler, J. 1985. "Insurance Issues and Superfund", Hearing Before the Committee on Environment and Public Works, U.S. Senate, 99th Cong, 1st. Sess., 115-37.

Chelius, James R. 1976. "Liability for Industrial Accidents: A Comparison of Negligence and Strict Liability Systems." *J. Legal Stud.* 5: 293. Journal of Legal Studies.

Colborn, Theodora E., Alex Davidson, Sharon N. Green, R. A. Hodge, C. Ian Jackson, and Richard A. Liroff. 1990. *Great Lakes Great Legacy?* Baltimore, MD: Conservation Foundation.

"Developments in the Law: Toxic Waste Litigation." 1986. *Harvard Law Review* 99: 1458.

Dewees, Donald N. 1986. "Economic Incentives for Controlling Industrial Disease: The Asbestos Case." *J. L. Stud.* 15: 289. Journal of Legal Studies.

——. 1990. "The Effect of Environmental Regulation: Two Case Studies." In *Securing Compliance: Seven Case Studies*, edited by M. Friedland. Toronto: University of Toronto Press.

Dewees, Donald N., and R. Daniels. 1988. "Prevention and Compensation of Industrial Disease." *International Review of Law and Economics* 8: 51.

Environmental Law Institute. 1980. *Six Case Studies of Compensation for Toxic Substances Pollution: Alabama, California, Michigan, Missouri, New Jersey, and Texas*. Washington, DC: Congressional Research Service, Library of Congress.

Environmental Protection Agency. 1989. "Asbestos Ban and Phaseout Rule, 40 CRF 763.160-179." *Federal Register* (July 12).

——. 1990. *Cancer Risk from Outdoor Exposure to Air Toxics*. Vol. 1. Research Triangle Park: U.S. EPA, EPA Report 100 450/1-90-004a.

Fishback, Price V. 1987. "Liability Rules and Accident Prevention in the Workplace: Empirical Evidence from the Early Twentieth Century." *J. Legal Stud.* 15: 305. Journal of Legal Studies.

Fleming, John. 1977. *The Law of Torts*. 5th ed. Sydney: Law Book Co.

Foster, Michele A. 1990. Public Policy Aspects of Enviromental Liability Insurance. Ph.D. diss., University of California at Riverside, Department of Economics. Ann Arbor, MI: University Microfilms International.

Friedman, Lawrence M., and Jack Ladinsky. 1967. "Social Change and the Law of Industrial Accidents." *Columbia Law Review* 7: 50.

Grad, Frank P. 1985. *Environmental Law*. 3rd. ed. New York: Matthew Bender.

Graebner, William. 1976. *Coal-Mining Safety in the Progressive Period*. Lexington: University Press of Kentucky.

Hensler et al. 1991. *Compensation for Accidental Injuries in the United States*. Santa Monica, CA: Institute for Civil Justice.

Huber, Peter. 1988. "Environmental Hazards and Liability Law." In *Liability: Perspectives and Policy*, edited by Robert Litan and Clifford Winston. Washington, DC: Brookings Institution.

IJC (Virtual Elimination Task Force, International Joint Commission). 1991. "Persistent Toxic Substances: Virtually Eliminating Inputs to the Great Lakes." Canada: IJC (July).

Inter-Agency Task Force on Products Liability Contractors Study. 1978. *Final Report on the Insurance Industry*. Washington, DC: U.S. Department of Commerce (cited as Inter-Agency Task Force).

Johnson, Rollin. 1991. "The Impact of Liability and Safety in General Aviation." In *The Liability Maze: The Impact of Liability Law on Safety and Innovation*, edited by Peter Huber and Robert Litan. Washington, DC: Brookings Institution.

Kakalik, J., P. Ebener, W. Felstiner, and M. Shanley. 1984. *Variation in Litigation Compensation and Expenses*. Santa Monica, CA: Institute for Civil Justice.

Keeton, W. P., D. B. Dobbs, R. E. Keeton, and D. G. Owen. 1984. *Prosser and Keeton on Torts*. 5th ed. St.Paul, MN: West Publishing Co.

Larson, A. 1988. "Tensions for the Next Decade." In *New Perspectives in Workers' Compensation*, edited by J. R. Burton. Ithaca, NY: ILR Press.

Mathews, Jessica T. 1991. *Preserving the Global Environment*. New York: Norton.

McGuire, Patrick. 1988. *The Impact of Product Liability*. New York: The Conference Board.

Menell, Peter. 1991. "The Limitations of Legal Institutions for Addressing Environmental Risks." *Journal of Economic Perspectives* 5 (Summer): 93-113.

Moore, Michael J., and W. Kip Viscusi. 1990. *Compensation Mechanisms for Job Risks: Wages, Workers' Compensation and Product Liability*. Princeton, NJ: Princeton University Press.

Mossman, et al. 1990. "Asbestos: Scientific Developments and Implications for Public Policy." *Science* 247 (January 19): 294. Mossman, Brooke T.

Ohshita, Stephanie B., and Christian Seigneur. 1993. "Risk and Technology in Air Toxics Control: California and the Clean Air Act Amendments." *Air and Waste* 43 (May): 723-28.

ORCA (Ontario Royal Commission on Asbestos). 1984. *Report of the Royal Commission on Matters of Health and Safety Arising from the Use of Asbestos in Ontario*. Toronto: Queen's Printer.

OSHA (U.S. Occupational Safety and Health Administration). 1986. "Occupational Exposure to Asbestos, Tremolite, Anthophyllite, and Actionlite; Final Rules." *Federal Register* (June 20): 22612-790.

Paustenbach, D. J. 1989. "Health Assessments: Opportunities and Pitfalls." *Columbia Journal of Environmental Law* 14: 379.

Poch, Harry. 1989. *Corporate and Municipal Environmental Law*. Toronto, Carswell.

Polinsky, A. Mitchell. 1989. *An Introduction to Law and Economics*. 2nd ed. Boston: Little, Brown.

Portney, P. R. 1990. "Air Pollution Policy." In *Public Policies for Environmental Protection*, edited by P. R. Portney. Washington: Resources for the Future.

Posner, Richard A. 1986. *Economic Analysis of Law*. 3rd ed. Boston: Little Brown.

Priest, George. 1989. The Insurance Antitrust Suits and the Public Understandings of Insurance. New Haven, CT: Yale Law School, Program in Civil Liability, Centre for Studies in Law, Economics & Public Policy. Manuscript.

Rea, Samuel A. 1982. "Non-Pecuniary Loss and Breach of Contract." *J. Leg. Stud.* 11: 35. Journal of Legal Studies.

Reuter, Peter. 1988. "The Economic Consequences of Expanded Corporate Liability: An Exploratory Study." Santa Monica, CA: Rand Corporation, N-2807-ICJ.

Schuck, Peter. 1986. *Agent Orange on Trial: Mass Toxic Disasters in the Courts.* Cambridge, MA: Belknap Press.

Schwartz, G. 1981. "Tort Law and the Economy in Nineteenth-Century America: A Reinterpretation." *Yale Law Journal* 90: 1717.

Shavell, Steven. 1987. *Economic Analysis of Accident Law.* Cambridge, MA: Harvard University Press.

U.S. National Academy of Sciences. 1984. *Toxicity Testing: Strategies to Determine Needs and Priorities.* Washington, DC: National Academy of Sciences.

U.S. GAO. 1992. *Superfund: Current Progress and Issues Needing Further Attention.* Washington, DC: U.S. Government Accounting Office.

Viscusi, W. Kip. 1991. *Reforming Products Liability.* Cambridge, MA: Harvard University Press.

Viscusi, W. K., and M. J. Moore. 1991. "Rationalizing the Relationship Between Product Liability and Innovation." In *Tort Law and the Public Interest*, edited by P. Schuck. New York: W. W. Norton & Company.

WCED (World Commission on Environment and Development). 1987. *Our Common Future.* New York: Oxford.

Weinrib, Ernest. 1989. "Understanding Tort Law." *Valparaiso Law Review* 23: 485.

Yandle, Bruce. 1993. "Superfund and Risky Risk Reduction." Paper presented to the Independent Institute Conference, June. See chapter 1, this volume.

8

Contracting for Health and Safety: Risk Perception and Rational Choice

Daniel K. Benjamin[1]

> *In matters touching on health and safety, contract [is] as dead as a doornail.[2]*

Life remains nasty, brutish, and short, albeit less so than when Hobbes first advanced the notion. Within the brief span allotted us, we must assess the hazards we face, and choose among them—an inference that does not resolve how and by whom that process shall be guided. For most of recorded history, both assessment and choice chiefly have been the prerogative of the individual; society and its institutions—most notably the law—have exhibited great faith in the ability of individuals to regulate the risks of their personal environments through the appropriate choice of contractual relationships.[3]

Over the past thirty years, however, contract gradually has been replaced by decisions of the courts and government regulatory agencies.[4] While there are many reasons for this development, there is little doubt that one important factor is the emergent belief that individuals are unable to make accurate risk assessments on their own. Indeed, there exists a body of literature arguing that individuals are subject to systematic errors when assessing the hazards they confront.[5] Specifically, it seems, individuals regularly tend to overestimate the risks of relatively safe activities and underestimate the hazards of relatively dangerous activities. As a result, consumers will fail to demand, and producers to supply, the optimal amount of safety. The alleged manifestations of these failures are excess accident rates and adverse health effects. This makes it sensible—it would seem—to redirect the process of assessing and choosing among risks from individuals to the experts at government agencies and in the judicial system.[6]

Yet there also is evidence that the experts are prone to error—or at least prone to assess hazards in such a way that subsequent risk management decisions cannot help but be in error.[7] Within the regulatory agencies, the drive for prudence in risk assessment has evolved into a degree of conservatism that can only be viewed as extremism: At every point in the process, the most conservative (i.e., worst case) assumptions are made, so that at the end of the process, the compound effect may be to overstate the true risks by a factor of ten or even one hundred —without saying so. At the same time, the judicial system has increasingly dismissed the consumer's ability to assess risk and willingness to bear it, and simultaneously denigrated producers' willingness to sell whatever level of safety and health that may be demanded. The result is not merely inappropriate (and excessively expensive) judicial and regulatory attempts to reduce health and safety hazards that are far less dangerous than alleged; along the way, the relative hazards of different threats may be so severely distorted that we end up attacking the less serious risks, leaving the more deadly to wreak havoc.[8]

If it is true that perfect judgment is unlikely on the part either of the individuals exposed to risks, or of the courts and agencies that loom so large in regulating risks, it becomes prudent to ask the question: How serious are the errors that individuals make when they assess hazards? In search of an answer to this question, I address three issues. First, I reexamine the risk perception data that has been the springboard for the literature arguing that individuals are prone to serious biases in hazard assessment. I find that the actual biases in hazard assessment are far smaller than they appear to be; indeed, when I examine hazards most relevant to the individuals involved, their judgments are strikingly accurate. Second, I examine a growing body of literature that deals with the ability of individuals to learn about the risks present in their environment. I find that there is increasingly strong evidence that as the rewards to being unbiased rise, the extent of bias diminishes. In effect, accuracy (of which bias is a component) is a matter of choice for individuals; their refusal to choose perfection in risk assessment is thus little different from their refusal to constrain their jewelry portfolios to flawless blue diamonds—which is to say, worthy of lament only in a world without scarcity. Third, and only briefly, I touch upon the growing dissatisfaction with the science and politics of risk assessment as undertaken by the experts. My overall assessment from this exercise thus comprises three elements: (1) individuals are much better at assessing risks than might be expected; (2) when the rewards to being right are high, people become better at assessing risk; and (3) there is little reason to believe the experts can assess risk better than individuals acting alone or through markets.

It is possible—as suggested by the opening quote—that the issues I am addressing are moot, or at best of interest only to a pathologist. But if it is true that the demise of contract has been prompted in part by a conviction that the individual is wrong and the experts right, and if this set of beliefs has things exactly backwards, then surely there is hope that the process is reversible. Ultimately, the health and safety choices we make are no better than the risk assessment process that precedes those choices. If we err in the risk assessment process, then the chances increase that our choices will make us worse off rather than better off— and may ultimately lead to less safety and poorer health than would otherwise have been the case. My results suggest that a greater reliance upon individuals as the assessors and managers of the risks in their personal environments, and thus a greater reliance upon contracting for health and safety, will enhance the quality of the choices that are made and the level of health and safety that results.

Background

Over the past twenty-five years researchers in both economics and psychology have identified numerous seeming anomalies in the way individuals make assessments about uncertain outcomes.[9] There are many phenomena that are not yet well understood. For example, most individuals appear to be grossly overconfident in their ability to judge the likelihood of uncertain outcomes; people rely too heavily on small samples and on the "availability" of evidence (i.e., on their ability to readily recall events); and they tend to "anchor" their estimates, under-adjusting probabilities in response to new information. I do not attempt to resolve all of these behavior patterns. Instead I seek to understand a narrow but important type of apparent error on the part of individuals: their inclination to overestimate the probability of low-risk hazards and underestimate the probability of high-risk hazards.

FIGURE 8.1

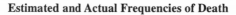

Estimated and Actual Frequencies of Death

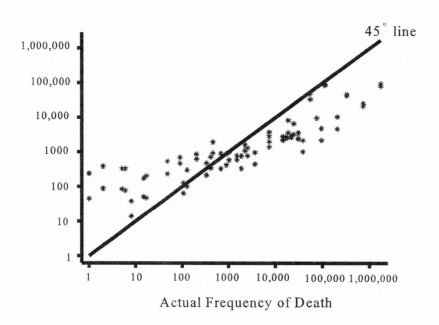

One example of the propensity for such biases is evident in Figure 8.1, prepared with data presented in Lichtenstein et al. (1978). Two separate groups of college students were told the annual death toll from one of two causes (motor vehicle accidents or electrocutions) in the United States, and then asked to estimate the frequency of forty other causes, ranging from smallpox to heart disease. The actual and estimated death rates are plotted on logarithmic scales in Figure 8.1, together with a 45-degree line, along which the data would fall if estimated and actual frequencies were equal. It is apparent from this Figure (and readily may be confirmed statistically) that the student responses do not match the actual frequencies. Indeed, the student responses are biased in a specific manner: They markedly overestimate the frequency of death for infrequent causes of death (such as botulism), and underestimate frequencies for common causes of death (such as heart disease). The original discoverers of this apparent bias offered no explanation for its existence, but were led to conclude that "People do not have accurate knowledge of the risks they face."[10] I begin by showing that despite its existence, this bias has little do with the risks people face, and—by inference—little to do with the decisions they make. I then inquire into the relevance of my findings for environmental policy.

Salience

The numbers behind Figure 8.1 were stimulated by a request that experimental subjects (college students) estimate "the frequency of deaths in the U.S. due to ... [these] lethal events...." It is apparent that the students were not very good at answering the question that was asked. Yet it is not obvious why they (or anyone else except a Public Health Service statistician) might have the slightest interest in knowing what the true numbers might be. This seems a peculiar contention, so I must be clear about what is meant.

Lichtenstein et al. interpret the data on frequencies of death by cause as being "the risks [people] face." In effect, these frequencies are viewed as being the outcome of multiplying the U.S. population by the true probabilities of dying from each of those causes. Thus, with a population of 205 million at the time of the study, and 1,025 deaths from electrocution, Lichtenstein et al. treat (1025/205,000,000) as being the probability that their respondents will die of electrocution. Thus, accidents involving motor vehicles and trains (which caused 1,517 deaths) are viewed as being approximately 50 percent more hazardous than electrocution for the survey respondents, while lightning strikes (which killed 107 people) are only about one-tenth as dangerous. Oddly, Lichtenstein et al. adopt this interpretation despite the fact that they did *not* ask any of the following questions:

1. What is the probability that a representative person will die during the next year due to each of these causes?
2. What is the probability that you (the respondent) will die next year due to each of these causes?
3. What is the probability that you or a representative person will die over the course of a lifetime due to each of these causes?

Just as strangely, the entire literature that has expanded upon Lichtenstein et al. has followed their original interpretation of the numbers, despite the fact that the data itself may have (almost) nothing to do with such judgments by individuals.[11]

To see the distinction, consider the following. Imagine asking a group of people how many people die as a result of (1) gunshot wounds and (2) lung cancer. Now imagine *interpreting* the replies as being responses to these questions: (1) What is the probability that you will die of a gunshot wound if you put a loaded handgun to your temple and pull the trigger? (2) What is the probability that you will die of lung cancer if you smoke a cigarette? For any remotely plausible set of answers to the questions actually asked, one would surely conclude—given the interpretations of the questions—that the respondents had no clear idea of the hazards of life that they faced. As absurd as this example might seem, I hope to show that it is nevertheless a useful approximation of the approach and subsequent interpretation of the work by Lichtenstein et al.

The causes of death examined by Lichtenstein et al., and the frequencies of death for each cause, are shown in Table 8.1. There are two attributes of these causes that are worthy of note. First, the latency of the causes—the delay between onset and death—varies enormously. For example, emphysema, syphilis, and lung cancer are all long latency causes, for the delay between the action that initiated the death and the death itself is typically measured in terms of decades. Conversely, measles, drowning, and falls are all short latency causes, for the delay between onset and death usually only ranges from minutes to a few days at most. The second significant attribute of these causes is that the age at which they kill varies enormously. The youngest killer is measles (at 8.7 years), while the oldest killer is strokes (at 78.4 years).[12]

The average death rates for these causes are the ex post outcomes of a variety of processes, some of which go back fifty years or more. The fatalities produced by long latency causes, such as cancer, stroke, and heart disease, are the result of past exposure rates, latency, lethality, and so on that are surely relevant for the parents and grandparents of the respondent students but may have little to do with the future likelihood of death for the students. Even if one were buying or selling life insurance, for example, the gross mortality statistics utilized by Lichtenstein et al. might do little more than establish the orders of magnitudes of the deaths due to various causes. Thus, between 1973 and

TABLE 8.1

Frequencies of Death, 1973

Cause	Deaths
Smallpox	0
Vitamin poisoning	1
Botulism	2
Measles	5
Fireworks	6
Smallpox vaccine	8
Whooping cough	15
Polio	17
Venomous bite	48
Tornado	90
Lightning	107
Nonvenomous bite	129
Flood	205
Hypothermia	334
Syphilis	410
Pregnancy	451
Hepatitis	677
Appendicitis	902
Electrocution	1025
Motor vehicle/train	1517
Asthma	1886
Firearms	2255
Poisoning (other)	2563
Tuberculosis	3690
Drowning	7380
Fire and flames	7380
Leukemia	14555
Falls	17425
Homicide	18860
Emphysema	21730
Suicide	24600
Breast cancer	31160
Diabetes	38950
Motor vehicle crash	55350
Lung cancer	75850
Stomach cancer	95120
All accidents	112750
Stroke	209100
All cancer	328000
Heart disease	738000
All disease	1740450

1988 the fatality rates for both emphysema and automobile accidents fell sharply (by 45 and 16 percent, respectively), while the rates for lung cancer and suicide both rose sharply (by 70 and 21 percent, respectively). Because an insurance contract covers *future* hazards, the fatality rates for 1988 would seem to be at least as relevant to students in 1973 as were 1973 fatality rates. So, I would argue, it is unsurprising that the students would know little about 1973 fatality frequencies for many of the causes in Table 8.1, simply because such numbers would be of little relevance to them.[13]

Is All Ignorance Bad?

Should we be concerned that the students (or any respondents) have no idea how many people died of a particular cause, such as lung cancer? For individuals to make correct decisions, they should know something about the hazard rates of the activities *they* will be undertaking. Consider bungee jumping and cigarette smoking. The adverse consequences of bungee jumping are immediate, and the observed frequency of jumping deaths will equal the probability of dying from it (*conditional* upon engaging in it) multiplied by the number of people who engage in the activity. We would expect bungee jumpers to be able to correctly assess the probability of dying or being injured while engaging in their sport. We might even question their rationality (or at least their intelligence) if they erred seriously in making this assessment. Yet, even if a bungee jumper has an unbiased estimate of the probability of dying from her activity, she may still not be very good at estimating the incidence of bungee deaths in the population as a whole, unless she also is good at estimating the population exposure rate for bungee jumping.

The contrast between individual hazard rates and population death frequencies is even more striking in the case of cigarette smoking. Most of the life-threatening adverse effects of smoking emerge only after about twenty pack-years of exposure.[14] So I ask, what is the hazard associated with smoking? "The" hazard depends on what is meant by "smoking." If we mean smoking one cigarette (the equivalent of one bungee jump?), then the answer is approximately zero. If we mean smoking a pack per day for a year then the risk remains about zero. And even if we mean the risk of smoking for twenty or thirty pack-years, the answer still may bear only tangentially on the observed death rate in the population as a whole—which depends on the actual exposure rate (how many people smoke and how long they smoke), the adverse effects of that amount of smoking, and the probability that something *else* will kill them before they die of smoking-related causes. In general, then, for long latency causes, I would expect an even lower ability of the individual to predict death rates in the population as a whole, and rightly so, because death rates for the population as a whole are even further removed from anything relevant to the individual.

The Relevance of Hazard Rates

For any potential cause of death, the actual probability of death (what I call the *hazard rate*) is equal to the probability of death conditional on exposure (or *lethality*), multiplied by the probability of exposure: $p = p_d\, p_e$. Thus, causes of death may differ from one another both in terms of their exposure rates (p_e) and their lethality (p_d). To take but one simple example: People of all ages fall down, yet falling is rarely lethal, *except* for the elderly. As people age, protective reflexes slow, bones become more brittle, and so on, so that the probability of dying from a fall begins to rise sharply at around age 65, because it is at that point that p_d begins to rise. As a result, the average age of death conditional upon dying from a fall is approximately 73 years.[15] Toward the other end of the spectrum, driving an automobile into a concrete bridge abutment at 85 miles an hour is almost always fatal (pd$\approx$ 1) to the occupants of the vehicle. Yet deaths by this cause are relatively rare, except among the young, who are prone to drive fast and to drive under the influence of alcohol and drugs. Because the probability of an 85 mph exposure to a bridge abutment (p_e) is relatively high for young persons, the average age of death due to automobile accidents is but 38.

Thus, due to variations in both p_d and p_e, the age profiles of the hazard rate can differ dramatically from one cause of death to another. Neither Lichtenstein et al. nor any of the subsequent work on risk assessment has accounted for this possibility. Instead, the standard method has been to treat the population death rate as the true hazard rate relevant for all individuals. In general, however, the hazard rate for any given segment of the population will not equal the population death rate, a fact that readily may be illustrated by simply examining the data.

In Figure 8.2, I show the distribution of deaths by age of death for two specific causes of death: stroke and homicide. The pattern for strokes

FIGURE 8.2

Incidence of Death by Age of Death

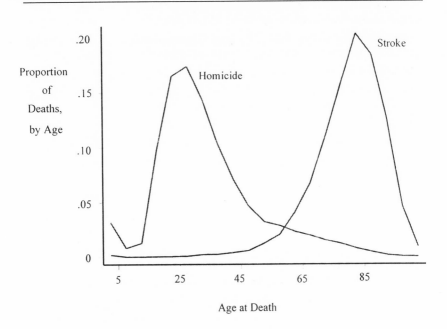

Age at Death

is very much like that observed for other long latency killers such as heart disease, most cancers, and so forth. The pattern for homicides is also representative of many other sudden killers, including motor vehicle accidents, electrocution, firearm accidents, and the like. By definition, $p_d = 1$ for homicides, so that the time pattern of deaths for this cause also illustrates the fact that, in general, rates of exposure to those activities that cause sudden death are not fully independent of age. In fact, for such causes of death, the hazard rate for young people is typically substantially greater than the population death rate, while for older people, the hazard rate is well below the population death rate.[16] Clearly, the deviation of a person's hazard rate from the population death rate for a given cause of death will depend on the age profile of the hazard rate and the age of the person in question. In the case of college students (the source of the data used in Figure 8.1), the hazard rate of things that tend to kill the young (such as motor vehicle accidents) will be greater than the population death rate for that cause, while the hazard rate for things that kill the elderly will be substantially lower than the population death rate.

Thus, if respondents base their estimates of *population death frequencies* on their own *hazard rates* (data with which they have some reason to be familiar), there will be a systematic effect of the age profile of a given cause of death on the deviation of the respondents' estimates from the actual population death rate: Respondents will offer up higher estimates (relative to population frequencies) for those causes that tend to kill earlier in life, and lower estimates for those causes that tend to kill later in life. If the age at which death occurs is not taken into account, the result will be exactly the sort of bias reported by Lichtenstein et al.: Low frequency events will be overestimated, while high frequency events will be underestimated.

This inference can be tested by explicitly allowing the effect of the population death rate to depend on the age of death. Doing so yields some striking conclusions.[17] First, as predicted by the preceding discussion, the age at which a cause kills has a strong negative effect on students' estimates of the hazard rate; for any given population death rate, causes that kill earlier are viewed as more hazardous than causes that kill late in life. In effect, the student-respondents are (quite sensibly) discounting the hazards associated with causes of death that kill chiefly the elderly.

Second, accounting for age of death dramatically increases predictability of the students' responses. Following the Lichtenstein et al. method, which considers only the population death rate, only about 20 percent of the students' responses can be explained statistically. Once the age of death is accounted for, nearly 80 percent of the student responses can be explained. Third, even for causes of death that are the least frequent, and thus most overestimated according to Lichtenstein et al., accounting for age of death brings the students' responses into close conformity with the facts. For example, there were no fatalities from smallpox in the Lichtenstein et al. sample; the student estimates that take age into account imply a fatality rate of 0.82 deaths per million. That is, when the true probability is zero, the students respond "the probability is one in a million."

Finally, the estimates I have developed enable me to infer the accuracy of the students in estimating the hazard rates most relevant to them—those that apply to causes that kill people their age. Thus, consider a hypothetical cause with an average age of death equal to the age of the respondent students. One might ask, what is the implied difference between the estimated death rate and the true death rate? If I assume the average age of the students to be 20, this difference is about 4 percent, a number that is small arithmetically, and is also not statistically different from zero. Of course the assumption that the students are 20 years old is arbitrary. One might alternatively ask, what age would the students have to be for their estimated death rate to be *exactly* equal to the true death rate? The answer is 17.5—which is not far from the average age of incoming college freshman.

Taken together, these findings suggest that for causes with an average age of death equal to the age of the respondents there is no tendency for individuals to overestimate low-probability events or underestimate high-probability events. Apparent biases such as those reported by Lichtenstein et al. are due to the fact that in the United States, causes that kill large numbers of people also tend to kill older people disproportionately.

My interpretation of these results may be summarized quite simply: Lichtenstein et al. asked the students an irrelevant question. For most causes of death, the observed population death frequency is at most only slightly related to the hazard rate facing the respondents to the question. My findings strongly suggest that the students responded to this question with answers that look remarkably like the hazard rates relevant for individuals in their age bracket. As it turns out, there is additional evidence that enables me to subject this inference to even more demanding tests.

More on Hazard Rates

The same study that produced the numbers I have been using thus far also produced another set of numbers that confirm the basic hypotheses. Lichtenstein et al. (1978) surveyed two additional groups of subjects—one again comprising students, the other consisting of members of the League of Women Voters. The latter group was chosen as being "representative of the best-informed citizens in the community." The researchers utilized the same forty-one causes of death discussed above, but placed them together in 106 pairs of causes. Subjects were asked, regarding each pair:

> Which cause of death is more likely? We do not mean more likely *for you*, we mean more likely *in general*, in the United States.

Consider all the people living in the United States—children, adults, *every-one*. Now supposing we randomly picked just one of those people. Will the person more likely die next year from cause A or cause B? (Lichtenstein et al. [1978, 554], emphasis in the original)

There are several key aspects of this experiment. First, the subjects were asked explicitly to estimate ratios of *hazard rates* (not population frequencies); this eliminates any possible ambiguities in interpreting the subjects' responses. Second, two different subject groups were asked to evaluate hazards for the same *representative* individual; if both groups independently make the same assessment, this would add to our confidence that the nature of their responses is not merely the result of chance. Finally, the fact that the respondents were asked to make pairwise *comparisons* of hazard rates across causes means that this experiment provides a slightly different specification against which to test hypotheses.

Despite having explicitly asked about hazard rates, Lichtenstein et al. once again used population frequencies of death when computing "true" ratios against which the subjects' responses were compared. Once again, the researchers concluded that the respondents were biased in their hazard assessments, particularly in underestimating the hazards associated with the most common causes of death. Yet the age at which a cause kills is crucial in assessing how hazardous that cause is to the *representative* individual. Consider falls and homicides. Both killed about the same number of people in the sample period, yet the people who died from homicides were thirty-five years closer to the median age than were the people who died from falls. Thus, for the representative individual, homicides presumably should have been viewed by the respondents as more hazardous. In effect, any failure to control for the ages of death of the causes will once again distort one's conclusions, with the magnitude and direction of the distortion depending upon (1) the correlation between average ages of death and frequencies of death, and (2) the correlation between ages of death and hazard rates.

For the United States, the first of these correlations is positive: Causes that kill many people (such as strokes, cancer, and heart disease) tend to kill them late in life. Table 8.2 illustrates the number of fatalities per cause and the average age of death for the least and most common causes of death. The average age of death among the ten biggest killers is twenty years greater than the average age of death among the ten smallest killers.

TABLE 8.2

Average ages of Death for the Least and Most Common Causes of Death

---

Least Common Causes:
    Fatalities per cause   690.5
    Average age of death   42.3

Most Common Causes:
    Fatalities per cause   156,830
    Average age of death   63.2

The second determinant of potential distortion is the correlation between ages of death and hazard rates. As I noted in the preceding section, this correlation is negative: Causes that kill later tend to be less hazardous to an individual of any given age. The reason for this negative correlation is simple: As the age at which a cause kills increases, the probability rises that some *other* cause of death will intercede. Thus, the probability of dying from the first cause decreases. For example, although falls (age of death = 73.8) and homicides (age of death 37.9) kill about the same number of people each year, falls are much less hazardous to the representative individual, because of the higher probability that something else will kill that individual before he or she becomes old enough to die of a fall.

Once the ages at which the causes kill are taken into account, I find that the seeming bias reported by Lichtenstein et al. completely disappears for all practical purposes.[18] Thus, for example, if we consider two causes of death that have the same age of death, both the League of Women Voters and the students estimate hazard rate ratios that differ only trivially from population death frequencies. Not only do the estimates of the two groups accord with the facts, they accord with each other. Moreover, for pairs of causes for which the ages of death differ, the voters and the students agree that, holding population death frequencies constant, each year sooner that a cause kills raises the implied hazard of that cause by about 5 percent.

Taken in conjunction with the earlier results, these findings suggest three general conclusions. First, population death frequencies and hazard rates are conceptually different, and at the level of the individual may bear little relationship to one another; any failure to recognize these differences is likely to lead to nonsensical conclusions. Second, the same data that have been interpreted as showing that people are poor at assessing the hazards they face actually reveal that people are remarkably accurate at estimating hazards. Third, two activities that have the same impact on life expectancies may pose markedly different hazards to individuals, and thus produce markedly different behavioral responses by individuals.

A Bayesian View

Based on the evidence assembled thus far, it appears that when individuals are asked the right question, they reply with remarkably accurate responses. Of course, these responses are arguably less than perfect, and so one might ask what determines their proximity to the truth? There are two strands of research that bear on this question.

The first of these, which was also the first systematic attempt to explain the existence of the apparent bias in risk assessment, has been pioneered by Viscusi (1984; 1985).[19] He argues that the existence and direction of any apparent bias in risk assessment is consistent with a Bayesian view of the world in which individuals have a set of prior beliefs about risks, beliefs that are updated in response to new information. The general nature of Viscusi's argument can be illustrated with a simple example.

Imagine asking an interplanetary visitor to estimate the risk of injury associated with activities one might engage in while sailing. For concreteness, let these activities be called jibing, tacking, and luffing. Suppose the visitor has never been sailing and has no knowledge of sailing terminology. Upon observing a variety of sailboats in action, the number of individuals on the boats, and the number of injuries inflicted upon them, the visitor would be able to formulate an estimate of the overall risk of sailing, but distinguishing among the hazards associated with each of the activities would be impossible. If compelled to produce estimates of these hazards, one approach for the visitor would be to attribute the average injury rate to each of the three activities, even if he knew that there were in fact differences among them.[20] If the visitor's judgments (perceived risk) were plotted as a function of the truth (actual risk), the result would be something like that shown in Figure 8.3.

FIGURE 8.3

Risk Estimates Made by an Uninformed Observer

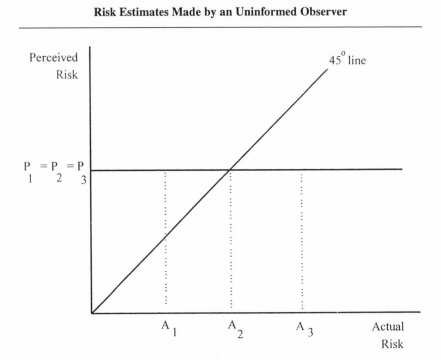

Here the actual risks of the activities are shown as A_1, A_2, and A_3, where $EA_i/n = \mu$ is the (average) true risk of sailing.[21] The visitor's estimates of these risks are given by $P_1 = P_2 = P_3 = \mu$. It is evident that the visitor's lack of knowledge about sailing produces a large and systematic bias in his estimates (perceptions) of risk: Despite being correct on average ($E_i/n = E_i/n$), the visitor overestimates the hazards associated with relatively safe activities ($P_1 > A_1$), while he underestimates the risks of relatively hazardous activities ($P_3 < A_3$).

Clearly, through some combination of sailing lessons and careful observation, the visitor could improve on these estimates in a Bayesian manner. As learning occurred and the visitor's prior beliefs were revised, P_1 would presumably decline toward A_1 and P_3 would rise toward A_3. But to engage in such learning would require that the visitor devote some resources to the task—resources that would be worth expending only if the rewards were great enough. The general points are: (1) as long as information is costly, hazards will be estimated only with error; and (2) the existence of this error will produce systematic biases in perceived risks. I would add a third point: The extent of the error and thus the magnitude of the bias will be negatively related to the rewards to being accurate.

To investigate these issues, Viscusi and O'Connor (1984) examine the responses of workers to job-related risks, seeking to determine whether workers behave in this sort of Bayesian manner. There are three key findings. First, the compensating wage differentials earned by workers are consistent with the hypothesis that workers are aware of and respond to risk differentials across different environments.[22] Second, when workers are given new information about risks, they revise their prior beliefs accordingly, demanding higher risk premia, for example, when on-the-job risks rise. Third, the arrival of new information does *not* cause workers to simply discard their prior information; instead they formulate revised estimates in a manner that reflects the relative precision and importance of the information sets. The net effect is that workers behave in a distinctly Bayesian manner, apparently incorporating a broad array of information into their decision-making—information that *includes* their prior beliefs. In such a setting, any attempt to determine whether individuals "accurately" assess environmental risks will almost surely conclude (for spurious reasons) that they do not.

To see this, suppose that one obtained a measure of "true" job risks and also estimated worker estimates of those risks. Further suppose that this data suggests the existence of bias. Viscusi and O'Conner's results imply that, even though Bayesian workers are distinguishable from non-Bayesians, they are observationally equivalent to biased workers. To successfully disentangle (1) biased workers from (2) Bayesian workers with non-trivial priors, one must know the unknowable—an exact enumeration of the information taken into account by workers.[23] Absent such knowledge, one might well argue that there are potential benefits from providing workers (or consumers) with new information about the risks they face, but there is no basis for claiming that the decisions they make based on existing information are somehow inferior.

The Viscusi and O'Connor results ignore a point subsequently raised by Harrison (1989; 1990) in a different context. In general, the finders of bias have relied on the results of surveys, in which respondents have little incentive to tell the truth, nor to learn the truth if they do not initially know it. Following the Bayesian approach suggested by Viscusi, one might imagine that survey respondents begin by simply classifying the causes of death presented by researchers as being nothing more than that—which is to say, clearly distinct from things that do *not* cause death, but otherwise indistinguishable from one another. Clearly, this is an overstatement of any plausible starting point for respondents, because the actual number of deaths associated with the various causes on Lichtenstein et al.'s list range from 0 to 1.7 million; one must surely presume that (almost) everyone knows that the fatality rate from heart attacks is higher than from measles. Nevertheless, by the earlier discussion it should be clear that even people who know *exactly* the conditional probability of being killed in a bungee jump may know almost nothing about the unconditional probability, simply because they lack the (largely irrelevant to them) information about how many people engage in bungee jumping each year. Thus, the mere fact that these "things" are lumped together as "causes of death" would be expected to upgrade the estimated frequency of low frequency causes and downgrade that of high frequency causes.[24]

If there is a tendency for people to behave in this way, then the survey venue—in which the penalty for such behavior is nil—is where it should be exhibited most clearly. Conversely, when the rewards for discriminating more finely are greater, there should be less of this behavior, and thus less bias in ranking the relative frequencies. Direct evidence on the role of rewards in reducing decision-making errors dates back at least to Siegel and Fouraker (1960). They found that in multiperson bargaining situations, increasing the opportunity cost of missing the optimum induced a tighter clustering of the observed outcomes in the neighborhood of the optimum outcome.

Siegel et al. (1964) obtained similar results in the context of individual decision making: Raising the monetary rewards to correct decision making moved individual decisions closer to the optimum, and also reduced mean square decision errors. Indeed, in an experimental setting well known (from prior research) to induce extreme bias in decision making, Siegel et al. found that modest monetary payoffs were sufficient to eliminate entirely all biases among many of the experimental subjects. More recently, Bull et al. (1987) and Drago and Heywood (1989) find that when payoff functions are more sharply peaked, the variance of the outcomes in tournaments is reduced markedly—exactly what one would expect if high rewards reduce decision making errors. Even more to the point, Smith and Walker (1992a) directly test for the influence of reward structure on decision making. They find that as the rewards to being correct increase, decision error decreases.[25] Smith and Walker also find that, holding payoffs constant, experience on the part of the experimental subjects also decreases decision errors.

These findings reinforce the work of Viscusi and O'Connor, for they not only mean that the experimental subjects were Bayesian, they also imply that the weight assigned to prior information in arriving at a posterior assessment depends on the rewards to being correct. Thus, while Viscusi and O'Conner note that the decisions of individuals may be improved by providing them with new information about their environment, the findings of Smith and Walker imply that individuals *will themselves* seek this information out if given the incentive to do so. The point, of course, is not that everyone will always correctly assess risk. It is rather that when the rewards for being right are smaller, people will invest little in being correct and thus will likely be wrong. On the other hand, as the rewards for being correct rise, so too will accuracy. Hand-in-hand with this result is that ultimately it is incentives that are the key to efficient risk assessment.

Expertise

It is tempting to conclude, as some have done, that because individuals may be prone to error in risk assessment, experts should make assessments on behalf of the people who would otherwise err. In my view, such a conclusion not only has things wrong; it has things backwards. If anything, the possibility of bias suggests that, to the extent that experts play a role in the process of risk assessment, they should appear as tools rather than arbiters in the decision-making process. Two recent but entirely characteristic episodes illustrate this point.

At the micro level, consider cadmium, a toxic industrial metal used to coat metals and to make batteries and pigments. There is general agreement that at high doses, cadmium can damage kidneys and possibly cause lung cancer. There is less agreement on what to do as a result.

During the early 1980s, the Occupational Safety and Health Administration (OSHA) began investigating the possibility of imposing stringent controls on workplace exposure to cadmium. OSHA's review of the evidence led it to propose rules that would have reduced exposure levels 100-fold, thereby (in the expert view of OSHA) saving the lives of fourteen workers each year. The Office of Management and Budget (OMB) subsequently reviewed the same body of evidence relied upon by OSHA and concluded (in its expert view) that the proposed regulations—far from saving fourteen lives each year—would actually kill at least twenty-five additional individuals because the regulations would reduce our wealth and so reduce the demand for safety throughout the country.[26] On the one hand, say the experts, the proposed regulations are an improvement; on the other hand, say the other experts, they are not. I do not know who is correct, and indeed that point may be unknowable. The issue is that the experts disagree, as is perhaps inevitable in risk assessment, for uncertainty is inherent in the process.[27] Despite this disagreement, a decision must be made. If the expert is to make the decision, we must still decide: *Which expert?*

At the macro level, consider the "Greenhouse Effect"—the apparent tendency of carbon dioxide (CO_2) and other gases to accumulate in the atmosphere, acting like a blanket that traps radiated heat, thereby increasing the earth's temperature. There seems little doubt that humankind is producing greenhouse gases at a record rate, and that they are steadily accumulating in the atmosphere. Airborne concentrations of CO_2, for example, are increasing at the rate of about 0.5 percent per year, and over the past fifty years, the amount of CO_2 in the atmosphere has risen a total of about 25 percent.[28] In principle, higher CO_2 levels should lead to higher global temperatures;[29] despite this, it is considerably less apparent what is actually happening to the earth's temperature, and why it is happening.

The National Academy of Sciences (1991) has suggested that by the middle of the twenty-first century, greenhouse gases could be double their levels of 1860, and that global temperatures could rise by as much as 2° to 9°—global warming on such a scale that much of today's temperate climes would become arid dust bowls. Such a view seems consistent with the fact that, on average over the past century, greenhouse gases have been rising and so has the average global temperature. Yet almost all of the temperature rise occurred *before* 1940, while most of the increase in greenhouse gases has occurred *after* 1940. In fact, global average temperatures fell about 0.5° between 1940 and 1970; this cooling actually led many prominent scientists during the 1970s to forecast a coming *ice age*! We thus find ourselves in the position that, less than twenty years after the experts assured us the evidence pointed directly at falling temperatures, we are now assured the evidence points toward rising temperatures. Moreover, the ground-based data that now point to warmer temperatures are directly contradicted by satellite-based data that show no change in global temperatures.[30] That the evidence may point in different directions is perhaps inevitable in risk assessment, for uncertainty is inherent in the process. Despite this, a decision must be made. If our decision is to be made on the basis of the best available evidence, we must still decide: *Which evidence*?

Once we admit the possibility that experts err (or at least disagree) and that evidence is sometimes ambiguous (or at least arguable), then we are forced to recognize that experts' judgments—and our choices among those judgments—depend on the incentives at hand. If it is true that rewards influence the quality of the decisions that people (expert or otherwise) make, then if we ignore this fact by simply abdicating to the experts, we are doomed to suffer higher costs and capture lower benefits than are actually feasible.

It is commonplace to think of "experts" as being as interchangeable as the units of labor or capital encountered in the standard textbook discourse on the theory of the firm: One can have more or less of them, but once one has one, one has all, save for a simple multiplicative factor. This mode of thought is both incorrect and dangerous, and the cadmium episode reveals quite clearly one dimension along which experts may differ substantively—and the issues thereby raised. As reflected in its name, OSHA is charged with protecting worker safety, and one must presume that the rewards facing its employees are structured accordingly. Similarly, both the name and mission statement of OMB emphasize managerial efficiency and the cost implications of government actions. The positive safety implications of the proposed cadmium regulations stem from the physical protections offered workers employed in handling cadmium; it was the evidence on these effects that OSHA's experts found most compelling. The adverse implications of the regulations—so compelling to the OMB experts—arise from their negative impact on our standard of living and the resulting decline in the demand for safety. It may be sheer coincidence that the respective experts' conclusions dovetailed so closely with the mandates of their respective agencies. In fact, I think it far more likely that these experts—like all experts—responded to the incentives they faced. They chose to evaluate the evidence not in a vacuum, but in accord with the perceived costs and benefits that would accrue to *them* as a result of their decisions. Given this, I am compelled to ask: *How shall the incentives facing the experts be determined?*

I would argue that only the people who bear the consequences of decisions can fully know the advantages and disadvantages of each expert decision, including those decisions that include, exclude, or assign weights to particular bodies of evidence. Consider the global warming (or cooling) issue: one body of evidence suggests that we act now to avoid further atmospheric warming; the other evidence suggests that we adopt a wait-and-see attitude. Does action now commit us to a major, irretrievable investment? Does inaction now commit us to inescapable losses in the future if we are wrong today? Absent knowledge of the loss function associated with different alternatives, the physics and chemistry of the issue are insufficient to determine which body of physical and chemical evidence should be acted upon. And the ultimate "experts" on that loss function are the individuals who will gain or lose as a result of the decisions that are made.

It will also do no good simply to suppose that we might instruct the experts to behave in a manner that is consistent with the wishes of those affected by their decisions. First, such a plan assumes that the experts are already *in situ*. Second, it assumes that the experts can know what other people know as well as those people know it. Although both assumptions are commonplace (at least implicitly), neither is tenable, for each amounts to assuming away a pivotal aspect of the risk assessment problem. To suppose—implicitly or explicitly—that we may rely on experts who are already "in place" requires that one of two conditions be satisfied: Either there must exist some *deus ex machina* who has anointed the experts on our behalf, or—in a world in which uncertainty is the essence of the entire discussion—we are somehow endowed with *certain* knowledge of the identity of the best expert in each case. The second assumption—that the appointed experts are endowed with full information of the preferences and endowments of the individuals affected by their decisions—is even more extraordinary, for it amounts to assuming that these experts are *endowed* with the information that is *produced* by the price system.[30] Given these inferences, I am driven to the conclusions that both the experts and the instructions under which they operate must be chosen by the people who will bear the costs and benefits of the experts' assessments.

The flip side of this, of course, is that those who are choosing the experts (or choosing to do without them) and are structuring the experts' incentives must themselves *bear* the full range of costs and benefits associated with their choices.

Ultimately, the uncertainty inherent in the process of risk assessment reinforces rather than obviates the need for individual resource owners to be the final arbiters of the risk assessment and decision-making process. There is simply no other way to choose (or choose to ignore) the experts, nor any way to weigh their findings. The consequences of risk assessment decisions will be borne by someone, whether that someone is the decision-maker or not. Those consequences will be the most advantageous possible only if those individuals that have the greatest incentive to decide among the unknowns—and the unknowable—are making those decisions.[32]

Conclusions

People err. But sometimes by less than we would have them. Frank Knight once observed, "We are so built that what seems reasonable to us is likely to be confirmed by experience or we could not live in the world at all."[31] The evidence I have found is consistent with Knight's view—at least as it applies to risk assessment undertaken by the individuals who must confront those risks.

Knight's observation was made at a time (1921) when the courts generally shared his opinion and when, except for the Food and Drug Administration, the regulatory agencies that so predominate hazard assessment today were not in existence. Whether an assertion similar to Knight's would apply to today's judicial and regulatory experts remains an unanswered question—but one that is surely worthy of further research.

Notes

1. This paper draws heavily on my work with William R. Dougan (1992, 1997, 2000), and I have benefited greatly from discussions with him and with Tim O. Ozenne. Bruce Yandle's comments on an earlier draft improved the present product. Financial support from the Center for Policy

Studies, Clemson University, and from PERC, Bozeman, MT, is grate-fully acknowledged. This paper was completed while I was on leave at Montana State University.

2. Huber (1988, 21).
3. See Huber (1988) and Posner (1986).
4. Cf. Morrall (1986); Huber (1988); Belzer (1991); and Viscusi (1992).
5. See Tversky et al. (1974); Lichtenstein et al. (1978); Arrow (1982); Kahneman et al. (1982); Viscusi and O'Connor (1984); Viscusi (1985); Slovic (1987); Zeckhauser and Viscusi (1990); and Viscusi (1993).
6. Throughout, I use the term "expert" in the broadest sense, to include all regulatory agency employees and all officers of the court system.
7. See National Academy of Sciences (1983); Nichols and Zeckhauser (1986); and Huber (1991).
8. See Nichols and Zeckhauser (1986) for a more complete development of this argument, particularly as it relates to the Environmental Protection Agency. See also Freudenberg (1988) and Slovic et al. (1991).
9. For an extensive compilation of the psychology literature, see Kahneman, Slovic, and Tversky (1982). Zeckhauser and Viscusi (1990) and Machina (1987) summarize much of the salient economics literature. See also Benjamin and Dougan (1997, 2000).
10. Lichtenstein et al. (1978, 578).
11. See, for example, Arrow (1982); Slovic et al. (1982); Viscusi (1985).
12. See below for a discussion of the sources and methods of constructing average age at death.
13. Although, as I shall show below, there are other numbers that might be relevant to them. It is about these that the students seemingly *do* know a great deal.
14. One pack-year of smoking is equivalent to smoking one pack of ciga-rettes per day for one year. Thus, twenty pack-years could consist of two packs per day for ten years or one pack per day for twenty years.
15. Average age of death computed based on Public Health Service (PHS) data for deaths by cause for various ages. PHS reports deaths by five-year age ranges. At the top end, the last PHS category includes everyone over the age of 100. I use the upper end of each range (and 105 for the >100 group) in computing average age of death by cause, thus probably over-stating by about 2.5 years the true average ages.
16. The exception, of course, is falls, mentioned earlier.
17. See Benjamin and Dougan (1992) for complete details.
18. Again, see Benjamin and Dougan (1992) for full details.
19. Slovic et al. (1982) offer *ad hoc* explanations for some of the outliers observed by Lichtenstein et al., but no systematic explanations. Arrow (1982) simply accepts the apparent biases as facts.
20. For example, if the visitor seeks to minimize the quadratic loss function of the form

$$L = E(\hat{\mu}\text{-}\mu)^2$$

where E denotes the expectations operator, μ is the true injury rate and $\hat{\mu}$

is the estimated injury rate, then as long as the visitor is willing to make *some* estimate of the risk of an activity, that estimate $\hat{\mu}$ will be set equal to μ.

21. For diagrammatic and expository convenience I let $A2 = \mu$, but nothing in the argument hinges on this assumption.

22. This finding is consistent with a broad body of literature on compensating wage differentials. See, for example, Thaler and Rosen (1976); Smith (1979); Viscusi (1979); and Brown (1980).

23. Alternatively, the observer might actually know the true risks, in which case *any* deviation—Bayesian or otherwise—is bias. One can only wonder how the observer obtained the data on true risk without market participants—arguably the highest-valued users of such information—being able to get their hands on it.

24. One implication of this argument is that the breadth of the frequencies presented to any given group will influence in a predictable way the extent of the apparent bias in the responses. The broader the range of frequencies, the greater will be the group's degree of overstatement of low frequency causes and understatement of high frequency causes.

25. These findings are consistent with the finding that weather forecasters are impressively unbiased in their forecasts of precipitation probabilities. See Camerer (1992) and Winkler and Murphy (1977). See also Harrison (1990).

26. See Davis (1992).

27. In terms of Figure 8.1, we may think of the measurements recorded along the horizontal axis as being the "truth," but to do so is to deceive ourselves: Experts A and B are likely to write down two completely different sets of numbers along that very same axis.

28. National Academy of Sciences (1991).

29. Laboratory analysis of glacial ice dating back at least 160,000 years indicates that global temperatures and CO_2 levels in the atmosphere do in fact tend to move together, suggesting that the impact of today's rising CO_2 levels will be higher global temperatures in the future. See National Academy of Sciences (1991).

30. See Spencer et al. (1990) and Christy and McNider (1994).

31. See Hayek (1945).

32. William Dougan and I have extended the empirical results discussed in this chapter in two directions. In Benjamin and Dougan (1997), we demonstrate that the biased and unbiased risk assessment hypotheses can be directly tested against one another using the data published in Lichtenstein et al. We perform these tests and find that the hypothesis of biased risk assessment is convincingly rejected in favor of the unbiased risk assessment hypothesis. In Benjamin and Dougan (2000), we show that when test subjects are asked about hazards faced by *their own age groups*, they display a remarkable lack of bias in estimating the risks of dying from alternative causes. These two papers update the developments in the literature that have taken place since Benjamin and Dougan (1992) was written.

33. Knight (1921, 227).

References

Arrow, Kenneth. 1982. "Risk Perception in Psychology and Economics." *Economic Inquiry* (January): 1-9.

Belzer, Richard B. 1991. "The Peril and Promise of Risk Assessment." *Regulation* (Fall): 40-49.

Benjamin, Daniel K., and William R. Dougan. 1992. "The Hazards of Risk Assessment." Clemson University Discussion Paper (November).

Benjamin, Daniel K., and William R. Dougan. 1997. "Individuals' Estimates of the Risks of Death: Part I—A Reassessment of the Previous Evidence," *Journal of Risk and Uncertainty* 15: 115-133.

Benjamin, Daniel K., William R. Dougan, and David Buschena. 2000. "Individuals' Estimates of the Risks of Death: Part II—New Evidence," *Journal of Risk and Uncertainty* (forthcoming).

Bull, Clive, Andrew Schotter, and Keith Weigalt. 1987. "Tournaments and Piece Rates: An Experimental Study." *Journal of Political Economy* (February): 1-33.

Camerer, Colin F. 1992. "Individual Decision Making." In *Handbook of Experimental Economics*, edited by J. Kagel and A. Roth. Princeton, NJ: Princeton University Press.

Christy, J. R., and R. T. McNider. 1994. "Satellite Greenhouse Signal." *Nature* 367: 325.

Davis, Bob. 1992. "Risk Analysis Measures Need for Regulation, But It's No Science." *Wall Street Journal* (August 6): A1.

Drago, Robert, and John S. Heywood. 1989. "Tournaments, Piece Rates, and the Shape of the Payoff Function." *Journal of Political Economy* (August): 992-998

Freudenberg, William R. 1988. "Perceived Risk, Real Risk: Social Science and the Art of Probabilistic Risk Assessment." *Science* 242 (October): 44-49.

Harrison, Glenn. 1989. "The Payoff Dominance Critique of Experimental Economics." University of New Mexico Discussion Paper.

Harrison, Glenn. 1990. "Expected Utility Theory and the Experimentalists." University of South Carolina Working Paper B-90-04.

Hayek, Friedrich A. 1945. "The Use of Knowledge in Society." *American Economic Review* (September): 519-30.

Huber, Peter W. 1988. *Liability.* New York: Basic Books.

———. 1991. *Galileo's Revenge: Junk Science in the Courtroom.* New York: Basic Books.

Knight, Frank H. 1921. *Risk, Uncertainty, and Profit.* New York: Houghton-Mifflin.

Lichtenstein, Sarah, Paul Slovic, Baruch Fischoff, Mark Layman, and Barbara Combs. 1978. "Judged Frequency of Lethal Events." *Journal of Experimental Psychology: Human Learning and Memory* 4, 6: 551-78.

Machina, Mark J. 1987. "Choice Under Uncertainty: Problems Solved and Unsolved." *Journal of Economic Perspective* (Summer): 121-54.

Morrall, John F. III. 1986. "A Review of the Record." *Regulation* (November/December): 25-34.

Murphy, A. H., and R. L. Winkler. 1977. "Can Weather Forecasters Formulate Reliable Probability Forecasts of Precipitation and Temperature?" *National Weather Digest* 2: 2-9.

National Academy of Sciences (NAS). 1983. *Risk Assessment in the Federal Government: Managing the Process.* Washington, DC: NAS.

———. 1991. *Policy Implications of Greenhouse Warming.* Washington, DC: NAS.

Nichols, Albert L., and Richard J. Zeckhauser. 1986. "The Perils of Prudence: How Conservative Risk Assessments Distort Regulation." *Regulation* (November/December): 13-24.

Posner, Richard A. 1986. *Economic Analysis of Law.* 3rd ed. Boston, MA: Little Brown and Co.

Siegel, Sidney, and Lawrence Fouraker. 1960. *Bargaining and Group Decision Making: Experiments in Bilateral Monopoly.* New York: McGraw Hill.

Siegel, Sidney, Alberta Siegel, and Julia Andrews. 1964. *Choice, Strategy, and Utility.* New York: McGraw Hill.

Slovic, Paul. 1987. "Perception of Risk." *Science* 236 (April): 280-85.

Slovic, Paul, James H. Flynn, and Mark Layman. 1991. "Perceived Risk, Trust, and the Politics of Nuclear Waste." *Science* 254 (December): 1603-07.

Slovic, Paul, Baruch Fischoff, and Sarah Lichtenstein. 1982. "Facts versus Fears: Understanding Perceived Risk." In *Judgment Under Uncertainty: Heuristics and Biases,* edited by D. Kahneman, P. Slovic, and A. Tversky. Cambridge: Cambridge University Press.

Smith, Vernon L., and James M. Walker. 1992a. "Monetary Rewards and Decision Costs." University of Arizona Discussion Paper.

———. 1992b. "Rewards, Experience and Decision Costs in First Price Auctions." University of Arizona Discussion Paper.

Spencer, R. W., J. R. Christy, and N. C. Grody. 1990. "Global Atmospheric Temperature Monitoring with Satellite Microwave Measurements: Methods and Results 1979-84." *Journal of Climate* 3: 1111-28.

Viscusi, W. Kip. 1979. *Employment Hazards: An Investigation of Market Performance.* Cambridge, MA: Harvard University Press.

———. 1985. "A Bayesian Perspective on Biases in Risk Perception." *Economics Letters* 17: 59-62.

———. 1989. "Prospective Reference Theory: Toward an Explanation of the Paradoxes." *Journal of Risk and Uncertainty* (September): 235-64.

———. 1991. "Risk Perceptions in Regulation, Tort Liability, and the Market." *Regulation* (Fall): 50-57.

———. 1992. *Fatal Tradeoffs: Public and Private Responsibilities for Risk.* New York: Oxford University Press.

———. 1993. "The Value of Risks to Life and Health." *Journal of Economic Literature* (December): 1912-46.

Viscusi, W. Kip, and Charles O'Connor. 1984. "Adaptive Responses to Chemical labeling: Are Workers Bayesian Decision Makers?" *American Economic Review* (December): 942-56.

Zeckhauser, Richard J., and W. Kip Viscusi. 1990. "Risk Within Reason." *Science* 248 (May): 559-64.

9

Regulation of Carcinogens:
Are Animal Tests a Sound Foundation?

Aaron Wildavsky[1]

Why does it matter if animal cancer studies are worthwhile or worthless or someplace in between? The answer is that regulation of exposure to chemicals, including the intermittent exposure to trace elements to which the general public is subject, is largely based on interpretations of animal cancer bioassays. If these tests are reasonably accurate in predicting the probability, sites, and severity of human cancers, then regulation of chemicals suspected of causing cancer (carcinogens) is on firm ground. But if these animal cancer tests are weak or worse, so that one cannot reasonably predict human cancers from them, then regulation rests on quicksand.

Animal studies have many important uses. Research into cancer mechanisms or problems of the immune system, for instance, may be furthered by introducing novel genes into small animals, such as transgenic mice, to discover better how life systems work (Hanahan 1989). There is no doubt that models based on research with animals have increased our understanding of metastasis, the spread of cancer (Schirrmacher 1989). None of the many invaluable animal cancer studies, however, tells us whether they are a valid source of evidence in predicting human cancer.

The Right Questions

We would like to know how much damage to human populations is caused by different types and quantities of exposure to various substances. That estimation requires answering subquestions about conditions. Routes of exposure may include breathing, eating, or through the skin (such as x-rays). Exposure may come from natural sources from medical uses or from man-made environmental conditions. Quantities may differ and the timing of exposure may differ. If a precise answer to the question of adverse effects cannot be given, we might be satisfied with a rough categorization of harm (see Hattis and Kennedy 1986, 66).

How reliable are these tests? If the tests were repeated on the same species, would we get nearly the same results? If they were repeated on different animal species, would we come up with similar results? If chemicals are carcinogenic in several animal species, it is more likely that they are carcinogenic to mammals in general, including human beings, than if they cause cancer in a single species. It is also important to distinguish rates and sites of cancer by age, because cancer is largely, though not entirely, a disease of old age, and by sex, because males and females are affected differently.

Dioxin in large and continuous doses appears unfriendly to mammals, but it is the dose that matters. Regulatory agencies assume that chemicals carcinogenic at some dose in any animal are also carcinogenic to human beings. But the assumption is questionable. As precisely as possible we wish to answer Freedman and Zeisel's question, "Are chemicals that have been shown to be carcinogenic through experimental animals also carcinogenic to humans?" (Freedman and Zeisel 1988, 14) The reason for their inclusion of the modifier "experimental animals" has to do with the particular conditions under which animals are tested. Therefore they also ask, "Do experimental animals (rodents, in particular) and humans have similar susceptibility to the carcinogenic effect of chemicals, or are rodents incomparably more susceptible than humans?" (14). The answer to the first question is: "sometimes," some rodent cancers are cancers in humans too but some are not. The same is true among different types of rodents. The answer to the second question is: "yes, mostly, but not always." If we know that a single LD50 dose for dioxin ranges from 2500 mg/kg in guinea pigs to 5000 in hamsters, a difference of 2500, does that give us confidence about rodent-to-human transfers?[2]

Suppose we were to find that there is a 10 percent probability that a substance causing cancer in a mouse or rat at a given dose will do the same in a human being. From one point of view, nine times out of ten the extrapolation from mouse to man would be wrong. From another point of view, why take chances with human health if the probability of getting a cancer is that high? Were we to find, however, that rodent cancers predict human cancers only 1 percent or .01 percent of the time then that might not be a reasonable approximation.

The Process of Animal Cancer Testing

Around 1915 or 1916 scientists learned that they could induce cancer in animals by treating them with certain chemicals. The methods of giving animals cancer vary greatly. Chemicals have been introduced into experimental animals by every orifice (orally, nasally, urethrally, vaginally, rectally), by various types of injections (intramuscular, intraperitoneal, intravenous, subcutaneous), by skin painting, by surgery, and by other methods (American Council on Science and Health 1984, 7).

Approximately 30 percent of the rodents get some form of cancer absent exposure to chemicals, though not all 30 percent die of it. This is one reason why a control group is essential. Because a chemical's effects at high doses may not show up at low doses, it is necessary to further subdivide the animals into different dose groups. Given that sex may play an important role in cancer, a further subdivision is between male and female. Usually there are three dose groups (0, 0.5, or 0.1 the maximum tolerated dose [MTD], and the MTD) and two species. Thus, there are at least twelve groups of animals. By convention and by statistical necessity, there are usually fifty animals, most often rodents, in each group.

Though only a few facts about the process of animal cancer testing have been given, we are already in a position to understand three of its most basic aspects: its short time compared to human epidemiological studies, its high cost, and its essentially statistical character. A great advantage of rodent testing is that these animals live only about two years. Therefore one doesn't have to wait too long to get results. One can also test any chemical, including new chemicals, for which epidemiological evidence may not be available. But the task is not easy or cheap. It is costly to keep these animals under controlled conditions for up to two years. The painstaking work of examining animals for tumors

requires pathologists. When each animal dies or, as the too-kind par-
lance states, is sacrificed, several pathologists must carefully examine
about forty sites within and around animal organs and tissues to search
for tumors, some of which are so small they can be discerned only with
high-powered microscopes.

A team of pathologists first work separately and then meet to resolve
differences before their findings are accepted for further evaluation (Chu,
Cueto, and Ward 1981). These pathologists consider whether the tumors or other abnormalities are actually induced by the chemical, an
opinion based on what they know about the normal incidence of tumors
and their experience. They ask themselves not only whether the incidence of tumors is higher but whether they are of a different size or
shape or color or contain any other signs that might show them to be
similar to or different from naturally occurring lesions (Chu, Cueto,
and Ward 1981, 256-57).

Now we are in a position to understand why rodent cancer tests are
so expensive. When one multiplies the time these tests take, roughly
three years, by the cost of keeping twelve groups of animals in controlled conditions, then adds the costs of killing and dissecting them,
preparing and examining forty slides per animal, and reconciling differences, one can see why the costs are substantial (see the discussion
in Rowan 1984). It is possible for a government regulator to conclude
that the tests are inadequate or that the substance being tested is a carcinogen. But it is not possible under the rules to say that the substance
is not, insofar as is known, a carcinogen; the closest government scientists are allowed to come is to say that "the compound has not been
shown to be carcinogenic" (Chu, Cueto, and Ward 1981, 252-53). What,
we may ask, is the meaning of classifying a substance as a suspected
carcinogen? It is worth attending closely to Chu and his colleagues'
discussion:

> If malignant tumors or a combination of malignant and benign tumors are
> produced, then the compound is considered carcinogenic to the animals. If
> the significant result is only the production of benign tumors, then the com-
> pound may pose a potential health hazard and is termed a suspected car-
> cinogen or a carcinogen, depending on the nature of the benign tumor. For
> example, 2,4-dinitrotoluene...was considered a suspected carcinogen since
> it induced only benign tumors (fibromas of the skin and subcutaneous tis-
> sue in male Fischer 344 rats and fibroadenomas of the mammary gland in
> females). Ideally, a distinction should be made between truly benign tu-

mors, which never progress to malignancy, and tumors that are in a benign state according to histopathologic criteria at the time of diagnosis. Scientific judgments in this area are limited by inability to predict the biological behavior of a lesion on the basis of morphological criteria, but it appears that there are few, if any, truly benign tumors in rodents. (Chu, Cueto, and Ward 1981, 257-58)

The "it appears" in the last sentence above reflects a judgment that any tumor might turn bad. Is it in the interests of public safety to treat all tumors, however benign in appearance, as if they might turn malignant, because we do not know they will not? Or is saying that they might turn malignant a way of prejudicing the outcome so that the chemicals will be found to induce cancers whether they do or don't?

In the Environmental Protection Agency (EPA) 1976 "Interim Procedures and Guidelines for Health Risk Assessments of Suspected Carcinogens," EPA Administrator Russell Train acknowledged that animal tests could not prove that a chemical would be carcinogenic in people, but that a substance would be considered a "presumptive cancer risk" if it "causes a statistically significant excess incidence of benign or malignant tumors in humans or animals" (U.S.EPA 1976, 21403). If benign is bad, what could be good?

Calculating Toxicity by the LD50 Test

In the field of pesticide regulation, lethality is calculated through the assignment of an LD50, the lethal dose for one-half of the test animals during the test period. The relevant number for aspirin would be 730 mg/kg, signifying that 50 percent of the test animals died when exposed to 730 mg of aspirin per kilogram of their body weight (Edwards n.d.). The larger the LD50, the more of a substance it takes to produce a toxic effect and the less harmful the chemical.

Among species most commonly used to carry out the LD50 test are fish, birds, rabbits, mice, and rats, although occasionally monkeys and dogs are used. Generally, about fifty or sixty animals of a particular species and a specific dosing method are used. The application is made by inserting a tube down the throat of the animal, by forcing injection of vapors, or by application to the skin (Paget 1970). The usual test lasts about two weeks, during which the animals either die or, at the end, are killed. The usual symptoms are bleeding from the mouth or eyes, convulsions, diarrhea, and what are exquisitely termed "unusual vocalizations." Rather than tolerate early death, according to the British Toxicological Society (1984), "There is pressure on the toxicologists to allow the study to continue, even when the animals are in distress since their premature killing may alter the end-point of the study, and so possibly affect the classification of the material being tested." Needless to say, animal rights advocates are not happy with this method.

Whether one believes that the LD50 test involves "a ritual mass execution of animals" (Rowan 1984, 207) or that "the main information they give is an indication of the size of dose required to commit suicide" (Baker 1969), or even whether most experts consider "the modern toxicological routine procedure a wasteful endeavor in which scientific inventiveness and common sense have been replaced by a thoughtless completion of standard protocols" (Zbinden 1976), there is ample scientific doubt about the value of the LD50 test for the purpose of predicting effects on humans.[3] The basic difficulty is that enormous differences between different (even closely related) species are reported, ranging from 5 to 75 times, which renders findings suspect.[4] If LD50 tests are useful in providing evidence to save human life from suffering, there would still have to be a debate over whether the animal suffering entailed can be justified. If, however, the observations are too unreliable to be useful, no such question arises.

Toxicity

The best explanation for laymen I have heard of the differences with which we have been concerned between very large and very small exposures in different kinds of species comes from reporter Richard Harris of National Public Radio (NPR 1992, 8-10), together with a number of cancer researchers and government officials. Their dialogue is instructive:

PENELOPE FENNER CRISP (Environmental Protection Agency): We're coming to discover that there are more differences between species than we had expected or, frankly, hoped that existed.

HARRIS: It turns out that a great many chemicals that can cause cancer in one species don't seem to do anything at all in another species. Here's an analogy.

[Excerpt from music from a CD]

HARRIS: The difference between rodents and people can be as dramatic as the difference between this CD and an LP. You could drop this CD, get it dusty, even scratch it, you wouldn't necessarily hurt it.

[Sound of record being scratched by a stereo needle and music played from an LP]

HARRIS: But try the same thing with a record, and you can just hear the damage. To be sure, some things will damage either a CD or a record album say a hot windowsill. Likewise, John Doull from the University of Kansas says some chemicals do cause cancer in all sorts of animals. Now, nobody's suggesting that these chemicals are harmless, but in some cases scientists believe that the standards may be vastly overstating the health risks. Again, this comes down to a necessary but flawed shortcut the EPA uses to size up a chemical. Scientists give a huge dose of chemicals to rats and then estimate the effects of that chemical at lower doses. By way of analogy, if you drop a bottle from 10 feet off the ground, it's pretty obvious what's going to happen.

[Sound of glass shattering]

HARRIS: This large drop is equivalent to a large dose of a chemical, and it can be deadly. But what if, instead of taking one bottle and dropping it from 10 feet, you take 10 bottles and drop them from one foot? It's like giving many people a smaller dose of that toxic chemical. Here's what the EPA assumes will happen.

[Sound of several bottles hitting the ground and one of them shattering]

HARRIS: They figure one of the 10 bottles will break. The reasoning is that one-tenth the dose, or one-tenth the drop distance, will do one-tenth the damage. In reality, though, this is what happens.

[Sound of several bottles hitting the ground]

HARRIS: There is, in fact, a safe height you can drop a bottle from without breaking it, and John Doull from the University of Kansas says the same idea holds for toxic chemicals.

DOULL: It is the dose, not the compound, that determines its adverse effects.

HARRIS: So, recently, researchers like Swenberg have started to dig deeper and ask why some chemicals trigger cancer in some animals. It's as though they're trying to understand the difference between turntables and CD players. And Swenberg says one especially interesting example is unleaded gasoline. You may have seen the sticker at the pump warning that gasoline causes cancer in laboratory animals. Well, here's the story with gasoline.

SWENBERG: It causes kidney cancer in male rats only, not in female rats and not in mice.

HARRIS: So what's going on? Swenberg decided to find out by studying those animals, and he discovered that a chemical in gasoline binds to a naturally occurring protein that's only found in the kidneys of male rats.

SWENBERG: And this results in a buildup of the protein and ultimately leads to the development of cancer. And since humans do not synthesize this protein, this is not likely to be a mechanism important to humans.

HARRIS: Swenberg says dozens of other chemicals besides gasoline cause this specific kidney cancer in male rats, including copy machine toner, a bathroom deodorizer, and even a natural chemical called D-limonene.

SWENBERG: It turns out that about two glasses of orange juice contains a carcinogenic amount of D-limonene for the male rat, but it has absolutely no effect on mice or on female rats, and I'm sure it has no effect on humans.

HARRIS: As a result of this research, the Environmental Protection Agency recently decided that if a chemical like gasoline only triggers this kind of kidney tumor in male rats and it doesn't do anything else bad, it's probably not going to cause cancer in people. So far there are just a handful of stories like this where scientists have actually figured out why a compound is causing tumors in certain animals. But there are a lot more studies in the works, including reassessments of dioxin, formaldehyde, and certain PCBs.

Knowledge of mechanisms yields far greater discriminatory power. With such knowledge scientists can determine whether there is a threshold below which there is no damage or whether harm occurs proportionate to the dose, however low that dose is. Without knowledge of the mechanism we cannot be sure of the dose-response relationship.

The War Over the Dose-Response Threshold

There is disagreement over whether there is a dose-response threshold, such that below a certain level no harm occurs, or whether the damage is linear, such that harm from a chemical increases or decreases as a proportion of the dose. It is important, to start with, to ask why such an apparently technical matter has occasioned so much dispute. Because the field of toxicology is built on the principle that the poison is in the dose, the opposing linear (or proportional) principle—that there is no threshold dose below which damage cannot occur—is a challenge to toxicological science.

A common statement about dose-response levels is that no one really understands what happens when people are exposed to very low levels of chemicals (see, for instance, Marx 1990). There is no difficulty in finding substances, such as the heart medicine digitalis, that are helpful at low doses but can be fatal at large doses. But that does not answer the question of whether there are substances for which no threshold exists (see Smith and Sharp 1984, 3). Given that there is a considerable range of sensitivity among human beings, it can always be said that some hypersensitive people might be adversely affected. The traditional response has been to use a margin of safety to take care of the supersusceptible. Given also that chemicals may interact with each other to create cancers that neither substance would alone, it cannot be said definitively that either is safe. By the same token, however, one chemical may render another harmless or less harmful (see Revision Without Revolution 1984). The regulatory response is that the dose-response relationship is linear. The rationale is that this provides a margin of safety for the public. The question is whether this assumption is true.

Going further into the "furious battle [that] rages around the 'threshold controversy'" (Rowan 1984, 234-35), it will be instructive to read a semiofficial account by high-ranking Environmental Protection Agency officials published in a major journal, *Risk Analysis*. The models the EPA uses attempt to establish an upper bound, nearly the worst that could happen, on the basis of a no-threshold linear response. Anderson et al. (1983) are quite open in saying that "recognition that the lower bound may be indistinguishable from zero stems from the uncertainties associated with mechanisms of carcinogenesis including the possibility of detoxification and repair mechanisms, metabolic pathways, and the role of the agent in the cancer process" (Anderson et al. 1983, 281). In short, for all the EPA knows, there may be no damage at low doses.

Furthermore, "[m]ost often there is no biological justification to support the choice of any one model to describe actual risk" (Anderson et al. 1983, 281). This task would be easy if there were data on actual environmental exposures in human beings, in which case an appropriate model could be fitted to the data. "In the absence of such data a variety of models can be used to fit the data in the observed range, but these models differ sharply [in the danger estimates they produce] at low dose" (Anderson et al. 1983, 281). If the choice of model determines the results, because they "differ sharply at low doses," why bother with the experiment? Employing the justification that nevertheless these models are the best available, Anderson et al. (1983, 281) state that "[i]t should be clear from the preceding discussion that the linear non-threshold model has been used by the EPA to place plausible upper bounds on risk, not to establish actual risk." This is a significant admission.

Use of the upper bound misleads people into thinking it is an actual estimate of hazard by an authoritative government agency when it is not. Use of "worst case" scenarios makes no sense, moreover, when the outcome may be zero and there is no biological sense in anticipating epidemic consequences.

Now we know that everything depends on which of the available statistical models are used and whether whichever one is chosen, in the absence of biological indications, tells us what we need to know. Does it?

The EPA claims that the linear, "no dose-response" model best fits knowledge about cancer causation. But its officials could not know this without knowledge of the mechanisms at work, in which case they would be able to choose a model they knew fit the causal relationship. At other times, they acknowledge the real basis for their choice of model, the desire to choose the most conservative estimate so as to, as the saying goes, act on the side of safety. But are they so acting?

If it is true, as Anderson et al. (1983, 289-90) say in their appendix, that "[t]here is no really solid scientific basis for any mathematical extrapolation model relating carcinogen exposure to cancer risks at the extremely low levels of concentration that must be dealt with in evaluating environmental hazards," then why make one? The answer must be that when going from rodents to people, most regulation of chemicals would lack a rationale that could be called scientific. No man-mouse extrapolation; no science; no regulation. The models that make this extrapolation plausible are the important thing. Extrapolations from animals dosed at very high levels to people exposed to far smaller levels make sense only in the context of the models of cancer causation into which they are meant to fit.

Multistage Models

Interpretation of cancer causation depends on models, which we can think of as representations of theories, with numbers attached, that give meaning to data. Actually, there have to be two models in one: first, a model of the biology underlying cancer causation, and second, the statistical approximation of that model. Getting accurate results depends both on the predictive power of the biological model of cancer causation and on whether the statistical approximation captures the causal structure of the model. If the model does not well describe cancer causation in human beings and if the statistical approximation does not well describe the model, the errors in both models multiply to give unsatisfactory results.

We can take the Armitage-Doll model as representative of those used by governmental agencies in regulation. It seeks to describe the relationship between exposures to chemicals and the incidence of cancer at various ages for men and women (see Moolgavkar 1986). The biological version portrays human cells as going through a number of stages that ultimately result in cancer. The hypothesis is that one or more cells receive an insult and then go through several changes that turn them into malignant cells, after which they proliferate. The times and the different stages are not specified. All these stages are probabilistic in that some cells under the same exposure will become cancerous and others will not. "Put another way, with multistage models," Richard Peto (1977, 1404) tells us, "when all the predisposing factors have been allowed for, luck has an essential role in determining who gets cancer and who does not." Thus the stages in the models are essentially probabilities, and the users do not know whether human cancer proceeds in those stages or according to those odds (Freedman and Navidi 1989, 72). In the field of economics, these would be called Markov chain models, which means essentially that every present stage depends on results of previous stages. The time spent in the various stages is assumed to be proportionate to the exposure of the affected individual.

The basic difficulty with multistage models, as the reader might imagine, is that there is little reason to believe they actually capture the biological process of cancer formation. At the same time, the statistical manipulations are very far from the causal requirements of the model, so that one has no idea what one has got when the result is cranked out (Moolgavkar 1986; Freedman and Navidi 1989). This brings us to the statistical interpretation of animal cancer studies, the most critical and least understood part of modeling cancer causation.

The existence of twelve different test groups shows that statistical inference is the essence of the matter: after all, conclusions are to be drawn by observing differences between the control group and other animals and between gender and dose groups. This is not something that is done by counting on the fingers of one hand. It requires methods based on statistical theory.

Biological Interpretation of Animal Cancer Tests

Fears et al. are wise in concluding that "[t]here is danger in relying solely on the finding of statistical significance without incorporating biological knowledge and corroborative evidence such as the presence of a dose-response relationship for experimentally consistent results in different species or sexes" (Fears, Tarone, and Chu 1977, 1941). But what if there is little or no biological knowledge?

In order to get accurate estimates of the probability that chemicals that cause cancer in animals also cause cancer in human beings, Salzburg (1983) recommends applying "the bioassay to a number of innocuous substances. There have to be some compounds that are not human carcinogens, or the whole exercise of looking for carcinogens makes no sense." Yet after examining the literature, he finds that "this was never done for the [rodent] lifetime feeding study" (Salzburg 1983, 65). His argument needs to be heard in full:

> Thus, it would appear that no attempt has ever been made to determine how well society can identify human carcinogens by feeding groups of 50 rats and mice, each, the suspect substance at maximum tolerated doses for their entire lives. Common scientific prudence would suggest that this assay be tried on a group of known human carcinogens and on a group of supposedly innocuous substances (such as sucrose or amino acids) before we either (1) believe that it provides some protection for society (sensitivity) or (2) believe it identifies mainly harmful substances (specificity). There is no substitute for such proper validation on any new bioassay. (Salzburg 1983, 63)

He believes "that we are confusing the effects of biological activity upon the old-age lesions of rodents with the thing we fear, cancer" (Salzburg 1983, 64-65; see also Salzburg 1980 and Tomatis, Agthe, and Bartsch et al. 1978). There is also the claim that cancers found in rat autopsies are induced by test procedures that feed the rats at maximum tolerated dose.

Mitogenesis: Is Cancer Caused by the Test and Not by the Chemicals?

Proof that the tumors observed in animal cancer tests are due to the huge doses delivered at the MTD would be fatal to the no-threshold idea, for then the animal cancer tests themselves would be taking out only what they put in: cancer in, cancer out. Bruce Ames and Lois Gold

(1990a; 1990b), among others, claim that the chronic wounding induced by delivering heavy doses of a chemical promotes cancer by inducing cell division, a process called mitogenesis. As the animal is effectively wounded or poisoned, it grows replacement cells, a process known to increase chances of mutation and hence of cancer.

The theory was prompted by finds that while cancer is thought to be accompanied by mutation, alteration, or damage of DNA, a large proportion of chemicals that cause cancers in animal tests do not in fact damage genes in other tests. There are, as a paper by Ames and Gold is titled, "Too Many Rodent Carcinogens," based on expectations flowing from knowledge of cancer and a belief there would be a great deal more cancer around if half the chemicals in the world caused this terrible sickness (Marsalis and Steinmetz 1990b, 10; Ames and Gold 1990a). Proliferation of cells with DNA damage is an important element in production of cancer in human beings. Cell proliferation is caused by chronic toxicity, by ionizing radiation, by chronic inflammation, and by hormones and viruses that cause infections that in turn lead to cells dying and hence to cell proliferation (Marsalis and Steinmetz 1990b, 10; see also Ames and Gold 1990b and 1991).

Support comes from Cohen and Ellwein (1990, 1007): "Chemicals that induce cancer at high doses in animal bioassays," they assert, "often fail to fit the traditional characterization of genetoxins. Many of these nongenetoxic compounds (such as sodium saccharin) have in common the property that they increase cell proliferation in the target organ." They argue that "the increase in cell proliferation can account for the carcinogenicity of the nongenetoxic compounds." Similarly, Daniel Krewski finds a "fairly strong" correlation between carcinogenicity and toxicity, which one would expect to find when test animals are being wounded by being fed the maximum tolerated dose (Marx 1990, 744). If the mitogenesis theory is correct, then rodent tests run at mitogenic doses are invalid as predictors of human cancer from exposures below toxic levels.

Ames (1990, A38) lays it right on the line: "We think the current approach to cancer risk assessment is bankrupt." A reply by Richard A. Griesemer, who is head of the Division of Toxicological Research and Testing at the National Institute of Environmental Health, is: "There are only two definitive ways to tell whether chemicals have potential to cause cancer. One is through epidemiological studies in humans. [T]he second way is to produce cancers in mammals" (Ames 1990, A38). In other words, we are not to experiment on people, so we have to do so on animals. But do we have to experiment on animals if the results are meaningless?

The most sustained attack on the Ames-Gold thesis (that the animal bioassay, by feeding animals the MTD, is itself causing the appearance of excessive rates of cancer) is by Perera (1990). Her first argument is that a variety of international agencies, including agencies in the United States, have "adopted the general assumption of low-dose linearity for carcinogensæ regardless of their presumed mechanism of action." The rationale is that there must be something to it since so many agencies have gone in the same direction. Why? Scientists usually do not argue from authority when they process a theory they can validate with evidence.

The reasons, according to Perera, are a general lack of understanding of the mechanisms of cancer causation, especially those termed nongenetoxic; a lack of agreement on a safe threshold level below which exposure would not be harmful to a diverse population; and "the desirability of preventing cancer through the use of testing and model systems, obviating the reliance on epidemiological data in humans."[5] The question is whether regulation should be undertaken as a replacement for the existing lack of knowledge. That is exactly what she advocates: "In the meantime, EPA cannot ignore its responsibility to evaluate and control synthetic chemicals since no one has yet devised an acceptable alternative" (Cogliano et al. 1991, 607). I disagree.

The argument that if we do not regulate we will count dead bodies is dead wrong. The predicted cancer rate at one in a million (even one in 100,000 or one in 10,000) is so low it will never be detected by epidemiology or any other method unless we know a lot more about the mechanism of cancer causation. "The problem with risk assessments based an animal tests," the Office of Technology Assessment's Michael Gough tells students, is that their theories cannot be tested (Gough, 1995).

False Positives or False Negatives

One way of looking at a test is to ask whether its error rate is low or whether there is a high proportion of false positives, i.e., those chemicals that do not cause cancer at the administered dose but are wrongly believed to do so (Fears, Tarone, and Chu 1977; Gart, Chu, and Tarone

1979). The opposite error, an overabundance of false negatives, i.e., those chemicals that cause cancer at the administered dose and are wrongly believe to be benign can also occur. Governmental agencies set their requirements so as to minimize the chances of false negatives.

Statistical phrases such as "low-dose extrapolation" and "low-dose linearity" determine regulation of chemicals in the United States. They are not sideshows; they are center ring. In 1985 an interagency committee working for the Executive Office of the President published thirty-one principles for conducting quantitative risk assessments (U.S. Office of Science and Technology Policy [OSTP] 1985). Appropriately, some of the principles emphasized the limitations of scientific knowledge about chemical carcinogenesis. For example, the committee acknowledged that existing knowledge was not up to determining whether chemicals that caused cancer at high doses have a carcinogenic effect at lower doses (OSTP 1985, 10376). The document accepted that "no single mathematical procedure is recognized as the most appropriate for low-dose extrapolation in carcinogenesis," but nevertheless endorsed linear extrapolation techniques: "uncertainties are involved in the use of any of the commonly employed extrapolation models, [but] models which incorporate low-dose linearity are preferred when compatible with the limited information [available]" (OSTP 1985, 10378). Why preferred? No doubt because they are conservative, though no one can calculate how conservative they are.

To extrapolate from rodents to people, a number of basic assumptions must be made, of which three are of primary importance: (1) the biology of these mammals is sufficiently similar to justify the extrapolation; (2) there must be an adjustment for the huge size of people compared to rodents; and (3) the vast differences in doses given to animals, which we have seen is essential to make rodent tests feasible, must be taken into account. After all, perhaps the greatest controversy surrounding animal cancer tests is whether a chemical given to rodents in huge doses would actually produce cancer at much lower exposure.

Where might one find false positives? Site is important. Research reveals that false positives are more likely to occur at sites with a high rather than a low number of spontaneous tumors (Fears, Tarone, and Chu 1977, 1941). It is also known that rare tumors are less likely to be false positives than are common ones. Thus knowledge of the spontaneous tumor rate is essential, especially if it is above 5 percent, because it then becomes difficult to tell the natural from the chemically caused tumor (Chu, Cueto, and Ward 1981, 259-62). The type of chemical also interacts with the type of animal; for instance, some rat organs, when exposed to chemicals, are pretty good predictors of tumors in mice, but mouse liver is a very poor predictor of tumors in rats. There is also a striking difference between chlorinated and nonchlorinated chemicals in regard to the sensitivity of mice versus other animals (Gold et al. 1989, 218; see also Marsalis and Steinmetz 1990a, 156).

The most important defect of animal cancer studies, as Freedman and Zeisel (1988) demonstrate, is that *the choice of statistical models overdetermines the results:* in speculation regarding the effects of the low doses to which human beings are subject in nonoccupational exposures, *the choices of statistical models produce outcomes that vary by hundreds, thousands, tens of thousands, and occasionally millions of times. Yet without knowledge of the biological mechanisms of cancer causation, there is no way of choosing among these models.*

In the study of the grain fumigant EDB, for instance, the probability that an individual would get cancer from eating food in which tiny amounts of EDB were present varies over a million times, depending on the model (Hattis and Kennedy 1986, 65). Using the same animal data but different statistical models in regard to saccharin, to take a well-known but extreme instance, led to differences of some five million times (National Academy of Sciences 1978). As Table 9.1 shows, statistical models that depend on different assumptions predict harms that vary by a factor of 200 for DDT, 800 for dioxin, and 40,000 for aflatoxin, a naturally occurring toxic substance found in various foods, especially peanut butter.

With results this far apart using different models, and without a plausible biological reason for preferring one model over another, the results are no better than guesswork. Would the reader accept the result in anything that mattered if that result was anywhere between 100 and 40,000 times off and the reader could not know which? Even if there were no alternative sources of information, we would do better ignoring such empty data.

Echoing lines from literature, my students often ask whether animal cancer tests could not be considered second best and therefore better than nothing. I ask them whether it would be second best if they want to go from Brooklyn to Queens and are advised to travel via Beijing. Some alternatives are so bad they should not be dignified by the designation "nth best," but rather should be recognized as too bad to be used.

Table 9.1

The Impact of the Model on Low-Dose Risk Estimates

Substance	One-hit	Multistage	Weibull	Multihit
Aflatoxin	1	30	1,000	40,000
Dioxin	1	400	400	800
DMNA	1	700	700	2,700
Dieldrin	1	3	200	1,000
DDT	1	2	70	200

Notes: From Food Safety Council (1980, Table 4). The virtually safe dose is estimated (from each of the four models) as that dose giving a risk of one in a million. The column for the multistage model shows the ratio of the estimated from one-hit model, for each of the five substances. Likewise for the Weibull and the multihit (Freedman and Zeisel 1988, 11).

In retrospect, to be sure, we can see that heading due west from Spain wasn't such a great route for Columbus to follow if he wanted to get to India. But with the more limited knowledge of Columbus's day, he was justified in sailing. Had he continued, more to the point, he would ultimately have gotten there. This much cannot be said of animal cancer tests. They can be performed forever without improving our ability to predict cancer in human beings. Had Columbus been even 200 times off, he would have been on another planet.

Alternatives

What are we to do to protect people against cancer? Eat fruits and vegetables? Stop smoking? These behavioral changes help a lot. But our question concerns the effects of chemicals. Epidemiology is the most accurate in predicting rates of cancers from different levels of exposure to various chemicals. It is also morally appropriate in that human beings are tested to protect human beings. But there are shortcomings. We could collectively decide to accept the lesser sensitivity of epidemiological studies to gain their accuracy and reliability, and at the same time seek the mechanisms of causation that alone hold promise of effective intervention. But that would leave many people worried.

Currently, the alternative to epidemiology is animal cancer tests at the MTD with interpretation based on multistage models dependent on administering the MTD. The advantages are better control over conditions in the laboratory and the ability to get short-term results. The defects are insuperable, however, for without knowledge of the biological mechanisms through which cancer is caused, there is no good way to interpret results. When all we know is that the potential link between exposure to a given chemical by rodents is dozens to tens of thousands of times away from human exposure, we know nothing of value. Animal cancer tests for the purpose of predicting the effects of small, intermittent, nonoccupational human exposures are not better than choosing at random. What, then, are we to do about the continuing stream of chemicals being used and concocted?

The demand either to use animal cancer studies or to find a substitute depends on the belief that very small amounts of chemicals cause significant amounts of damage to human beings and other living creatures. If large doses are required to cause large effects, epidemiology would do. The belief in the power of small doses is reinforced by the related view that there is no level below which damage does not occur.

If no level is low enough to be safe, then we would expect, with the introduction of so many new chemicals from the 1950s onward, that deleterious effects would show up in health statistics. But they do not. Neither a general cancer epidemic nor the once-feared asbestos epidemic has materialized. Once we control for cigarette smoking and hence lung cancer, for age (because cancer is largely a disease of old age), and for AIDS-related cancers, overall cancer rates either are falling or have leveled off. Life expectancy has gone up decade by decade from the turn of the century.

This should not be surprising. The chemicals regulated are so small in amount and so far from people that they could hardly do much damage unless, through unknown mechanisms, very small exposures are doing significant harm. In a seminal study, Michael Gough (1990) showed that if everything the EPA claimed for its regulation proved out, the most that could happen would amount to a 1 percent or smaller reduction in cancer rates. The EPA criterion for regulation limits exposure to some 374,000 times less than a dose shown to cause harm in experimental animals (Gaylor 1989). The health benefits from limiting exposure to such tiny amounts are likely to vary from insignificant to nonexistent.

In an article titled "Information Value of the Rodent Bioassay," Lester Lave and his co-authors (1988) conclude that

> [f]or almost all of the chemicals tested to date, rodent bioassays have not been cost-effective. They give limited and uncertain information on carcinogenicity, generally give no indication of mechanism of action, and re-

quire years to complete. Instead, some of these resources should be devoted to improving the sensitivity, specificity and cost-effectiveness of alternatives to the long-term bioassay, such as *in vitro* [literally, "in glass," i.e., in the tube] and *in vivo* short-term test schemes.[6]

I agree. I also agree with Salzburg that

> [p]resently the lifetime feeding study pre-empts the field. As long as it is considered to be useful in detecting human carcinogens this very expensive and time-consuming procedure will continue to drain the toxicological resources of society. This report questions its usefulness and suggests that it is time to seriously consider the alternatives. (Salzburg 1983, 66)

There are alternatives. Some *in vitro* tests seek to detect damage to DNA within the cell nucleus. The Ames bacterial test detects mutagens. None of these *in vitro* tests predict human cancers (Goldberg and Frazier 1989).

Another of many possibilities is pharmacokinetics, in which quantitative mathematical modeling is used to estimate such things as how much of a given chemical gets to a particular kind of tissue and the absorption, metabolism, and distribution of the chemical in the human body (Wilson 1991). This is a theory-building exercise. How it can be related to theory testing is not yet decided. The conclusion has to be that while alternatives are promising, they have yet to fulfill their promise.

If epidemiology is too insensitive and animal cancer tests are invalid, the questions remain: What should be done to reduce cancer rates? How should the multitudes of chemicals be treated until we possess the knowledge to eliminate or restrict those that cause human cancers at low doses? Should our collective decision be to cut through the complexity by severe regulation? Would such a policy actually improve our health? What type of strategy is suited for our current ignorance? What do we actually know about the sources of carcinogenic chemicals to which human beings are exposed? The two categories of interest are synthetic carcinogens produced by industry and natural carcinogens produced by plants to ward off their predators.

It is no small matter to read a report by *Washington Post* science writer Malcolm Gladwell beginning with the headline "New Panel Questions Traditional Carcinogen Testing: Cancer Experts Respond to Growing Doubts about Massive-Dose Experiments in Animals" (September 8, 1990, A5). The article starts out by saying: "The nation's top experts on assessing whether chemicals cause cancer say that traditional methods are sometimes misleading and that improvements or entirely new methods should be developed." At a meeting lasting three days, a National Academy of Sciences panel on risk assessment stated, in Gladwell's words: "The use of rats and mice to test potential carcinogens, a practice that has formed the basis for regulating chemicals in the United States for more than twenty-five years, should be brought under sharp scrutiny. Many scientists said the studies are too unreliable and too inaccurate to form the basis for evaluating risks to humans" (Gladwell 1990, A5). The panel was particularly critical of feeding animals at the maximum tolerated dose, because it was so much greater in proportion to weight than that to which human beings were subject.

Reform versus Revolution in Risk Assessment

In summary, rodent studies are more rapid but too inaccurate, while mechanistic studies are exceedingly accurate but very slow. Epidemiology lies in the middle on both counts: it is far more accurate than animal tests (being tried on humans at normal doses) but not accurate enough to detect effects at low doses, especially with smaller populations. Leo Levenson recommends a return to the traditional method of controlling the consequences of toxic substances without making special cases out of those that might conceivably cause cancer. For this purpose he would accept the results of rodent cancer tests but would then apply a 100-fold reduction from the level that caused cancer as determined by rodent bioassays. If there were knowledge or experience that led to greater concern, he would then multiply by 10 to reach a level 1,000 times below the animal test level. Were there less reason for concern a level only 10 times under might be applied.

The virtue Levenson sees in this traditional approach is, first and foremost, that it has worked well in the past. It also has other advantages: it is relatively speedy, so it can be applied to new chemicals, and it is relatively straightforward. This traditional "safety factors" approach would also end fascination with and stultification by vanishing small levels of chemicals.

Arguments Against the Use of Safety Factors
for Potential Carcinogens[7]

The 1985 federal interagency committee that published principles for conducting quantitative risk assessment gave four reasons for continuing to treat animal carcinogens differently from other chemicals and rejecting a "safety factors" approach:

1. The Baseline for Applying Safety Factors (NOAEL or LOAEL) is too Sensitive to the Particular Experimental Design.

2. Safety Factors Fail to Use All the Information from Dose-Response Curves.

3. Safety Factors Imply Absolute Safety.

4. Safety Factors Imply Thresholds.

The funny thing about these arguments is that they all apply equally well to the quantitative cancer risk extrapolation methods that the interagency committee endorsed. Let's look at how the interagency report presented each of the arguments in turn.

1. Sensitivity to Experimental Design: The interagency report argued that "[i]n spite of its common use, there are a number of potential problems associated with the safety factor approach. The observation of no treatment-related effects at a given dose may depend, at least in part, on the number of animals exposed at the particular level" (OSTP 1985, 10439). Thus the fewer animals you use, the harder it will be to see a statistically significant increase in tumors or other effects attributable to the chemical exposures. However, experimental design makes a great difference with the linear extrapolations as well. Regulatory agencies must establish protocols for the minimum number of animals and experimental conditions that can constitute an acceptable study no matter what technique is finally used to characterize health risks posed by the chemicals.

2. Safety Factors Do Not Make Use of Information from Dose-Response Curves: According to the interagency report, "[t]he determination of a NOAEL ignores the shape of the dose-response curve, even though it would seem that a curve that has a shallow slope in the experimental NOAEL region potentially represents a greater toxicological hazard than one that rises steeply in this region" (OSTP 1985, 10439). In other words, when you apply a safety factor to a dose that appears harmless, you are failing to make use of information from the relationship between higher doses and increased cancer rates. This is correct but there is also no way to know whether a linear risk extrapolation technique can make good use of the high-dose relationships either. Since there is no "adequate biological rationale" for any extrapolation technique, it is hard to see how quantitative acrobatics that incorporate extra-high-dose information are likely to improve low-dose risk estimates.

3. *Safety Factors Are Arbitrary:* The interagency report complains that "there is no biological justification for the general use of any specific safety factor" (OSTP 1985, 10439). Safety factors are always arbitrary but at least they are transparently arbitrary, and there can be an informed debate about what safety factor to choose without anyone maintaining the illusion that there is one "correct" margin of safety. By contrast, the linear extrapolation methods sanctioned by the committee contain a large number of equally arbitrary assumptions about how to use tests involving high chemical doses in animals to predict risks to people at much lower chemical doses, and the assumptions *are effectively concealed from lay people.* When a model result is announced stating, "Such and such a dose allows a maximum risk of 1–1 million," most people have no idea what they are hearing.

4. *Safety Factors Imply Thresholds:* "Another important consideration that would argue against the use of a safety factor approach in cancer risk assessment is the fact that this approach assumes the existence of a true population threshold below which no adverse effects can occur. Even if the concept of the individual thresholds could be supported, the well-recognized genetic variability in the human population would effectively prevent the estimation of a general population threshold value" (OSTP 1985, 10439). In other words, the use of a safety factor would give people a false sense of security about potential residual cancer risks for sensitive individuals. By implication, quantitative risk estimation procedures are more honest in admitting that there could always be a potential risk, no matter how low the exposure. Is this so?

We agree that the government should never promise, and should never accept the responsibility, to eliminate all risk. The EPA and FDA can respond to the unanswerable questions "Is this standard absolutely safe?" or "Is there still a risk of cancer from drinking this water?" with the honest reply, "We think that to the best existing knowledge the chemicals in the water protect human health more than would their diminution or removal." But there is no reason the regulators need to promise that the use of a safety factor implies absolute safety. In contrast, the use of seemingly precise quantitative risk estimates gives the illusion that regulators know more than they really do about cancer risks, if any, from low-dose exposures.

Conclusions

Here are Levenson's and my conclusions regarding human safety standards for chemicals:[8]

- The less of a potentially toxic chemical people are exposed to, the less likely they will get sick from the chemical. This includes chemicals that may cause cancer. It would be better if we talked about carcinogenic or toxic "doses" of chemicals, rather than calling the chemicals themselves "carcinogenic" or "toxic."

- If the only evidence about toxic effects in a chemical is from high doses, there is no good reason to apply the effects at lower doses to people or other animals. Numerical extrapolations will all be statistical games; they cannot provide insight into real risks. These ideas hold true for all types of chemicals both carcinogenic and noncarcinogenic. There is no guarantee that any chemical dose will be "absolutely safe." But we can make good guesses that a particular dose will be small compared to other potential disease factors.

- Congress's attempts to ban chemical carcinogens in the nation's food supply appear to stem from beliefs that chemicals could be easily divided into those that "cause cancer" and those that do not, and that the public health benefits of eliminating "cancer-causing" chemicals entirely had to be greater than the expense. In fact, the categories of "carcinogen" and "not carcinogen" are fundamentally flawed; many chemicals may cause cancer at very high doses but would not cause cancer at lower doses. The cost of reducing chemical residues all the way down to nondetectable amounts turns out to be increasingly high as scientific technology for detecting ever smaller amounts of chemicals improves. And the public health benefits of eliminating tiny amounts of synthetic chemical residues have been called into question given the defense systems that people have, which work to prevent damage from low levels of natural chemical carcinogens to which people are ordinarily exposed in their diets. Cost, however, is not the strongest argument against the criteria used to regulate chemical exposures in the United States today; the strongest argument is that there are no health benefits.

The result of adopting safety-factor criteria would be a reasonable level of protection in the light of existing knowledge, while greatly facilitating regulation. Our expectation (Levenson is a former EPA project director) would be far fewer Superfund sites, greater willingness to clean up the remaining sites, and greater capacity to clean up at the stipulated levels. Lawyers' fees and waiting lists would decline precipitously, and accomplishments would rise.

Cut the Gordian Knot: Reject Regulation
Based on Weak Causes and Weaker Effects

As theories and evidence against the validity and predictive value of animal tests accumulate, a return to the traditional safety-level rule-of-thumb approach will become much more politically feasible. Nevertheless, given a choice, I would recommend a rejection of the existing risk-assessment approach. While acknowledging the importance of political feasibility, I would rather advocate the approach I believe best enhances human (and, for that matter, animal) health and safety.

My objection to a return to the time-tested safety-factor method is that it would be based, as Levenson stipulates, on rodent cancer tests, which I believe to be worthless in predicting cancer in human beings (or, indeed, other species).

Instead, I propose that the government of the United States use its resources and those in the private sector it regulates to enhance two approaches that show promise of developing a knowledge-based policy of cancer control: epidemiology and discovery of actual cancer-causing mechanisms in humans and other species.

Let's take another look at the weaknesses of epidemiology. Studies of human populations do not reveal possible harm from small doses of chemicals even if they exist. This weakness could be diminished by putting resources into doing larger studies and developing better statistics, but as long as regulatory action is conducted at very small risk levels, such as one in a million, the weakness would remain. Over time, mechanistic understanding is the only way of distinguishing those small insults that are harmful from those that the human body successfully defends against. What should be done in the meantime?

I propose that small harms from small causes be ignored until we learn how to identify them more reliably so that the harm done by generating so many false positives is exceeded by the health gains from discovering true positives. We do not give up much by ignoring small harms for two reasons: one is that they are small (or epidemiology would pick them up); the other is that there is no good reason, given the invalidity of animal studies, to believe that the models actually pick up real dangers except by accident. In fact, we are suffering those harms now, if they exist, because we do not know their causes and, hence, are unable to take effective preventive measures. A desire to prevent cancers (even more, a desire to show the public that their government is trying to protect its citizens) is not the same as actually providing protection. The pretense of protection, however, is expensive not only in its loss of money but in its loss of the very health and safety it is supposed to defend. I will not be a party to a method of risk assessment and regulation that makes people sicker in the name of keeping them healthier, and that is exactly what is happening.

The other shortcoming of epidemiology is that it takes a long time, because the latency periods of cancers may be decades long. True, but not, I think, conclusive. A long time is a long time; if a disease takes so long, it cannot be striking people down at early ages, because the evidence would already have shown up. We could argue over whether preventing deaths when people are already quite old should be part of governmental policy. There is no need to do that, however, because short-term harms perpetrated by preventive measures are palpable, while long-term gains, in the absence of knowledge about cancer causation, are dubious.

While the long-latency-period argument is cogent for occupational exposures, it has much less force for the general citizenry exposed only to intermittent small doses. For the population at large, moreover, shorter-term evidence from epidemiological studies has value in that if workers exposed to comparatively large doses, even for only a few years or a decade, show no ill results, the probability that small and sporadic exposures would be harmful is low. In the same way, when we learn that symptoms decline or disappear when doses are reduced, this dose-response relationship gives us a pretty good idea that we can identify the cause.

My advice is to cut the Gordian knot of chemical regulation by requiring a standard parallel to that set for medical drugs; where evidence of efficacy in promoting health is properly demanded of medical drugs, so should evidence of harm to health be demanded for regulation of chemicals. Unless and until the existing reversal of causality (negative evidence that a chemical does no harm) is replaced by the evidence of harm, regulation will continue to harm health in the name of protecting it.

Notes

1. The late Aaron Wildavsky was Class of 1940 Professor of Political Science and Public Policy at the University of California, Berkeley, and a Research Fellow for The Independent Institute.

2. The LD50 dose is the dose which is lethal to 50 percent of the population (lethal dose 50).
3. See Sharpe (1988) for a list of authorities with negative verdicts.
4. See the numerous examples in Rowan (1984).
5. Perera (1990, 1644). See Ames and Gold's response to Perera in the same issue (Ames and Gold 1990b).
6. Lave et al. (1988, 633). As the NMAS (National Medical Advisory Service) asserts: "After more than 15 years of utilizing the B6C3F1 mouse as a mainstay animal on which to perform cancer risk assessment studies, many in the scientific community are calling for a methodology review. At issue is whether or not using this particular test mouse results in inaccurate conclusions. It was specifically bred to be sensitive to cancer causing agents, and it has a high rate of spontaneous tumors (25 to 30 percent). The theory behind the creation of this type of mouse was that if the tests were being performed on a very sensitive animal, the data produced would be conservative, therefore setting very cautious levels of exposure. However, there is a groundswell of opinion today which recognizes that this test mouse has produced results that are overly cautious, and perhaps an inaccurate base upon which misleading risk assessments are being conducted" (NMAS 1992, 1).
7. This section was written by Leo Levenson.
8. All of these principles relate to standards designed to protect *human* health. Additional factors must be considered if concerns have been raised regarding effects of chemicals on animals, plants, or ecosystems.

References

American Council on Science and Health. 1984. Pamphlet, *Of Mice and Men: The Benefits and Limitations of Animal Cancer Tests.*

Ames, Bruce N. 1990. "More on Carcinogen Test Processes." *Los Angeles Times* (August 31): A38.

Ames, Bruce N., and Lois Swirsky Gold. 1990a. "Too Many Rodent Carcinogens: Mitogenesis Increases Mutagenesis." *Science* 249 (August 31): 970-71.

———. 1990b. Response to Perera. Letter to the Editor. *Science* 250 (December 21): 1645-46.

———. 1991. "Endogenous Mutagens and the Causes of Aging and Cancer." *Mutation Research* 250, 1-2: 3-16.

Anderson, Elizabeth A., and the Carcinogenic Assessment Group of the United States Environmental Protection Agency. 1983. "Quantitative Approaches in Use to Assess Cancer Risk." *Risk Analysis* 3, 4: 277-94.

Baker. 1969. Quoted in Rowan.

British Toxicological Society. 1984. *Human Toxicology* 3:85-92. Quoted in Robert Sharpe, *The Cruel Deception.* London: Thorson's Publishing, 1988, 94-95.

Chu, Kenneth C., Cipriano Cueto, Jr., and Jerrold M. Ward. 1981. "Factors in the Evaluation of 200 National Cancer Institute Carcinogen Bioassays." *Journal of Toxicology and Environmental Health* 8, 1-2: 251-80.

Cogliano, Vincent James, et al. 1991. Carcinogens and Human Health: Part III. Letter to the Editor. *Science* 251 (February 8): 606-7.

Cohen, Samuel M., and Leon B. Ellwein. 1990. "Cell Proliferation in Carcinogenesis." *Science* 249 (August 31): 1007-11.

Edwards, Gordon J. N.d. Worried about Pesticides in Food and Water? Here Are the Facts. Pamphlet distributed by National Council for Environmental Balance, Inc.

Fears, Thomas R., Robert E. Tarone, and Kenneth C. Chu. 1977. "False-Positive and False-Negative Rates for Carcinogenicity Screens." *Cancer Research* 37 (July): 1941-45.

Freedman, D. A., and William C. Navidi. 1989. "Multistage Models for Carcinogenesis." *Environmental Health Perspectives* 81: 169-88.

Freedman, David A., and H. Zeisel. 1988. "From Mouse to Man: The Quantitative Assessment of Cancer Risks." *Statistical Science* 3, 11: 3-56.

Gart, J. J., K. C. Chu, and R. E. Tarone. 1979. "Statistical Issues in the Interpretation of Chronic Bioassay Tests for Carcinogenicity." *Journal of the National Cancer Institute* 62: 957-74.

Gaylor, D. W. 1989. "Preliminary Estimates of the Virtually Safe Dose of Tumors Obtained from the Maximum Tolerated Dose." *Regulatory Toxicology and Pharmacology* 9: 101-8.

Gold, Lois Swirsky, Leslie Berstein, Renae Magaw, and Thomas H. Sloane. 1989. "Interspecies Extrapolation in Carcinogenesis: Prediction between Rats and Mice." *Environmental Health Perspectives* 81: 211-19.

Goldberg, Alan M., and John M. Frazier. 1989. "Alternatives to Animals in Toxicity Testing." *Scientific American* 261, 2: 24-30.

———. 1990. "How Much Cancer Can EPA Regulate Away?" *Risk Analysis* 10, 1: 1-6.

Gough, Michael. 1995. "It's not science. What can science do about it?" *Health Physics* 71:275-278

Hanahan, Douglas. 1989. "Transgenic Mice as Probes into Complex Systems." *Science* 246 (December): 1265 ff.

Hattis, Dale, and David Kennedy. 1986. "Assessing Risks from Health Hazards: An Imperfect Science." *Technology Review* 89, 4: 60-71.

Lave, Lester B., Fanny K. Ennever, Herbert S. Rosenkranz, and Gilbert S. Omen. 1988. "Information Value of the Rodent Bioassay." *Nature* 336 (December 15): 631-33.

Marsalis, Jon C., and Karen L. Steinmetz. 1990a. "The Role of Hyperplasia in Liver Carcinogenesis." In *Mouse Liver Carcinogenesis: Mechanisms and Species Comparisons,* edited by D. S. Stevenson, J. A. Popp, J. M. Ward, R. M. McClain, T. J. Slaga, and H. C. Pitot. New York: Wiley-Liss.

———. 1990b. "Scientists Question Use of Rodent Tests in Risk Assessment." *Chemecology* (October): 10.

Marx, Jean. 1990. "Animal Carcinogen Testing Challenged." *Science* 250 (November 9): 743-45.

Moolgavkar, Suresh H. 1986. "Carcinogenesis Modeling: From Molecular Biology to Epidemiology." *Annual Review of Public Health* 7: 151-69.

National Academy of Sciences. 1978. *Saccharin: Technical Assessment of Risks and Benefits.* National Research Council, Committee for Study on Saccharin and Food Safety Policy, Washington, DC, chapter 3., pp. 72 and 61 ff.

National Medical Advisory Service (NMAS). 1992. *NMAS Advisor IV* (4).

National Public Radio. *Morning Edition.* Transcript. Washington, DC. Thursday, March 12, 1992, 10:00 A.M. EDT.

Paget, G. E., ed. 1970. *Methods in Toxicology.* Boston: Blackwell Scientific Publications.

Perera, Frederica P. 1990. Carcinogens and Human Health: Part 1. Letter to the Editor. *Science* 250 (December 21): 1644-45.

Peto, Richard. 1977. "Epidemiology, Multistage Models, and Short-Term Mutagenicity Tests." In *Human Risk Assessment.* Book C of *Origins of Human Cancer,* edited by H. H. Hiatt, J. D. Watson, J. A. Winsten. Cold Spring Harbor (N.Y.) Laboratory.

"Revision Without Revolution in Carcinogen Policy." 1984. *Regulation* 8, 4: 5-7.

Rowan, Andrew H. 1984. *Of Mice, Models, and Men: A Critical Evaluation of Animal Research.* Albany: State University of New York Press.

Salzburg, David. 1980. "The Effects of Lifetime Feeding Studies on Patterns of Senile Lesions in Mice And Rats." *Drug & Chemical Toxicology* 3: 1-33.

——. 1983. "The Lifetime Feeding Study in Mice and Rats—An Examination of Its Validity as a Bioassay for Human Carcinogens." *Fundamental and Applied Toxicology* 3: 63-67.

Schirrmacher, Volker. 1989. "Immunobiology and Immunotherapy of Cancer Metastasis: Ten-Year Studies in an Animal Model Resulting in the Design of an Immunotherapy Procedure Now under Clinical Testing." *Interdisciplinary Science Reviews* 14, 3: 291-303.

Sharpe, Robert. 1988. *The Cruel Deception.* London: Thorson's Publishing.

Smith, Allen H., and Dan S. Sharp. 1984. A Standardized Benchmark Approach to the Use of Cancer Epidemiology Data for Risk Assessment. Typescript presented at the Environmental Protection Agency Symposium on Advances in Health Risk Assessment for Systemic Toxicants and Chemical Mixtures, Cincinnati, Ohio, October 23-25.

Tomatis, L., C. Agthe, and H. Bartsch et al. 1978. "Evaluation of the Carcinogenicity of Chemicals: A Review of the Monograph Program of the International Agency for Research on Cancer (1971 to 1977)." *Cancer Research* 38: 877-85.

U.S. Environmental Protection Agency (EPA). 1976. Interim Procedures and Guidelines for Health Risk Assessments of Suspected Carcinogens. *Federal Register* 41:21402-21405.

U.S. Office of Science and Technology Policy (OSTP). 1985. Chemical Carcinogens: A Review of the Science and its Associated Principles, February 1985. *Federal Register* 50:10372-10442.

Wilson, Alan E. G. 1991. "Encouraging the Development and Application of Pharmacokinetic Modeling in Risk Assessment." *American Industrial Health Council Quarterly* (Summer): 6-9.

Zbinden. 1976. Quoted in Rowan.

10

Protecting Groundwater and the Unintended Transport of Toxic Substances by Water

David T. Fractor

1. Introduction

The greatest portion of current and potential cost to firms and individuals emanating from toxic pollutants concerns groundwater. The very large liabilities imposed by Superfund and the other laws requiring risk-preventive actions, including cleanups, are primarily aimed at reducing the pollution and the risk of pollution to underground water supplies. Policy makers and civil servants who carry out the policies are understandably determined to protect this precious resource. But the cost is often very great, so that well-informed decisions are crucially important. To understand the available policy options, and thus to make rational policies and to promulgate sensible rules, requires some knowledge about not only the role and importance of the groundwater resource, but also about how and when groundwater may be threatened, and about what can be done to protect or to rehabilitate the resource. This chapter provides an overview of some critical hydrologic and engineering facts faced by those making decisions affecting liability and groundwater supplies.

Groundwater is important because it constitutes the earth's primary source of freshwater. Groundwater supplies contain enormous stocks built up over even thousands of years. Only 3 percent of our freshwater stock comes from reservoirs, streams and lakes; groundwater supplies the remaining 97 percent. In the United States, 86 percent of the fluid fresh water available comes from groundwater (Travis and Eltnier 1984, 9). In addition to being the primary source of drinking water, groundwater has also enabled agriculture to flourish in many regions. It is the source of one-third of industrial water usage and one-half of the nation's agricultural irrigation. The development of groundwater resources has increased as the potential development of surface supplies has diminished. Groundwater is a unique resource; a reliable supply of potable water subject to less variability than surface supplies.

Groundwater supplies are often a safer and more reliable supply than is surface water, which is more readily exposed to pollutants than is groundwater. Groundwater pollution can be unique compared with other polluted resources because of the slow rate of natural replenishment. Groundwater movement tends to be very slow, so a contaminated aquifer could take centuries to flush contaminants out. Contaminated groundwater may pose a threat to surface systems when it provides the base flow for streams.

Many groundwater contaminants do not biodegrade readily. The complex hydrological and geological nature of groundwater aquifers can make identification of pollution sources very difficult. Those responsible for polluting the groundwater environment have often avoided responsibility for their actions, since it has often been unclear where responsibility lies. The scope of groundwater pollution is potentially vast; the EPA estimates that at least half of the nation's land area has geologic factors that would allow for groundwater contamination.

This paper will explore methods by which groundwater quality can be protected through individual decisions that encourage cost effective and creative responses to environmental protection. The nature and importance of the groundwater resource is discussed in section 2. Section 3 discusses groundwater pollution problems, including sources and means of contamination. Methods of remediation are discussed in section 4. Section 5 discusses market-oriented alternatives for protecting groundwater quality.

2. The Groundwater Resource

Groundwater is essentially any water found beneath the surface that can be collected, or water that flows naturally to the earth's surface.

Groundwater is pumped by wells or flows out through springs. The uses to which groundwater is put have historically been about 65 percent for irrigation, 15 percent for industry, and the remainder for urban and rural drinking water. The importance of groundwater in the nation's water budget is summarized in Table 10.1, while the importance of groundwater to each state is reported in Table 10.2.

Groundwater is particularly desirable because of its widespread availability, in contrast to surface supplies that may be available on a seasonal basis at best in the arid West and Southwest. Groundwater supplies also eliminate the need for widespread transmission networks. Groundwater tends to be of high quality naturally because of limited exposure to surface activities. Aquifers store large quantities of water with little evaporation and without the expense of surface storage facilities.

Aquifers are water-bearing strata characterized by openings or pores through which water can move. These geologic formations are capable of transmitting and yielding significant quantities of water. Aquifers are typically composed of relatively porous materials such as sand and gravel. Generally, coarse-grained sedimentary material constitute the best types of aquifers, as these more readily collect and release water than, say, finer materials such as clay.

A distinction is made between unconfined and confined aquifers according to the presence or absence of a water table. Simply put, an unconfined aquifer does not have an overlying confining layer. Water that percolates into the ground directly enters the unconfined aquifer, which is in direct contact with the atmosphere through the open pores of the material between the aquifer and the land surface. Thus, the unconfined aquifer is in balance with atmospheric pressure and water movement is directly related to gravity. The water table acts as the upper surface of the zone of saturation. Unconfined aquifers can be thought of as underground lakes within porous materials.

In contrast, a confined aquifer (also known as an artesian aquifer) is characterized by an overlying layer that is less permeable than the aquifer itself and may not be in direct contact with the atmosphere. They are generally bounded by impervious or semi-pervious materials. Water in these aquifers is under pressure and may flow directly to the surface when a well is sunk. Because of their relatively high pressure, confined aquifers yield water through wells at a level above the upper confining layer of the aquifer.

TABLES 10.1

Water Usage in the United States

Year	(Billions of Gallons Per Day) Groundwater	Surface	Percent Groundwater
1950	34.0	150	18.5
1955	47.6	198	19.4
1960	50.4	221	18.6
1965	60.5	253	19.3
1970	69.0	303	18.5
1975	83.0	329	20.1
1980	83.9	361	18.9
1985	73.7	320	18.7
1990	80.6	327	19.8

Source: Table 29, Environmental Quality, 23rd Annual Report, 1992

TABLES 10.2

Importance of Groundwater by State

	State	Ground	Surface	% Ground
1	Kansas	4,800	866	84.7
2	Arkansas	3,810	2,100	64.5
3	Mississippi	1,580	933	62.9
4	Nebraska	5,590	4,450	55.7
5	Arizona	3,100	3,330	48.2
6	New Mexico	1,510	1,780	45.9
7	Oklahoma	568	707	44.5
8	South Dakota	249	425	36.9
9	Hawaii	655	1,490	30.5
10	California	15,100	34,600	30.4
11	Vermont	37	89	29.4
12	Texas	7,410	17,900	29.3
13	Iowa	671	2,090	24.3
14	Nevada	908	2,830	24.3
15	Minnesota	685	2,150	24.2
16	Florida	4,050	12,930	23.9
17	Idaho	4,800	17,500	21.5
18	Utah	815	3,500	18.9
19	Georgia	1,000	4,440	18.4

20	Alaska	72	334	17.7
21	Washington	1,220	5,810	17.4
22	Colorado	2,340	11,200	17.3
23	Louisiana	1,440	8,980	13.8
24	North Dakota	127	1,040	10.9
25	Missouri	640	5,470	10.5
26	Oregon	660	5,880	10.1
27	New Jersey	668	6,270	9.6
28	New Hampshire	84	810	9.4
29	Wisconsin	570	6,170	8.5
30	Wyoming	526	5,700	8.4
31	Indiana	635	7,400	7.9
32	New York	1,100	14,100	7.2
33	Illinois	968	13,500	6.7
34	Rhode Island	27	381	6.6
35	Ohio	730	12,000	5.7
36	Pennsylvania	799	13,500	5.6
37	Michigan	600	10,800	5.3
38	Tennessee	444	8,010	5.3
39	North Carolina	435	8,320	5.0
40	Kentucky	205	3,990	4.9
41	Delaware	79	1,580	4.8
42	Virginia	341	6,910	4.7
43	Maine	66	1,460	4.3
44	West Virginia	227	5,210	4.2
45	Alabama	347	8,250	4.0
46	Connecticut	144	3,640	3.8
47	Maryland	219	6,490	3.3
48	Massachusetts	315	9,340	3.3
49	South Carolina	214	6,610	3.1
50	Montana	203	8,450	2.3
		--------	--------	
	Sum	73,783	321,715	18.7
Source:	Statistical Abstract of the United States, 1992, Table 348.			
	Data is for 1985.			

Perched aquifers are an intermediate classification. In some areas over an unconfined aquifer, material with less permeability than the surrounding materials may be present (such as clay or silt beds). Water which would naturally percolate through the unconfined aquifer may be intercepted, thereby creating a perched water table that is above the remainder of the water table of the unconfined aquifer.

Water below the surface is typically classified in two zones, the zone of aeration and the zone of saturation. Between the land surface and the water table is found the aeration zone. Water within this zone is under lower pressure than the atmosphere, so that a well sunk in this zone will not yield water (as gravitational forces lead the water down towards the zone of saturation). Water that percolates below the vegetation root zone reaches a level at which all of the openings or voids are filled with water. This is known as the saturation zone. If not restricted by an impermeable layer, the upper surface of this zone is known as the water table. When the ground formation over the saturation zone keeps the groundwater at a pressure greater than atmospheric pressure, the groundwater is said to be under artesian pressure.

The zone of saturation is characterized by matter wherein all space between soil materials are filled with water (in the zone of aeration these interstices are filled with water or air). Water within the saturation zone can move freely through wells, other water bodies, or spring naturally to the surface. The upper layer of this zone is known as the water table. Because of the presence of air in this zone, through which the majority of groundwater pollutants must pass, many contaminants have an opportunity to biologically degrade, adsorb onto soil particles, or chemically react to form more innocuous products.

The movement of water within the aquifer and to the surface is measured according to the hydraulic head, hydraulic conductivity, storage coefficient, and specific yield (Travis and Eltnier 1984, 17). The hydraulic head is the vertical distance from the water table to a reference point such as sea level. Hydraulic conductivity is the quantity of water that will move through a given area in a specific amount of time, which is affected by how porous the medium is and the fluid flowing through that medium. The storage coefficient measures the volume of water an aquifer takes into or releases from storage for a unit change in the hydraulic head. Finally, specific yield is the ratio of the volume of water released from an aquifer, under the influence of gravity alone, to the total volume of the aquifer. The specific yield provides information on the change in storage volume as the aquifer volume changes.

The availability of groundwater is determined from the basic groundwater equation; the change in groundwater storage is equal to recharge minus discharge. Recharge to an aquifer helps replenish the available supply. Sources of recharge include deep percolation from precipitation, seepage from surface water bodies (such as lakes and streams), underground flow from a connected aquifer, and artificial recharge. Artificial recharge can occur from planned actions such as the development of spreading grounds, which helps reduce runoff of precipitation in urban areas, or injection from recharge wells. Artificial recharge also occurs incidentally from irrigation and sewage effluent spreading grounds, seepage from reservoirs, canals and drainage ditches, and from other sources.

Groundwater discharge occurs primarily four ways. First, during certain seasons of the year, groundwater may naturally discharge to hydraulically-connected streams. Second, water may flow from springs and seeps. Third, if the water table is near enough to the land surface, evaporation and transpiration may occur. Fourth, and most familiarly, artificial discharge occurs from wells and drains.

3. Sources of Pollution

Groundwater contamination is not a new problem (Barcelona et al. 1988, 1). Studies performed in the mid-1800s linked the contamination of wells by cholera to seepage from earth privy vaults even before the discovery of the microorganisms that caused the disease.

Because groundwater tends towards chemical equilibrium with the minerals it comes into contact with, some groundwater supplies are naturally contaminated. Deep wells can have high concentrations of minerals, sulfates, and chlorides. A number of chemicals are commonly found in ground water, including calcium, magnesium, sodium, potassium, iron, bicarbonate, sulphate, chloride, nitrate, fluoride, and silica. Also, salt water intrusion can be a problem in coastal aquifers. Anaerobic decomposition of buried organic matter can occur, thus contaminating groundwater with gases such as ammonia and methane.

Nevertheless, groundwater found near the land surface tends to be of high quality, as natural cleansing and filtration occurs while the water moves vertically and horizontally towards the zone of saturation within the aquifer. Soils and soil microbes naturally break down contaminants. Groundwater contamination occurs from organic and inorganic chemicals. Any chemicals that are easily soluble and penetrate the soil are prime candidates for groundwater pollutants. Groundwater pollution emanates from a variety of sources:

1. Agricultural runoff of pesticides, herbicides, and salts percolate into groundwater supplies;
2. Accidental spills and leaks;
3. Salts leach into the soil from road de-icing;
4. Toxic organic chemicals are dumped into industrial landfills, pits and lagoons;
5. Toxic chemicals dumped into municipal landfills have leached into drinking water supplies. According to a 1984 EPA report, land disposal of hazardous waste in 1981 included 59% by deep well injection, 35% by surface impoundment, 5% by landfill, and 1% land treatment (Brown and Daniel, 1985). The EPA estimates more than 75% of 75,000 sanitary landfills are polluting ground waters with leachate;
6. In some densely populated residential areas, groundwater contamination problems has occurred from the toxic chemicals used to clean septic tanks;
7. Illegal dumping of toxic wastes, also referred to as "midnight dumping" (Travis and Eltnier 1984, 50). Wastes have been illegally dumped (sometimes into urban sewer systems) to avoid the high cost of proper disposal. Some toxic waste generators may unknowingly hire firms posing as hazardous waste disposers. Instead of properly disposing of the dangerous substances, they drive tankers around back roads at night releasing toxics on the roadway;
8. Leaching of pollutants occurs from underground petroleum storage tanks.[1]

Means of Contamination

Whatever the specific source, the primary means of contamination are through infiltration, direct migration, interaquifer exchange, and recharge from contaminated surface water (Barcelona 1988, 2-4). Infiltration is the most common mechanism of groundwater contamination, occurring as contaminated water moves slowly through the spaces of the aquifer soil matrix. Moving downward under gravity, it dissolves materials with which it comes into contact. Leachate is formed when the water dissolves contaminants as it percolates downward through a contaminated zone. This leachate continues to migrate downward until the saturated zone is reached and then spreads in the direction of groundwater flow.

Contamination also occurs in an aquifer when pollution migrates directly into the groundwater from below-ground sources (e.g., storage tanks, pipelines) which lie within the saturated zone. Direct contact of contaminants with ground water may also come from storage sites and landfills excavated to a depth near the water table. Interaquifer exchange occurs when contaminated ground water mixes with uncontaminated ground water when one water-bearing unit is hydraulically "communicating" with another, such as when a well penetrates more than one water-bearing formation to provide increased yield. Finally, contamination can result from surface water recharge. Under normal condition, groundwater moves toward surface water bodies. Sometimes, the hydraulic gradient is such that a reversal of flow occurs (such as when a water body is in flood stage) and surface water contaminants can then enter the groundwater.

How Big a Problem?

The primary threat of groundwater contamination is to drinking supplies, especially to rural America, which relies almost exclusively on groundwater. Additional important threats are related to the presence of synthetic chemicals and saline water intrusion. Predicting the spatial extent of groundwater contamination is difficult, however, because of a number of factors that influence the extent of contamination: the location of disposal, the pattern of water development within the aquifer, the hydrological characteristics of the aquifer, the behavior of each contaminant in the soil, rock and water environment, the impact of the gaseous aspect of the unsaturated zone, the density of the contaminant compared to that of water, and the interaction between contaminants. Moreover, many contaminants are odorless or tasteless and require sophisticated sampling technology to detect. The slow movement of groundwater favors chemical and biochemical reactions approaching equilibrium.

As of 1984, the presence of over 200 chemical substances in groundwater had been documented and well over 50,000 synthetic chemicals were in use in the United States. Chlorinated compounds found in drinking water include TCE, PCE, and vinyl chloride (Nyer 1993, 136). As we previously noted, however, many of these materials occur naturally, especially minerals dissolved from geologic materials in contact with water. Water quality is not a clearly defined concept and varies considerably depending on use and even local habits and circumstances. Water quality can be evaluated in numerous dimensions, including temperature, chemical species present, biological components, radioactive substances, color, taste, odor, and so on.

Although percolating water may be cleansed by filtration and adsorption, once water reaches the saturated water table, little further cleansing takes place. Depending on the hydraulic gradient and the permeability of the aquifer, water movement may be very slow. Combined with the abiotic conditions of the saturated water table, contaminated sites may naturally remain so for centuries. On the positive side, a plume of contaminants may move so slowly that the water a short distance away may remain unaffected.

In stagnant groundwater, solutes slowly disperse and affect a greater although less concentrated volume. When groundwater flows, solutes also move with it, but they do not necessarily move at the same speed and in the same direction as the water. Ultimately, analysis of groundwater pollution involves a multitude of tasks. These include determination of the source of contamination, identification of the contaminants, their likely movement through the aquifer, predicted concentrations throughout the aquifer, the potential hazard posed by these concentrations, and the viability and cost of remediation (Hosseinipour 1991).

Groundwater Tracers

Because of the "out of sight, out of mind" nature of groundwater aquifers, one key to designing mechanisms that will encourage decisionmakers to act in a way that is environmentally responsible and economically efficient is to identify the process by which a given dose of contaminant impacts groundwater quality. The use of groundwater tracers represents a means to identifying potential pollution problems and, where appropriate, perhaps economically rational alternatives to remediation.

"[A] tracer is matter or energy carried by ground water which will give information concerning the direction of movement and/or velocity of the water and potential contaminants which might be transported by the water" (Davis et al. 1985, 1). A tracer can be natural (for instance, heat carried by hot-spring waters), accidentally introduced (such as from a fuel oil spill), or introduced intentionally. Tracers are introduced into groundwater to simulate a contaminant, with concentrations analyzed to provide data on how a contaminant plume migrates under heterogeneous conditions (Waldrop 1991).

Typically, a series of test wells are used to measure the flow of the tracer through the aquifer, including speed of transmission, direction, and concentration. Hydraulic connection was studied in Europe in the 1860s using dyes and salts. Fluorescent dyes experiments were performed in the early 1900s in France. In the 1950s, radioactive tracers were developed, allowing very precise and selective tracer measurement. However, their use as been curtailed for public health reasons. In the 1960s, naturally-occurring radioisotopes and stable isotopes came into use. In the last two decades, extremely sensitive tracers have been developed such as fluorinated organic acids and halocarbons.

Ideally, one would use a nontoxic tracer that travels with the same velocity and direction as the water and does not interact with solid material. But no ideal tracer exists. Because of the complexities of groundwater aquifers, it is appropriate to use different types of matter for various tracing purposes. Indeed, possible groundwater tracers number in the thousands. In addition to naturally-occurring tracers, researchers have utilized solid matter, ions, dyes, isotopes, gases, and radionuclides.

Water temperature is a potentially useful natural tracer because water temperature changes slowly as water migrates subsurface. Temperature changes can help detect recharge from surface sources such as rivers. Solid material in suspension can be used as a tracer in areas where water flows in large conduits, such as limestone or dolomite aquifers. Solid particle tracers include paper, yeast, bacteria, viruses, and spores. Bacteria are the most commonly used microbial tracers, because of their ease of growth and detection.

Ionic compounds such as common salts have been used extensively as groundwater tracers. Ionic tracers have been employed for a wide range of hydrologic problems concerning the determination of flow paths and residence time and the measurement of aquifer properties. In most situations anions, negatively charged ions, are not affected by the aquifer medium and do not decompose. Ionic tracers are monitored by changes in electrical conductance in the aquifer. Tracer movement from the injection well to the observation well is observed by an electric circuit that utilizes the conductivity of the groundwater. Organic dyes have been used for surface water and ground water tracing since the late 1800s. Dyes, such as fluorescent dyes, are inexpensive, simple to use, and effective.

4. Treatment of Groundwater Pollution

The basic approach for addressing groundwater pollution problems involves one of three methods (Hosseninipour 1991; Nolan and Boardman 1990). The first method seeks to contain the plume of pollutants using natural or human-made barriers to stop seepage through the aquifer. The second is to extract the polluted water, treat it, and return the treated water back into the aquifer to dilute the remaining pollutants. The third method involves in-situ (on-site) treatment involving chemical or biological processes. Application of these methods has been limited because of the expense and uncertainties about effectiveness. Instead, groundwater modeling has been used to assist in choosing remediation procedures.

Relevant parameters for determining appropriate containment and treatment include the initial contaminant concentration, desired effluent quality, treatment duration, and groundwater flow rate to the treatment system. Length of treatment is related to groundwater recovery rates and the number of pore volumes required to flush contaminants adequately. The last issue is affected by the hydraulic conductivity of the aquifer, the decay rate of the contaminant, and the adsorbability of the contaminant (the solubility of the contaminant and whether it strongly adsorbs to the organic matter in the aquifer). The latter reason explains why pump and treat strategies are not very effective for petroleum-hydrocarbon contamination.

Physical Containment Techniques[2]

Impermeable Barriers. Barriers used to contain, capture or redirect groundwater flow include slurry walls, grout curtains, sheet piling, and hydraulic barriers. Slurry walls are an effective, inexpensive way to control groundwater flow in unconsolidated earth material. The walls are usually made of a bentonite-soil mixture, but concrete may be used. Slurry walls must generally be sunk into the confining layer of the aquifer to be effective. They are placed upgradient of contaminated areas so that groundwater flows around the area, downgradient to catch groundwater after it has flowed through the contaminated area, and circumferential to wastes so that contaminated groundwater are trapped while uncontaminated waters migrate around the walls.

A grout curtain is a somewhat less effective barrier consisting of particulates (such as cement) or chemicals that are injected through wells into soil mass or rock to reduce water flow and strengthen the geologic formation. The primary use of grouting is to seal voids in rocks.

Sheet piles are relatively primitive barriers consisting of wood, pre-cast concrete, or steel. Although relatively inexpensive, sheet piles are not very reliable. Wall integrity is difficult to predict as the underlying rock formation may damage the sheet pile when installed in the aquifer.

Hydraulic barriers are potentially effective if sufficient information exists about the hydrology and geology of an aquifer. The principle of this approach is to modify the hydraulic gradients around contaminated groundwater through depressions and recharge of the water table. Through pumping and recharge strategies, including the use of inter-ceptor ditches and withdrawal wells, the movement and size of a plume of contaminants may be controlled, eventually leading to removal and cleanup (Zheng et al.). Hydraulic barriers are most effective in high permeability soils and rock formations, which more effectively allow the contaminated groundwaters to be pumped and treated.

Surface Water and Leachate Controls

The most common controls in this category are diversion and collec-tion systems, grading, capping, and revegetation. Diversion and collec-tion systems are used primarily to control surface water flows at waste disposal sites. Dikes, berms, terraces, channels, and waterways are used to prevent or intercept run-on and runoff. Sedimentation basins and ponds are used to reduce the quantity of suspended matter in surface runoff. Storage or discharge of water is accomplished through seepage basins and drainage ditches. Treated groundwater is discharged back into the aquifer. These systems work best in areas where the groundwa-ter table is shallow and the soils are highly permeable.

Grading involves reshaping of the land surface to alter the flow of runoff, rate of erosion, and infiltration. Capping is used to prevent rain-water from infiltrating into groundwater, perhaps to prevent contami-nants from being further leached into the aquifer. Caps may be multi-layer or single layer structures, requiring a layer of low permeability, and perhaps top layers of vegetation on topsoil and a layer of sand for drainage. The key lower layer is often comprised of a synthetic liner combined with low-permeability soil. In some instance, caps are made of concrete or asphalt. Revegetation is often used in conjunction with grading or capping to provide stability to the surface, decreasing erosion, and to enhance aesthetics.

Subsurface Drains. If groundwater contamination is limited to the shallow part of an aquifer, subsurface drains may be used to reduce the size of a contaminant plume or to lower the water table to prevent contact with buried wastes. Conduits are buried to collect and transport leachate.

Groundwater Treatment Technologies

Groundwater treatment technologies consist of physical, chemical, and biological processes.

Physical Treatment. Physical treatment of groundwater typically utilizes the volatile, adsorptive, and insoluble properties of the contaminant. These proven technologies are applicable to both recovered groundwater and to wastewater.

Phase-generated hydrocarbon recovery involves extraction and separation of contaminants originating from leaking underground storage tanks and associated transfer operations. This technique is appropriate for contaminants that are lighter than water but does not treat residual material that adsorb to aquifer solids. Extraction may involve the use of dual pumps, one to create a cone of depression in the water table and the other to collect the free hydrocarbons from within the cone. In addition, oil-water separators are used to recover additional free product from recovered groundwater.

Air stripping is one of the most widely used treatments for removal of volatile contaminants from groundwater. This technique is used for treating TCE, benzene, toluene, and other aromatic hydrocarbons. Contaminated groundwater is pumped into the top of a treatment tower while air is introduced into the bottom. This method allows for the mass transfer of contaminant from the liquid phase to the vapor phase in the center part of the tower which is the aeration basin.

Adsorption is a natural process in which molecules of a liquid or gas are attracted to and then held at the surface of a solid (Nyer 1993, 127). Adsorption onto activated carbon is a physical process whereby the attraction is caused by the surface tension of the carbon. The organic compounds are removed from the water and are transferred to the surface of the carbon. The molecules are not changed or destroyed, and must be properly disposed of. Therefore, as with any treatment technique, the ultimate disposal of any remaining contaminant must be accounted for.

Activated carbon adsorption is used to remove high weight, low-solubility constituents from groundwater, such as phenols and naphthalene. Basically, this system involves feeding the contaminated water into vessels of activated carbon, leaving spent carbon and treated effluent. This method is considered to be relatively expensive, but allows for the removal of a greater level of contaminants than, say, by air stripping alone (Nolan and Boardman 1990, 34).

Filtration is used to remove solids from groundwater, often in conjunction with other technologies. A typical application removes the suspended solids before treatment with carbon adsorption to prevent fouling of the carbon bed. Typically, sand and anthracite beds are used to filter the contaminated groundwater.

Chemical Treatment. Chemical treatment of groundwater, an alternative pump-and-treat approach to remediation, includes coagulation-precipitation, reduction/oxidation, and neutralization. Coagulation-precipitation is used in the treatment of wastewater. Coagulation is promoted by the addition of chemicals that allow the particles in the water to more readily act with one another. The soluble agents are transformed into insoluble materials by the process of precipitation.

Reduction/oxidation (redox) reactions are those in which the oxidation state of one reactant is increased while that of another is decreased. For example, chemical oxidation can make a compound more amenable to biological treatment. Chemical reduction reduces the oxidation state of a substance. Reduction is used to treat lead and mercury in water.

Neutralization involves adjusting the pH of water. It is used to control chemical reactions, pretreat water prior to biological treatment, control the rates of chemical reactions, and other purposes. Compounds used to neutralize waters include sulfuric acid, hydrochloric acid, sodium hydroxide, and lime.

Biological Treatment. Biological treatment, to achieve the microbial degradation of organics in wastes, is routinely used to remove organics from municipal and industrial wastewaters, and is applicable to groundwaters containing moderate levels of organics. Moreover, biodegradation may allow for the in-situ treatment of contaminated materials.

Most commonly, treatment takes place in an aerobic environment. Conventional biological treatment is with an activated sludge process, yielding low effluent concentrations. The typical treatment combines microorganisms, oxygen, and the organic contaminants to form carbon dioxide, water, and biomass (in the form of sludge). Toxic organics amenable to treatment include aromatic hydrocarbons, phenols, and naphthalene. Contaminants less amenable to treatment include TCE, chloroform, and carbon tetrachloride. Conventional biological treatment is typically less expensive than physical and chemical treatments.

An alternative treatment is in-situ biodegradation, which involves modification of the subsurface environment to promote microbial growth and biodegradation of organic contaminants. Oxygen (often in the form of hydrogen peroxide) and nutrients are introduced in the subsurface either through surface infiltration or injection wells to stimulate indigenous microorganisms. The effectiveness of in-situ biodegradation is affected by many factors, including the hydrologic characteristics of the aquifer, the type and distribution of contaminants in the aquifer, the nutrient concentration of the groundwater, and so on.

Harker (1991) discusses the possibility of using microorganisms to treat hazardous waste. He notes that care must be taken to supply appropriate amounts of oxygen and nutrient material for the organism to survive and break down the contaminant. He believes the use of microorganisms is likely to be most successful in a pump and treat environment that can be altered to allow the microorganism to flourish.

As a practical matter, one has to consider the many factors that can affect the efficacy of any particular treatment system. For example, there may be material other than the contaminant in the groundwater that can affect the treatment system that is installed.

Examples of Containment and Remediation

Hydraulic Barriers. Zheng et al. (1989) describe a study performed in the vicinity of a chemical plant in the Northeast. Field investigations indicated that groundwater beneath the plant site had been contaminated by wastes released from manufacturing processes and that a plume of contaminated groundwater was migrating towards a river. A network of injection and extraction wells was proposed to flush contaminations out of the aquifer and prevent them from discharging into the river. The researchers were able to determine that the "dividing surface" would effectively contain the contaminant sources.

Pump and Treat. Jackson et al. (1989) studied an aquifer near Ottawa, Canada that was polluted with toxic organic chemicals from the disposal of laboratory solvents in shallow trenches of a landfill immediately above the aquifer. Alternative remediation options were considered. Impermeable barrier walls were rejected as unsuitable, given the permeable nature of the underlying bedrock. Much of the plume can be removed hydraulically over a period of time by the operation of purge wells pumping to an on-site treatment plant from which the purified water is returned to the aquifer by recharge wells, thereby diluting remaining contaminants. A final phase of decontamination might involve in-situ biorestoration.

One proposed cleanup solution was to install an impermeable slurry wall or grout curtain around the zone of highest contamination. This method is effective if there exists an impermeable flow beneath the contamination zone. In this case, the limestone bedrock underlying the plume is very permeable and hydraulically connected to the overlying aquifer. A test of the pump and treat approach was conducted by injecting uncontaminated groundwater and two non-reactive tracers into the contaminated aquifer via an injection well and the withdrawal of a similar amount of water from another well five meters away. This allowed a determination to be made of the rate that contaminants within the aquifer would become sufficiently diluted to achieve targeted cleanup goals.

Bioremediation of Styrene Spill. Kuhlmeier (1989) studied a tanker truck spill in Ohio that shows how bacteria can be used for site restoration. Within the first ten weeks of operation, a 90 percent reduction of the styrene spilled was accomplished. Injection and recovery wells were driven into the subsurface. The recovery wells controlled the ground water flow within the aquifer. Acclimated sludge was mixed with hydrogen peroxide (for added oxygen) and injected into the aquifer. Applications of bacteria throughout the remainder of the cleanup were based on lab analysis of soil microbe concentrations. The styrene was completely oxidized to carbon dioxide and water. Kuhlmeier believes that in-situ treatment of hazardous waste spills, leaking underground tanks, and landfills has been shown to be a technically viable and a cost effective method of remediation.

Landfills and the Protection of Groundwater Quality. Landfills pose a particular concern in the public's mind, because of the extensive use of landfills for disposal of wastes and potential threats to public drinking supplies. Lee and Jones (1991) argue that current landfill containment techniques, including the use of membrane liners and soil-clay liners in "dry tomb" landfills will ultimately not keep moisture out of the landfill. At best, these systems postpone the pollution of groundwater from leachate, as even the best system fails. Once leachate has penetrated the uppermost liner, the waste must be removed to prevent groundwater pollution.

Possible approaches to cleanup include more expensive techniques which will not impose these pollution costs or to dispose of waste in locations where groundwater pollution is not possible. With that in mind, some have proposed establishing a solid waste management site in South Dakota. As long as those generating the waste are not internalizing all costs, these other options will fail as they are much more expensive than current practices.

One suggested alternative is to treat municipal solid waste to produce nonpolluting residues that may be safely disposed of by land burial. This would involve procedures such as incineration, or biochemical fermentation, a process that chemically leaches wastes, leaving a stable residue. Lee and Jones advocate stabilizing fermentable organics in a landfill. First, potentially hazardous components of the wastes, such as lead, cadmium, and other heavy metals would be removed. Thereafter, the decomposable components of shredded solid waste would be fermented, accompanied by a practice of leachate recycling. Leaching the nonfermentable residues will reduce long-term threats to the underlying groundwater. Eventual treatment could render the leachate of the fermentable residues virtually harmless. Moreover, recoverable (though not necessarily economically viable) methane gas would be produced during this process.

5. Protecting Groundwater: An Economic Analysis

Groundwater is complex. It differs from most other resources because time often has little effect on the level of contamination because of the slow movement of water within most aquifers. Any economically-rational approach to protecting groundwater needs to first consider the likely relationship between emissions and environmental damage, which can be complicated by the hydrological and geological factors discussed earlier. To establish property rights in groundwater quality, and thus to make possible a market approach to groundwater policy, the hurdle of defining the source of emissions needs to be overcome. While some have suggested branding chemicals so that the source of emissions can be identified precisely (Stroup 1989, 873), if such technology is not on the horizon, efforts to better understand the dynamics of particular aquifers will help us to at least identify with greater accuracy the location of the disposal that caused the damages.

Establishing property rights to water quality provides an incentive for decisionmakers to consider the effects of their actions upon others and to provide for an efficient amount of water quality, not a zero level of pollution. A possible protocol would be to establish concentration limits for various pollutants with those responsible for exceeding these concentrations being held liable for damages. Also, the interactive effects between toxics needs to be identified, since a number of toxic substances may be present together (particularly in landfills).

Since the relevant physical damage measure is the concentration of contaminants at the receptor site, rather than emissions from the polluting source, aquifer identification may help identify the maximum level of source pollution that will not create harm to potential parties. Moreover, as the ability to code toxic chemicals improves over time, it will ease the identification of the actual polluter. In the meantime, groundwater tracing will help expand knowledge of aquifer characteristics and the dynamic effects associated with the emission of groundwater pollutants.

It is important to consider the costs of improved definition of property rights to groundwater quality. Information is costly, so the benefits of more complete aquifer modeling should be compared with relevant marginal modeling costs.

It would be in polluters' interest to be informed about the hydrological characteristics of the aquifer, so they would know the likely impact of their actions on others (assuming they are held responsible for their actions). For example, a potential polluter might find it in his interest to install barriers around the plume of pollution, arrange to treat the water at the well head for affected basin residents, pretreat the contaminants prior to disposal to lower its toxicity, compensate affected parties to shut down their wells, arrange to dispose of pollutants at locations over the aquifer that will create less harm (or off-site completely), and so on. One can envision potentially affected parties agreeing to compensation, the provision of alternative supplies by polluting parties, or both, if that represents a cheaper alternative than maintaining water quality at their wells. In times of drought, when recharge rates are reduced and pollutant concentrations are likely to increase, polluters may pay to artificially recharge specific areas of the aquifer to reduce those concentrations to legal limits.

Midnight dumping could be treated as a criminal activity, analogous to theft.[3] Illegal dumpers are stealing environmental quality from others.

With respect to disposal of wastes in landfills, current practices postpone groundwater pollution (by delaying when the underlying groundwater is affected) so that future generations pay the costs of cheaper garbage disposal today, including compromising public health, loss of water resources, and the costs of remediating polluted groundwaters and aquifers (Lee and Jones 1991). Policies could be implemented to make the landfill operators responsible for any contamination. This would cause dumping charges to reflect more accurately the actual marginal costs of disposal and treatment.

It might be economically desirable to establish a toxic waste facility in a groundwater polluting-resistant locale (such as South Dakota [Lee and Jones 1991]), although it is important to account for all relevant costs, including the expense (and likely increase in air pollution) of transporting wastes to the disposal site, the increased likelihood of an accidental spill given the increase in transport miles, and so on.

A key to designing an efficient institutional response to groundwater pollution is to focus on the ultimate goal of guaranteeing the availability of groundwater at a given quality level. By focusing decisionmaking on participants, the development of techniques to achieve this minimum quality level at the lowest possible cost will be encouraged. An ideal market-oriented system for controlling groundwater pollution might incorporate the following:

1. Identify maximum allowable water pollutant concentration levels, for both individual toxics and multiple emissions. Associated emission limits could be modified over time as hydrological conditions change (due to drought or excessive recharge, for example).
2. Identify individual emission sources (if feasible) or emission locations. A well monitoring program that periodically tests wells for pollutant concentrations could be used to effectively monitor pollutant flows.
3. Identify likely damage locations and the rate of potential pollution exchange between portions of the aquifer.

4. Identify physical and economic damages at each location.
5. Establish tradeable rights to maximum pollutant concentrations. These could be granted de facto by making polluters subject to appropriate standards of liability, which could be "traded" to the polluter (or some other party) by potentially affected parties waiving liability claims for excessive pollution, either permanently or for a given time period.[4]
6. Establish criminal liability for illegal dumping.

A possible approach for implementing a property rights scheme is to establish a pollution bidding district that auctions rights to pollute measured by receptor concentrations, not emissions. Rights could be geographically limited such that pollution limits are not exceeded anywhere within the aquifer. If this is considered to be too restrictive, pollution in excess of allowable limits would be permitted when affected parties waive liability, as previously suggested. Rights could be auctioned on an appropriate periodic basis (one year, five years, etc.) with proceeds used to defray monitoring costs, provide excess coverage for clean-up, and so on. As aquifer conditions change, the quantity of rights auctioned could be modified accordingly.

Potential Problems to Overcome

For a complex resource such as groundwater, defining complete property rights to environmental quality for all relevant dimensions of time and space may be difficult to accomplish. Potential problems include the following:

1. The cost of aquifer identification to establish all relevant water flows. Someone will have to bear these costs and the monitoring costs. The cost of identification must justify the potential gains from doing so.
2. Identifying relevant interactive effects from disposal of different types of pollutants. This may be particularly critical in establishing liability amongst multiple emitters. For example, suppose party A dumps a pollutant that is within concentration limits at any relevant receptor well in the aquifer. Meanwhile, party B dumps a different type of pollutant that is also within concentration limits at the receptor wells. The interaction between the two pollutants, however, creates a third pollutant that exceeds concentration limits. Who is liable for damages? A priority scheme could be established based on seniority (or some other criteria) so that the polluter whose dumping results in the excessive interactive pollution be held liable, even though their own emissions may be far lower than other polluters. This last problem may be more apparent than real, as the polluter with less senior dump-

ing rights could pay the senior polluter to limit their emissions to such a level that the interactive effects do not cause pollution concentration limits to be exceeded. Nevertheless, it must be decided if it is appropriate to bestow on past polluters the initial rights to pollute.
3. All parties could be required to reveal the substances they are emitting into an aquifer. Information on toxic releases could help other parties determine whether their own emissions could create toxic interactive effects.
4. Polluting parties could be required to possess insurance (or alternatively to post a bond) sufficient to pay for potential pollution damages.

These complex questions typically have no easy, cheap, or uniform answers; yet effective and cost-efficient policy institutions must consider them. Policies that ignore any of these key questions are unlikely to minimize the costs of dealing with pollution. However it is also true that gathering detailed answers may cost more than the answers are worth. When an aquifer is both highly valuable and vulnerable to releases of pollution from the ground above, and the relevant information about the aquifer and the geology is also very costly relative to the likely benefits of detailed planning or of market trades in rights to water quality, then a relatively simple rule might serve well. Such a rule might be of this form: "Certain wastes must not be stored in the regulated area unless they are contained and protected by specified precautionary construction measures." The same sort of rule might be overly protective, at excessive cost, if the aquifer is less valuable, or if some or all of it is protected by natural geological barriers against potential pollution.

Conclusion

We conclude by noting again that groundwater is important, and increasingly so as demands for it increase. Some groundwater aquifers are more valuable and important than others, given their location, quality, and accessibility. Some are better protected, either by geology or by effective precautions. When a valuable aquifer becomes polluted by toxic chemicals, the cost of cleanup or of lost uses will often generate large financial liabilities that reflect these costs. When property rights in groundwater quality are fully specified and protected, damages and damage risks will be properly managed to minimize the sum of all costs. The information requirements for this, or any other efficient management system can be costly, however, so that in some cases simpler, less sophisticated regulatory schemes may be preferred. As developing technologies and better science reduce the cost of knowledge, we can hope that the benefits of information and incentives produced by a system of property rights and market trades might become attainable.

Notes

1. Recent examples include an underground storage tank at Delta Airlines in Atlanta found leaking a petroleum-based solvent, Varsol (*Atlanta Constitution*, August 10, 1991), and one million gallons of jet fuel found in the ground at Williams Air Force Base in Mesa, Arizona, threatening Phoenix-area water supplies (*USA Today*, July 30, 1991).
2. The following sections on groundwater treatment are based on the excellent survey presented by Nolan and Boardman (1990).
3. This has been proposed by Stroup (1989, 873).
4. For a discussion of liability and negligence rules for regulating pesticide contamination of groundwater, see Wetzstein and Centner (1992).

References

Abriola, L., ed. 1989. *Groundwater Contamination*. IAHS Publication No. 185. Walingford, U.K., International Association of Hydrological Sciences.

Anderson, T., and D. Leal. 1988. "Going With The Flow: Expanding The Water Markets." *Cato Institute Policy Analysis*, No. 104 (April 26).

Barcelona, M., J. Keely, W. Pettyjohn and A. Wehrmann. 1988. *Handbook of Groundwater Protection*. New York: Hemisphere.

Bradbury, K. 1991. "Tritium as an Indicator of Ground-Water Age in Central Wisconsin." *Ground Water* 29, 3 (May-June): 398-404.

Brown, K., and D. Daniel. 1985. "Potential Groundwater Implications of Land Disposal of Toxic Substances." In *Issues in Groundwater Management*, edited by E. Smerdon and W. Jordan. Austin: Center for Research in Water Resources, University of Texas.

Bruggink, T. 1992a. "Third Party Effects of Groundwater Law in the United States: Private Versus Common Property." *American Journal of Economics and Sociology* 51, 1 (January): 1-17.

———. 1992b. "Privatization Versus Groundwater Central Management: Public Policy Choices in the 1990s." *American Journal of Economics and Sociology* 51, 2 (April): 205-22.

CH2M Hill. 1990a. *Public Review Draft Basinwide Technical Plan Report: Volume One - San Gabriel Basin, Los Angeles, California*. EPA Contract No. 68-01-7251 (April 17).

———. 1990b. *Public Review Draft Basinwide Technical Plan Report: Volume Two - San Gabriel Basin, Los Angeles, California*. EPA Contract No. 68-01-7251 (April 17).

Council of Environmental Quality. 1992. *Environmental Quality: 22nd Annual Report* (March).

———. 1993. *Environmental Quality: 23rd Annual Report* (January).

Cross, B. 1993. "Groundwater Safety is a Public Charge." *Environmental Protection*, 4, 3 (March): 44-47.

Custodio, E. 1989. "The Role of Groundwater Quality in the Decision-Making Process for Water Resources." In *Groundwater Management: Quantity and Quality*, edited by A. Sahuquillo, J. Andreu, and T. O'Donnell. IAHS Publication No. 188. Walingford, U.K.: International Association of Hydrological Sciences.

Davis, S., D. Campbell, H. Bentley, and T. Flynn. 1985. *Ground Water Tracers*. Dublin, Ohio :National Water Well Association.

Feldman, D. 1991. *Water Resources Management: In Search of an Environmental Ethic*. Baltimore, MD: Johns Hopkins University.

Fractor, D. 1982. A Property Rights Approach to Groundwater Management, Ph.D. diss. University of Oregon.

Harker, A. 1991. "The Potential Use of Genetically Engineered Microorganisms in the Remediation of Environmental Pollution." In *Ground Water: Proceedings of the International Symposium*, edited by G. Lennon. New York: American Society of Civil Engineers.

Hosseinipour, E. 1991. "Groundwater Models and Aquifer Restoration: An Overview." In *Ground Water: Proceedings of the International Symposium*, edited by G. Lennon.

Houghton, M. 1993. "EPA-State Programs Protect Groundwater." *Environmental Protection*, 4, 1 (January): 24-32.

Jackson, R. et al. 1989. "Aquifer Contamination and Restoration at the Gloucester Landfill, Ontario, Canada." In *Groundwater Contamination*, edited by L. Abriola. IAHS Publication No. 185. Walingford, U.K. : International Association of Hydrological Sciences.

Kamler, K. 1993. "Natural Restoration." *Environmental Protection* 4, 2 (February): 36-42.

Kuhlmeier, P. 1989. "Enhanced Microbial Degradation of Styrene in Shallow Soils and Ground Water." In *Groundwater Contamination*, edited by L. Abriola.

Lee, G., and R. Jones. 1991. "Landfills and Ground-Water Quality." *Ground Water* 29, 4 (July-August): 482-86.

Lennon, G., ed. 1991. *Ground Water: Proceedings of the International Symposium*. New York: American Society of Civil Engineers.

McTernan W., and E. Kaplan, eds. 1990. *Risk Assessment For Groundwater Pollution Control*. New York: American Society of Civil Engineers.

Nolan, B. III, and G. Boardman. 1990. "Aquifer Restoration Techniques - A Status Report." In *Risk Assessment For Groundwater Pollution Control*, edited by W. McTernan and E. Kaplan. New York: American Society of Civil Engineers.

Nyer, N. 1993. *Practical Techniques for Groundwater and Soil Remediation*. Boca Rotan, Florida: Lewis.

———. 1992. *Groundwater Treatment Technology.* 2nd ed. New York: Van Nostrand Reinhold.

Ostrom, E. 1990. *Governing the Commons: The Evolution of Institutions for Collective Action.* Cambridge: Cambridge University Press.

Portney, P., ed. 1990. *Public Policies for Environmental Protection.* Washington, DC: Resources for the Future.

Sahuquillo, A., J. Andreu, and T. O'Donnell, eds. 1989. *Groundwater Management: Quantity and Quality.* IAHS Publication No. 188. Walingford, U.K.: International Association of Hydrological Sciences.

Smerdon, E., and W. Jordan, eds. 1985. *Issues in Groundwater Management.* Austin: Center for Research in Water Resources, University of Texas.

Stephenson, F. 1992. "Chemical Oxidizers Treat Wastewater." *Environmental Protection* 3, 10 (December): 23-27.

Stroup, R. 1989. "Hazardous Waste Policy: A Property Rights Perspective." *Environment Reporter* 20, 21 (September 22): 868-73.

Travis, C., and E. Eltnier, eds. 1984. *Groundwater Pollution: Environmental and Legal Problems.* Boulder, CO: Westview.

Uhlman, K. 1992. "Groundwater Dating Locates VOC Source." *Environmental Protection* 3, 9 (November): 56-62.

U.S. Advisory Commission on Intergovernmental Relations. 1991. *Coordinating Water Resources in the Federal System: The Groundwater-Surface Water Connection.* A-118 (October). Washington, D.C.

U.S. Department of Interior. 1985. *Ground Water Manual.* Denver, CO: Government Printing Office.

U.S. Environmental Protection Agency. 1989. *The Toxics-Release Inventory: Executive Summary.* EPA 560/4-89-006. Washington, D.C.

Waldrop, W. 1991. "Overview and Objectives of the Macrodispersion Experiment." In *Ground Water: Proceedings of the International Symposium,* edited by G. Lennon.

Wetzstein, M., and T. Centner. 1992. "Regulating Agricultural Contamination of Groundwater Through Strict Liability and Negligence Legislation." *Journal of Environmental Economics and Management* 22: 1-11.

World Resources Institute. 1993. *The 1993 Information Please Environmental Almanac.* Boston, MA: Houghton Mifflin.

Zheng, C. et al. 1989. "Effectiveness of Hydraulic Methods for Controlling Groundwater Contamination." In *Groundwater Contamination,* edited by L. Abriola.

11

Improving the Reliability of Scientific Testimony in Court

David E. Bernstein, Kenneth R. Foster and Peter W. Huber

Introduction: Shopping for Experts

The scientific and legal issues raised when science enters the courtroom have been discussed many times before, from varying perspectives.[1] We focus here on assessing the reliability of scientific testimony, which is a core issue in toxic tort suits, as well as many other kinds of litigation that rely on scientific evidence.

Science and the law handle technical evidence in fundamentally different ways. The fabled reliability of science arises in the long term, as scientists question scientific theories, confirm and extend novel experimental results, and incorporate new findings into the body of accepted scientific knowledge. This process, which one author has called a "knowledge filter,"[2] is a collective phenomenon that transcends the contributions of individual scientists.

The American legal system, which is based on an adversarial process, handles scientific data in a way that is the very opposite to that of the "knowledge filter." Lawyers are allowed (indeed, encouraged as a way to carry out their professional obligations to be effective advocates) to construct theories to explain "what happened" and to spin facts to further their clients' cases. This process extends to scientific controversies, with opposing lawyers choosing experts who will present views favorable to their cases.

● Moreover, while attorneys are stuck with the testimonial limitations of available fact witnesses, experts with a range of opinions and abilities can be found on almost any issue. Thus, for example, there might exist only a few eyewitnesses to an automobile accident, but a lawyer can choose among an almost unlimited pool of experts to testify about the nature and extent of the victim's injuries. This gives an attorney almost unlimited opportunity to "shop" for an expert with a pleasing courtroom manner who will testify in support of the attorney's theory of the case.[3]

At its worst, the high fees (in some cases in excess of $1,000 per hour) create incentives for venal experts to shape their testimony to benefit the lawyer who hires them.[4] Even absent venality, however wide the latitude given to a lawyer's choice of expert witness, distorts the truth-seeking process. As Judge Jack Weinstein has noted, "[a]n expert can be found to testify to the truth of almost any factual theory, no matter how frivolous." [5]

It is not surprising, therefore, that much proffered expert testimony is problematic. Problems include:

● Medical diagnoses based on questionable tests. Clinical ecology, for example, is a fringe medical specialty that purports to treat immune-system problems created by trace environmental pollutants. Some clinical ecologists have been prominent in litigation, testifying about the cause of plaintiffs' illnesses, on the basis of immune-system tests that mainstream medical organizations find unpersuasive for diagnosis of disease. [6]

● Erroneous and slanted review of scientific data. For example, one witness who has been prominent in Bendectin litigation (alleging birth defects caused by the morning sickness drug) was criticized by a judge for his slanted and inaccurate review of the scientific literature. The witness argued that use of the drug was associated with birth defects—despite a consensus of scientific expert groups that no such connection has been shown in the literature. [7] A similar scenario has occurred in numerous silicone breast implant trials. [8]

● Legally irrelevant testimony couched in impressive-sounding scientific jargon. For example, one frequent witness (an epidemiologist) for plaintiffs in Bendectin litigation testified that the drug had not been proven safe to her satisfaction; in effect, the burden of proof has shifted the plaintiff to prove causation of injury to the defendant to disprove causation. [9]

Some critics argue that calls for "good science" in the courtroom are really surrogates for a legal position that would raise the standard of proof, to the (presumably very high) levels that the scientific community would require, from the (presumably lower) standards of legal proof. But the problems listed above have nothing to do with scientific excellence (as a scientist would regard it), but rather with deficiencies in expert testimony for use in legal proceedings. A witness who presents naked opinion couched in scientific-sounding jargon is not engaged in bad science (in fact, is not engaged in science at all), but is presenting testimony that fails to meet legal requirements.

Admissibility of Expert Testimony

Courts have often dealt with the problems presented by expert testimony the same way they have dealt with much other problematic testimony—by letting the jury evaluate the testimony, no matter how flawed it might be. Courts traditionally rely on cross-examination to uncover ordinary biases, such as a witness's familial or friendly relationship to one of the parties, as well as to reveal contradictions or inconsistencies in a witness's testimony. [10]

Some authorities have argued that cross-examination will also reveal an expert witness's bias or errors to the jury. [11] Revealing "bias" is problematic: it is clearly difficult for opposing counsel to discredit an expert as a hired gun, when his own expert—who, perhaps, is scrupulously honest—is on his payroll. Relying on cross-examination to uncover scientific errors is even more problematic. It is not at all clear how a jury composed of laypersons—who must sit through days or even weeks of complex testimony, and who must generally rely solely on their memory of this testimony, without consulting any documentation—can consistently separate the scientific wheat from the chaff. The "let it all in" approach to scientific testimony, then, is inconsistent with the law's insistence that proffered testimony—expert or otherwise—help the fact-finder arrive at a correct decision.

Given the high direct and indirect costs of litigation, particularly mass tort suits that might include thousands of plaintiffs, there is great economic benefit to making sure that the proffered testimony meets appropriate legal standards, and excluding that which does not. *Phantom Risk* describes some of the problems with excessive litigation fueled by expert testimony of questionable legal value. [12]

Daubert and Joiner

In 1993, the U.S. Supreme Court handed down a landmark ruling on scientific evidence, Daubert v. Merrell Dow Pharmaceuticals.[13] Daubert attempted to alleviate the problem of unreliable scientific testimony by requiring judges to evaluate the scientific validity of proffered scientific testimony. The issue in Daubert was the proper interpretation of the Federal Rules of Evidence relating to scientific evidence. The Rules were codified in 1975 and are the rules of evidence that federal judges apply today. Many, though not all, state courts revised their rules along similar lines.

The most important rule with regard to expert testimony is Rule 702. This rule states that trial testimony is admissible from any qualified scientific expert who possesses "scientific, technical, or other specialized knowledge [that] will assist the trier of fact [the jury] to understand the evidence or to determine a fact in issue."

Before Daubert, some federal judges believed that Rule 702 allowed almost any testimony by a qualified scientific expert to be presented to a jury. In response to criticism that courts were issuing decisions based on questionable scientific evidence,[14] other courts engaged in stricter scrutiny of scientific evidence. [15]

Daubert resolved this growing split among the federal courts. The Supreme Court held that Rule 702 requires that district court judges serve as "gatekeepers," who exclude improper expert testimony. [16] Parsing the language of Rule 702, the Court held that, to be admissible, proffered scientific evidence must constitute "scientific knowledge."[17] This requirement, according to the Court, establishes a standard of evidentiary reliability. "Evidentiary reliability," the Court held, means "trustworthiness"; it depends on "scientific validity."[18] The Court added that Rule 702 requires that proposed expert scientific testimony must "assist the trier of fact to understand the evidence or to determine a fact in issue."[19] Proposed testimony must therefore have some scientific relevance to the issue at hand. [20]

The Court then enumerated four factors that may "bear on the inquiry" as to whether scientific evidence is admissible. [21] First, courts faced with challenged scientific evidence should determine whether the theory or technique at issue can be (or has been) tested. Peer review and publication, the Court added, are important, though not generally dispositive, factors. The Court also directed judges' attention to determining the known or potential rate of error of a technique in question, as well as to the existence and maintenance of standards controlling the technique's operation. A final consideration is the degree of the acceptance of the method or theory at issue. [22]

One final issue still needed to be resolved. In the toxic tort arena, some judges and legal scholars argued that Daubert required courts to limit themselves to determining whether a scientific expert witness was relying on studies that used a methodology appropriate for inquiry into the general subject at issue. Others maintained that courts should also review the expert's reasoning in extrapolating from those studies to causation. [23]

This debate was more than simply academic. One court adopted the "general methodologies only" approach, and upheld a multimillion dollar jury verdict with a poorly reasoned opinion, affirming the district court's decision to admit testimony that breast implants cause systemic disease.[24] Another court adopted the view that expert witnesses must not only use accepted methodologies, but must extrapolate from them in valid ways, and excluded similar evidence regarding implants. [25]

The more stringent interpretation of Daubert was vindicated by the Supreme Court's December 1997 opinion in General Electric Co. v. Joiner. [26] The Court held that scientific testimony should be rejected when there is "too great an analytical gap between the data and the opinion proffered." The Court ruled that an expert's testimony linking PCBs with cancer was properly excluded.

Even after Joiner, however, two significant problems with the admissibility of scientific evidence loom. First, while federal courts are bound by Supreme Court rulings, most state courts have not yet adopted Daubert, much less Joiner, leaving plenty of opportunities for scientifically invalid testimony to be admitted into legal proceedings. Within a few months after Joiner was decided, two state courts, one in Massachusetts and one in Oregon, held that highly questionable evidence that breast implants cause disease was admissible. [27] Both courts chose to ignore Joiner, and ruled that scientific evidence is admissible so long as it is based on methodologies that might be reliable. Challenges to how such methods are used by experts in a particular case, the courts held, go to the weight, not the admissibility, of the evidence.

Helping Judges Evaluate Scientific Evidence

The second post-Joiner problem with scientific evidence is that district court judges are not usually well qualified to engage in the type of inquiry into technical matters that Daubert and Joiner require. One mechanism for reducing this problem is to educate lawyers and judges about science through the process of continuing professional education. A manual on scientific evidence published by the Federal Judicial Center[28]—soon to go into its second edition—is valuable to judges in evaluating proffered scientific testimony. Judicial education programs, such as one run by George Mason University's Law and Economics Center, also help enlighten judges regarding scientific issues.

More important, judges could appoint expert witnesses to advise them on scientific issues. This idea has the support of many legal scholars, including Justice Stephen Breyer, [29] and seems sound if procedural obstacles can be overcome. The purpose is not to turn over decision making to scientists or to enforce some artificially high standard of scientific excellence, but rather to help judges understand what the scientists are saying. Scientific discourse, as other professional discourse, is characterized by unstated assumptions, use of tacit criteria, and use of specialized concepts that laypersons find difficult to follow. And scientific arguments often rely on statistical analysis whose validity may be difficult for a layperson to evaluate. A judge can be helped greatly in his or her assessment of proffered testimony by using a court-appointed expert.

Most trial lawyers vehemently oppose the use of court-appointed experts, arguing (correctly) that this will change the balance of the current system. Some lawyers and sociologists of science argue that there is no such thing as a neutral expert. But it is possible to find knowledgeable scientists of high principle, and a nonpartisan judge has a good prospect of finding a nonpartisan expert.

Courts faced with breast implant claims have actively used court-appointed experts as advisors. Federal District Judge Robert Jones appointed a panel of breast implant experts. In December 1996, he relied on the advice of these experts and delivered an opinion, discussed previously, [30] excluding the plaintiffs' evidence in the several dozen breast implant cases pending before him. [31]

In April 1996, two federal judges and one state judge in New York City issued a joint opinion appointing a neutral panel, which, in turn, was to appoint a scientific panel to advise the judges on the state of the scientific evidence regarding breast implants and immune-system disease. [32] The plaintiffs' steering committee for the federal class action went to Judge Samuel Pointer, who is in charge of the federal multidistrict breast implant litigation, and asked him to appoint his own panel of scientific experts. Judge Pointer agreed to do so, and the New York judges suspended their effort. [33]

The panel appointed by Pointer concluded that there was no evidence linking silicone gel breast implants and systemic disease. [34] The conclusions of the panel are bound to be influential in deciding the outcome of the immense amount of litigation on this issue. This successful approach--appointing panels of experts to advise judges about complex technical issues arising in litigation—is likely to be used by judges in other class action suits as well.

Another recent and very hopeful development is the announcement by the American Association for the Advancement of Science (AAAS) of a new project to provide neutral science advice to judges. [35] The $500,000 project, called the Court Appointed Scientific Experts (CASE) project, grew out of the joint panel of AAAS and the American Bar Association. It will use panels of scientists and scientific societies to identify neutral experts to respond to requests from a judge. The project is also developing guidelines to avoid conflicts of interest, which is clearly a crucial issue if the plan is to withstand challenge.

Conclusion

Daubert is not a magic wand that can let a judge separate science from pseudoscience, nor was that the intention of the High Court. Asking judges to listen carefully to what scientists and engineers have to say, and to judge admissibility using criteria that are not much different from those that an intelligent layperson would use to judge an empirical claim about the world is, on the one hand, common sense, and, on the other hand, an important legal development. Much hard work will be required of judges if the promise of Daubert is to be realized. But if judges continue to take their role as gatekeepers seriously, Daubert and its progeny will provide a practical way to reduce the volume of litigation and increase the reliability of science in the courtroom.

Notes

1. For example, P.W. Huber, *Galileo's Revenge: Junk Science in the Courtroom* (Basic Books, 1991); S. Jasanoff, *Science at the Bar: Law, Science, and Technology in America* (Harvard University Press, 1995).

2. H. H. Bauer, *Scientific Literacy and the Myth of the Scientific Method* (University of Illinois Press, 1992).

3. Samuel R. Gross, "Expert Evidence," 1991 *Wisc. L. Rev.* 1113, 1127.

4. Samuel R. Gross, "Expert Evidence," 1991 *Wisc. L. Rev.* 1113, 1127.

5. Jack B. Weinstein, "Improving Expert Testimony," 20 *U. Rich. L. Rev.* 473, 482 (1986).

6. L. J. White, "Clinical ecology," 111(2) Ann. Int. Med. 168-178, (1989).

7. K. R. Foster and P. W. Huber, *Judging Science* (Cambridge, MA: The MIT Press 1997).

8. For more information on the breast implant litigation, see David E. Bernstein, "The Breast Implant Fiasco," 87 *California Law Review* 457 (1999).

9. Foster and Huber, *Judging Science.*

10. Id.

11. Ronald J. Allen and Joseph S. Miller, "The Common Law Theory of Experts: Deference or Education?" 87 *Nw. U. L. Rev.* 1131, 1146 (1993).

12. Kenneth R. Foster, David E. Bernstein, and Peter W. Huber, eds., *Phantom Risk: Scientific Inference and the Law* (Cambridge, MA: The MIT Press, 1993).

13. 509 U.S. 579 (1993).

14. See Foster, Bernstein, and Huber, *Phantom Risk,* and P. W. Huber, *Galileo's Revenge: Junk Science in the Courtroom* (Basic Books, 1991). One notorious case was Wells v. Ortho Pharmaceutical Corp., 615 F. Supp. 262 (N.D. Ga. 1985), which resulted in a $4.7 million award for birth defects allegedly resulting from the mother's use of spermicide (*Phantom Risk,* pp. 137-138).

15. See, for example, Christopherson v. Allied Signal Corp., 939 F. 2d 1106 (5th Cir. 1991) (en banc).

16. Daubert, 509 U.S. at 597.

17. Id. at 590 n.9.

18. Id.

19. Id. at 591.

20. Id.

21. Id. at 593-94.

22. Id.

23. Compare David E. Bernstein, "The Admissibility of Scientific Evidence After Daubert v. Merrell Dow Pharmaceuticals, Inc.," 15 Cardozo L. Rev. 2139, 2161 (1994) (arguing that Daubert requires courts to assess reasoning), with Kenneth Chesebro, "Taking Daubert's 'Focus' Seriously: The Methodology/Conclusion Distinction," 15 Cardozo L. Rev. 1745 (1994).

24. Hopkins v. Dow Corning Corp., 33 F.3d 1116 (9th Cir.1994).

25. Hall v. Baxter Healthcare Corp., 947 F. Supp. 1387 (D. Ore.1996).

26. 118 S. Ct. 512 (1997).

27. Vassallo v. Baxter Healthcare Corp., 696 N.E.2d 909 (1998); Jennings v. Baxter Healthcare Corp., 1998 WL 55039 (Ore. App.).

28. The Federal Judicial Center, Reference Manual on Scientific Evidence (West, 1994).
29. For example, Joiner, 118 S. Ct. 512 (Breyer, J., concurring).
30. See earlier in this chapter.
31. Hall v. Baxter Health Care, 947 F. Supp. 1387 (D. Ore. 1996).
32. Eliot Marshall, "New York Courts Seek 'Neutral' Experts," 272 *Science* 189 (Apr. 12, 1996)..
33. In re Breast Implant Cases, 942 F. Supp. 958, 960 (E. & S.D.N.Y. 1996).
34. The Panel's report is available on the Internet at http://www.fjc.gov/ BREIMLIT/SCIENCE/summary.htm
35. "Project offers judges neutral science advice," *Science* 284:1600 (June 4, 1999).

About the Editors

Richard L. Stroup is professor of economics at Montana State University, senior associate at the Political Economy Research Center, and a member of the Board of Advisors of The Independent Institute. His Ph.D. in economics is from the University of Washington. Professor Stroup has served as director of the Office of Policy Analysis at the Department of the Interior and has published widely in professional journals and in popular media on the economics of resources and the environment. He has lectured on these topics throughout the U.S. and internationally to professional and general audiences. He is co-author with James Gwartney of the best-selling textbook, *Economics: Private and Public Choice*, now in its 9th edition, and of *What Everyone Should Know About Economics and Prosperity*. In addition, he is contributing co-editor with John Baden of *Bureaucracy vs. the Environment*, and *Natural Resources: Environmental Myths and Bureaucratic Management*.

Roger E. Meiners is professor of law and economics at the University of Texas at Arlington and a research fellow at The Independent Institute. Having received his Ph.D. in economics from Virginia Tech, he has served as director of the Center for Policy Studies at Clemson University, director of the Atlanta Regional Office of the Federal Trade Commission, associate director of the Law and Economics Center at Emory University, and a member of the South Carolina Insurance Commission. He is the author or editor of fourteen books including *Barriers to Corporate Growth* (with B. Baysinger and C. Zeithami), *Economic Consequences of Liability Rules* (with B. Yandle), *Federal Support of Higher Education* (with R. Amacher), *Managing in the Legal Environment* (with A. Ringleb and F. Edwards), *Taking the Environment Seriously* (with B. Yandle), *Victim Compensation*, and The Independent Institute book, *Regulation and the Reagan Era* (with B. Yandle).

About the Contributors

Daniel K. Benjamin is professor of economics at Clemson University and senior associate at the Political Economy Research Center. He is a graduate of the University of Virginia, where he was elected to Phi Beta Kappa, and he received his Ph.D. from the University of California at Los Angeles. He has served as chief of staff of the U.S. Department of Labor, Staff Economist at the President's Council of Economic Advisors, and deputy assistant secretary of labor. Professor Benjamin has also taught at the University of California at Santa Barbara, University of Washington, and Montana State University and he has been a National Fellow at the Hoover Institution, Visiting Scholar at the American Enterprise Institute, and Visiting Distinguished Scholar at the University of Liverpool. His books include *Undoing Drugs*, *U.S. and U.K. Unemployment Between the Wars*, *The Economics of Public Issues* (with R. Miller and D. North), *The Economics of Macro Issues* (with R. Miller), and *Economic Forces at Work* (editor), and he is formerly an associate editor of *Economic Inquiry*. His articles have appeared in the *Journal of Political Economy*, the *Journal of Law and Economics*, the *Journal of Risk and Uncertainty*, and elsewhere.

Bruce L. Benson is DeVoe Moore Distinguished Research Professor of Economics at Florida State University. He received his Ph.D. from Texas A & M University, and he has taught at Pennsylvania State University and Montana State University. Professor Benson has been an Earhart, F. Leroy Hill and Salvatori Fellow. His research interests focus on law and economics with emphasis on private alternatives to publicly provided law and legal services, the evolution of legal institutions, and the economics of crime. He has published over 100 articles in scholarly journals, contributed more than thirty book chapters, and authored four books: *The Enterprise of Law, The Economic Anatomy of a Drug War: Criminal Justice in the Commons* (with D. Rasmussen), *American Antitrust Law in Theory and in Practice* (with Melvin L. Greenhut), and The Independent Institute book, *To Serve and Protect: Privatization and Community in Criminal Justice.*

David E. Bernstein is associate professor at George Mason University School of Law. Professor Bernstein graduated from Yale Law School in 1991, and began teaching at George Mason in 1995. Bernstein is co-editor with Kenneth Foster and Peter Huber of *Phantom Risk: Scientific Inference and the Law,* (MIT Press, 1993), and has written about scientific evidence for *Science,* the *California Law Review,* the *Yale Journal of International Law,* the *Wall Street Journal,* and other academic and popular venues. He is the author of over forty articles, book chapters, and essays.

Jo-Christy Texas Brown is a founding partner of the firm of Brown & Carl, L.L.P. in Austin, Texas. She received her undergraduate degree from Texas Tech University in 1973, and her J.D. from the University of Texas School of Law in 1987. Over the past twelve years, she has been in private practice, focusing in the areas of environmental, local government and water law. She has served as the vice-chair of annual reports on environmental law for the Urban Lawyer, State and Local Government Law Section, American Bar Association, from 1995 to 1999. She is a frequent author, contributor and speaker on the topic of environmental law and law firm management. She co-authored the second edition of *What to Do When The Environmental Client Calls,* published by the American Bar Association's Section of Natural Resources Energy, and Environmental Law.

Donald N. Dewees is professor of economics and law at the University of Toronto. A graduate of Harvard Law School, he also received his Ph.D. in economics from Harvard University in 1971. He has served as director of research for the Ontario Royal Commission on Asbestos, as a member of the Science Advisory Board of the International Joint Commission, as vice-chair of the Ontario Market Design Committee (writing rules for electricity restructuring) and as vice-dean and acting dean of the Faculty of Arts and Science of the University of Toronto.

He has published extensively in academic journals and is the author, editor or co-author of seven books including *Economics and Public Policy: The Automobile Pollution Case*, *The Regulation of Quality*, *Controlling Asbestos in Buildings: An Economic Investigation*, *Exploring the Domain of Accident Law* (with David Duff and Michael Trebilcock), and *Reducing, Reusing and Recycling: Packaging Waste Policy in Canada* (with Michael Hare).

Kenneth R. Foster is professor of bioengineering at the University of Pennsylvania. Since receipt of his Ph.D. in 1971, Dr. Foster has been engaged in studies on the interaction of non-ionizing radiation and biological systems, including possible health risks of electromagnetic fields such as emitted by cellular phones and power lines. In addition he has written widely about scientific and technology and the law. He has published approximately ninety technical papers in peer reviewed journals, numerous other articles, and is the author of two books related to technological risk and the law. His latest book is *Judging Science*.

David T. Fractor is a member of Phillips & Fractor LLC, a Pasadena, California economics and statistics consulting firm. He received his Ph.D. in economics from the University of Oregon, and he has taught at the University of California at Los Angeles, California State University at Northridge, State University of New York at Cortland, University of Oregon, and Montana State University. Dr. Fractor has been chief operating officer of Findlay, Phillips and Associates, Managing Director of JurEcon, Inc., and an Earhart Fellow. He is a contributor to the books, *Water Rights* and *Natural Resources*, and his articles have appeared in *Defense Counsel Journal*, *Litigation Economic Digest*, and *Journal of Forensic Economics*, among others. He has also served as associate editor of *Litigation Economic Digest*.

David D. Haddock is professor of law and of economics at Northwestern University. He received his Ph.D. from the University of Chicago, and he has taught at Emory University, Yale University, Ohio State University, and the University of California at Los Angeles. Professor Haddock is a contributor to the books, *Economic Consequences of Liability Rules*, *Fortune Encyclopedia of Economics*, *The Stock Market*, *Economics of Corporate and Capital Markets*, and

Law and Economics and the Economics of Legal Regulation. His articles have appeared in the *American Economic Review, Journal of Law and Economics, Virginia Law Review, California Law Review, Federal Communications Law Journal, Journal of Economic Literature,* and many other journals.

Peter W. Huber is a lawyer and writer. He earned a doctorate in mechanical engineering from M.I.T., and served as an assistant and later associate professor at M.I.T. for six years. His law degree is from the Harvard Law School. He clerked on the D.C. Circuit Court of Appeals for Judge Ruth Bader Ginsburg, and then on the U.S. Supreme Court for Justice Sandra Day O'Connor. He is a senior fellow of the Manhattan Institute for Policy Research, partner in the Washington, D.C. law firm of Kellogg, Huber, Hansen, Todd & Evans, and chairman of Telecom Policy and Analysis: a Kellogg, Huber Consulting Group. Peter Huber is the author of seven books including *Liability: The Legal Revolution and Its Consequences, Galileo's Revenge: Junk Science in the Courtroom* and his most recent, *Hard Green: Saving the Environment from the Environmentalists.* He is also co-author of several books including *Judging Science: Scientific Knowledge and the Federal Courts* (with Kenneth R. Foster). His articles have appeared in scholarly journals including the *Harvard Law Review, Yale Law Journal,* and *Science,* among many others . In addition, he writes a regular column for *Forbes* magazine and has contributed to numerous other magazines and newspapers throughout the United States.

Daniel D. Polsby is associate dean of academic affairs and professor at George Mason School of Law. He previously taught at Northwestern University as Kirkland & Ellis Professor of Law. He has served as legal counsel to Commissioner Glen O. Robinson of the Federal Communications Commission and practiced law with Wilmer, Cutler and Pickering in Washington, D.C. In 1971-72 he served as Law Clerk to the late Harold Leventhal, U.S. Court of Appeals, District of Columbia Circuit. Professor Polsby was Chicago correspondent for the *Economist* from 1990-1994. Professor Polsby received his B.A. from Oakland University (1964) and earned his J.D. *magna cum laude* from the University of Minnesota (1971). He is the author of *Federalism and Retail Auto Markets,* a contributor to the books, *Communications for*

Tomorrow (G. Robinson, ed.) and *Coming Out of the Ice* (J. and D. Bast, eds.), and his articles and reviews have appeared in the *Journal of Legal Studies*, *Yale Law and Policy Review*, *Federal Communications Law Journal*, *Constitutional Commentary*, *Emory Law Review*, *Supreme Court Law Review*, *Michigan Law Review*, and many other journals.

W. Kip Viscusi is the John F. Cogan, Jr. Professor of Law and Economics and director of the Program on Empirical Legal Studies at Harvard Law School. Professor Viscusi received his Ph.D. in Economics from Harvard University. He has published 17 books and over 200 articles, most of which deal with different aspects of health and safety risks. His books were awarded the Kulp-Wright Award for Outstanding Book on Risk and Insurance from the American Risk & Insurance Association in 1992, 1993, and 1994, which also awarded him the Mehr Article Award in 1999. His latest book is *Calculating Risks? The Spatial and Political Dimensions of Hazardous Waste Policy* (with James T. Hamilton). Professor Viscusi is the founding editor of the *Journal of Risk and Uncertainty* and serves on the editorial boards of seven other journals, including the *American Economic Review* and the *Review of Economics and Statistics*.   He has consulted to the U.S. Office of Management and Budget, the Environmental Protection Agency, the Occupational Safety and Health Administration, the Federal Aviation Administration, and the U.S. Department of Justice on issues pertaining to the valuation of life and health. Professor Viscusi also served on the Science Advisory Board of the U.S. Environmental Protection Agency for seven years.

Aaron Wildavsky was the Class of 1940 Professor of Political Science and Public Policy at the University of California at Berkeley. He received his Ph.D. from Yale University and has served as president of the Russell Sage Foundation and dean of the Graduate School of Public Policy at the University of California. Professor Wildavsky has taught at City University of New York, Hebrew University, and Oberlin College. A founding member of the Board of Advisors of The Independent Institute, he has been president of the American Political Science Association, President of the Policy Studies Organization, director of the Survey Research Center, and senior editor of *Society*. A fellow of the Association of Public Policy Analysis and Management, American

Academy of Arts and Sciences, and the National Academy of Public Administration, he is the author of thirty books, including *Searching for Safety, Risk and Culture* (with M. Douglas), *The Politics of the Budgetary Process, Urban Outcomes* (with F. Levy and A. Melsner), and *Presidential Elections* (with N. Polsby).

Bruce Yandle is Alumni Professor of Economics at Clemson University. He received his Ph.D. from Georgia State University and has served as executive director of the Federal Trade Commission and senior economist for the President's Council on Wage and Price Stability. A consultant to the World Bank, Environmental Protection Agency, U.S. Department of Energy, and the Office of Management and Budget, Professor Yandle is the author of *Labor and Property Rights in California Agriculture, Alternative Systems for Allocating Air Quality, Economics of Environmental Quality, Establishing Markets for Water Quality, Common Sense and Common Law for the Environment, Managing Personal Finance,* and *Environmental Use and the Market.* He is also the editor of The Independent Institute book, *Regulation and the Reagan Era* (with R. Meiners) and the books, *Public Choice, The Strategic Use of Regulation for Anticompetitive Purposes, Benefit-Cost Analyses of Social Regulation,* and *Land Rights: The 1990s' Property Rights Rebellion.*

Index